Environmental Stress Physiology of Plants and Crop Productivity

Edited by

Tajinder Kaur

Department of Agriculture
Sri Guru Granth Sahib World University, Fatehgarh Sahib
Punjab 140407, India

&

Saroj Arora

Department of Botanical and Environmental Sciences
Guru Nanak Dev University, Amritsar
Punjab 143005, India

Environmental Stress Physiology of Plants and Crop Productivity

Editors: Tajinder Kaur and Saroj Arora

ISBN (Online): 978-1-68108-790-0

ISBN (Print): 978-1-68108-791-7

ISBN (Paperback): 978-1-68108-792-4

need for a court order if at any point you breach any terms of this License Agreement. In no event will any delay or failure by Bentham Science Publishers in enforcing your compliance with this License Agreement constitute a waiver of any of its rights.

3. You acknowledge that you have read this License Agreement, and agree to be bound by its terms and conditions. To the extent that any other terms and conditions presented on any website of Bentham Science Publishers conflict with, or are inconsistent with, the terms and conditions set out in this License Agreement, you acknowledge that the terms and conditions set out in this License Agreement shall prevail.

Bentham Science Publishers Ltd.
Executive Suite Y - 2
PO Box 7917, Saif Zone
Sharjah, U.A.E.
Email: subscriptions@benthamscience.net

BENTHAM SCIENCE

CONTENTS

FOREWORD

I am delighted to jot down the foreword for this book that deals with plants' response to numerous abiotic stresses and various approaches and concepts to mitigate their adverse effects. While the demand for food is increasing continuously, agriculture productivity is threatened by various environmental factors, often related to temperature change and warming. The most typical abiotic stresses affecting the plants are high temperature and drought, resulting in unbearable crop losses. Global temperature change further increases the frequency of warmth stress, flood, and drought, which have a negative impact on crop yield and poses a heavy challenge to global food security. Presently, the best priority to satisfy the worldwide food demand is to sustain and improve crop yield.

This book **"Environmental Stress Physiology of Plants and Crop Productivity"** edited by **Prof. Saroj Arora and Dr. Tajinder Kaur** presents a broad picture of plant responses to major abiotic stress factors, plants adaptations during stress conditions like the role of antioxidative system and plant growth regulators, biotechnological approaches for abiotic stress tolerance in plants for sustainable agriculture, and the role of secondary plant compounds and signaling molecules to modulate the oxidative stress in plants under stressful conditions. These topics, written by experts in their respective fields, make this book highly useful to the scientific community. It covers a good range of subjects that provide the eader a comprehensive overview of stress tolerance to varied environmental factors affecting plant growth and development.

I congratulate the editors for compiling this publication that provides the reader the simplest way forward in abiotic stress management to enable more productive agriculture. I am sure that the readers within the field of abiotic stress management, agriculture, and biotechnology will find this book very useful.

K. Muthuchelian
Dayananda Sagar University, Bangalore
Formerly Vice-Chancellor, Periyar University
Salem (TN)
India

PREFACE

The human population is increasing worldwide at a much faster rate and is expected to increase from ~7 billion to ~ 10 billion by the end of the year 2050. On the other hand, agricultural productivity is not increasing at the desired rate to feed all the people due to the negative impact of various environmental stresses. Stresses in plants include heat, cold, drought, flooding, salinity, radiations, heavy metal toxicity, and nutrient loss, limiting agricultural productivity. In the present scenario of global environmental change, increasing crop productivity and minimizing the losses in crop yield is a major concern for all nations to cope with increasing food requirements.

The present book **"Environmental Stress Physiology of Plants and Crop Productivity"** integrates a broad cross section of scientific knowledge and expertise about the mechanisms underlying plant responses against different environmental stressors. It is a timely contribution to a subject that is of eminent importance. Chapter *one* gives an overview of major abiotic stress factors affecting growth and development in plants. Chapter *two* and *three* focus on two major abiotic stressors, *i.e.*, drought and salinity, that adversely affect crop productivity and quality. Chapter *four* covers factors responsible for temperature variations around the globe, plants' response to temperature variations, and its impact on crop yield. Chapter *five* covers the effect of mineral deficiency on plant stress responses. Chapter *six* deals with the role of nitric oxide in providing tolerance to plants under salt stress. Chapter *seven* highlights the role of antioxidative defense system of plants in mitigating the harmful effects of excessive production of reactive oxygen species under various types of stress factors. Chapters *eight* and *nine* discuss the importance of melatonin and phenylpropanoid, respectively, in mitigating the adverse effects of abiotic stress in plants. Chapter *ten* comprehensively deals with the role of different plant growth regulators in abiotic stress tolerance. Chapter *eleven* present an overview of genomics, proteomics, and metabolic approaches, and *twelve* deals with the advent of new technologies like CRISPR gene technology to develop plant resistance against various environmental changes.

We express our thanks to all the contributors. We would like to thank Prof. (Dr) K. Muthuchelian for writing the foreword. Finally, it is a profound pleasure to thank Bentham Science for taking up the publication of this book. We hope that this book will provide current knowledge on abiotic stress in plants and will lead to new discussions and efforts to deal with various environmental stress factors.

Tajinder Kaur
Department of Agriculture
Sri Guru Granth Sahib World University, Fatehgarh Sahib
Punjab 140407, India

&

Saroj Arora
Department of Botanical & Environmental Sciences
Guru Nanak Dev University, Amritsar
Punjab 143005, India

Summary

The knowledge of plant responses to various abiotic stresses is crucial to understand their underlying mechanisms as well as the methods to develop new varieties of crops, which are better adapted. *Environmental Stress Physiology of Plants and Crop Productivity* will provide a timely update on plants' response to a variety of stresses such as salinity, temperature, drought, oxidative stress, and mineral deficiencies. The book will focus on how plants adapt to abiotic stress and methods of improving plants' tolerance to abiotic stresses. The book will be beneficial to scholars and researchers working in the field of botany, agriculture botany, crop physiology, soil science, and environment sciences.

List of Contributors

Ajay Kumar	Department of Botanical and Environmental Sciences, Guru Nanak Dev University, Amritsar, Punjab 143005, India
Arpna Kumari	Department of Botanical and Environmental Sciences, Guru Nanak Dev University, Amritsar, Punjab 143005, India
Atamjit Singh	Department of Pharmaceutical Sciences, Guru Nanak Dev University, Amritsar, Punjab 143005, India
Avinash Kaur Nagpal	Department of Botanical and Environmental Sciences, Guru Nanak Dev University, Amritsar, Punjab 143005, India
Balbir Singh	Department of Pharmaceutical Sciences, Guru Nanak Dev University, Amritsar, Punjab 143005, India
Drishtant Singh	Department of Molecular Biology and Biochemistry, Guru Nanak Dev University Amritsar, Amritsar, Punjab 143005, India
Harjit Kaur Bajwa	Department of Botany and Environmental Science, Sri Guru Granth Sahib World University, Fatehgarh Sahib, Punjab 140407, India
Hina Khan	Department of Botany and Environmental Science, Sri Guru Granth Sahib World University, Fatehgarh Sahib, Punjab 140407, India
Inderpreet Kaur	Department of Chemistry, Centre for Advanced Studies, Guru Nanak Dev University, Amritsar, Punjab 143005, India
Jaskaran Kaur	Department of Botanical and Environmental Sciences, Guru Nanak Dev University, Amritsar, Punjab 143005, India
Jatinder Kaur Katnoria	Department of Botanical and Environmental Sciences, Guru Nanak Dev University, Amritsar, Punjab 143005, India
Jyoti Mahajan	PG Department of Agriculture, Khalsa College, Amritsar, Punjab 143005, India
Kirandeep Kaur	Department of Pharmaceutical Sciences, Guru Nanak Dev University, Amritsar, Punjab 143005, India
Kritika Pandit	Department of Botanical and Environmental Sciences, Guru Nanak Dev University, Amritsar, Punjab 143005, India
Manik Sharma	PG Department of Agriculture, Khalsa College, Amritsar, Punjab 143005, India Department of Botanical and Environmental Sciences, Guru Nanak Dev University, Amritsar, Punjab 143005, India
Neena Bedi	Department of Pharmaceutical Sciences, Guru Nanak Dev University, Amritsar, Punjab 143005, India
Neha Sharma	Department of Botanical and Environmental Sciences, Guru Nanak Dev University, Amritsar, Punjab 143005, India
Preet Mohinder Singh Bedi	Department of Pharmaceutical Sciences, Guru Nanak Dev University, Amritsar, Punjab 143005, India
Rajani Gupta	Department of Botanical & Envrionmental Sciences, Hindu College, Amritsar, Punjab 144001, India

Rajinder Kaur	Department of Botanical and Environmental Sciences, Guru Nanak Dev University, Amritsar, Punjab 143005, India
Rattandeep Singh	School of Bioengineering and Biosciences, Lovely Professional University, Jalandar, Punjab 144001, India
Ravinder Singh	Department of Botanical and Environmental Sciences, Guru Nanak Dev University, Amritsar, Punjab 143005, India
Renu Bhardwaj	Department of Botanical and Environmental Sciences, Guru Nanak Dev University, Amritsar, Punjab 143005, India
Sakshi Sharma	Department of Botany, DAV College, Amritsar, Punjab 143005, India
Samiksha	Department of Zoology, Guru Nanak Dev University, Amritsar, Punjab 143005, India
Sandeep Kaur	Department of Botanical and Environmental Sciences, Guru Nanak Dev University, Amritsar, Punjab 143005, India PG Department of Botany, Khalsa College, Amritsar- 143005, Punjab, India
Saroj Arora	Department of Botanical and Environmental Sciences, Guru Nanak Dev University, Amritsar, Punjab 143005, India
Satwinderjeet Kaur	Department of Botanical and Environmental Sciences, Guru Nanak Dev University, Amritsar, Punjab 143005, India
Satwinder Kaur Sohal	Department of Zoology, Guru Nanak Dev University, Amritsar, Punjab 143005, India
Shagun Verma	Department of Botanical and Environmental Sciences, Guru Nanak Dev University, Amritsar, Punjab 143005, India
Sharad Thakur	CSIR-Institute of Himalayan Bioresource Technology, Palampur, H.P. 176061, India PG Department of Agriculture, Khalsa College, Amritsar, Punjab 143005, India
Sneh Rajput	Department of Botanical and Environmental Sciences, Guru Nanak Dev University, Amritsar, Punjab 143005, India
Tajinder Kaur	Department of Agriculture, Sri Guru Granth Sahib World University, Fategarh Sahib- 140407, Punjab, India

CHAPTER 1

Abiotic Stress in Plants: An Overview

Sharad Thakur[1,2], Ravinder Singh[3], Jaskaran Kaur[3], Manik Sharma[2,3], Kritika Pandit[3], Satwinderjeet Kaur[4] and Sandeep Kaur[3,4,*]

[1] *CSIR-Institute of Himalayan Bioresource Technology, Palampur, H.P. 176061, India*

[2] *PG Department of Agriculture, Khalsa College, Amritsar, Punjab 143005, India*

[3] *Department of Botanical and Environmental Sciences, Guru Nanak Dev University, Amritsar, Punjab 143005, India*

[4] *PG Department of Botany, Khalsa College, Amritsar, Punjab 143005, India*

Abstract: Environmental stress is one of the major limiting factors for agricultural productivity worldwide. Plants are closely associated with the environment where they grow and adapt to the varying conditions brought about by the huge number of environmental factors resulting in abiotic stress. Abiotic factors or stressors include high or low temperature, drought, flooding, salinity, mineral nutrient deficiency, radiation, gaseous pollutants, and heavy metals. High salinity, drought, cold, and heat are the major factors influencing crop productivity and yield. The negative impact of various abiotic stress factors is the alteration in the plant metabolism, growth, and development and, in severe cases, plant death. Abiotic stress has been becoming a specific concern in agriculture leading to unbearable economic loss to the breeders. Thus, understanding these stresses will help in achieving the long-term goal of crop improvement, therefore, minimizing the loss in crop yield to cope with increasing food requirements. With this chapter, an attempt has been made to present an overview of various environmental factors that are hostile to plant growth and development, thereby leading to great loss in crop yield.

Keywords: Abiotic stress, Crop yield, Drought, Heat, Salinity.

INTRODUCTION

Plants continuously face unfavorable environmental conditions that affect their growth and development by altering their metabolic activities, eventually leading to plant death. Abiotic stress is defined as environmental conditions that cause alterations in plant growth and development and limit yield below optimum levels [1 - 3] and can result in unacceptable economic losses. Abiotic stress factors include water stress (drought, flooding, waterlogging), extreme temperatures

---

* **Corresponding Author Sandeep Kaur:** P.G. Department of Botany, Khalsa College, Amritsar, Punjab 143005, India; Tel: +91-9877168954; E-mail: soniasandeep4@gmail.com

(heat, cold, and freezing), too high or too low irradiation, insufficient mineral nutrients in the soil, excessive soil salinity, gaseous pollutants, and heavy metals. However, drought and salt stress pose severe threats to agriculture resulting in loss of crop yield. These threats may become more intense by global climate change and population growth [4]. The ability of crops to achieve their yield potential (maximum possible yield) if all inputs are non-limiting can be affected by a number of stress events. In most agricultural systems, crops hardly reach their yield potential due to stress events, be they abiotic or biotic. Water is often the largest restraint to limiting crop yield. Water deficiency reduces the rate of transpiration and photosynthesis in addition to the absorption of nutrients and water by the root surface, resulting in lower crop yield as compared to yield potential [5]. The average yield of major crop plants is reduced worldwide by more than 50% due to abiotic stress [6, 7]. It is predicted to become even more severe as desertification increases. The present yearly loss of arable area may double due to global warming by the end of the century [8, 9]. Crop production may be reduced to about 70%, with the majority of the crops performing at only 30% of their genetic potential with regard to yield [10]. The problem will become more intense by a simultaneous increase in population growth, creating more pressure on existing cultivated land and other resources [11]. Given the increasing human interference by humans with nature, only 2.75% of the global land area is not affected by some environmental limitations, according to its yearly report by Food and Agriculture Organization (FAO), 2018. It has been projected that more than 90% of the land in rural areas is distressed by various environmental stress factors at some point during the growing season [1].

Abiotic stress exhibits a huge challenge in our goal for sustainable food generation. Drought and rising temperature are two of the main abiotic stressors around the world that reduce crop productivity and influence the ability to meet the food demands of the rising global population, particularly given the current and growing impacts of climate change. Several studies have reported that increased temperature and drought can decrease crop yields by 50% [12]. Salt stress is also one of the major severe limiting factors for crop growth and production, which has been elevated mostly by agricultural practices such as irrigation [13]. More than 6% of the world's land is affected by salinity. Of the present 230 million hectares of irrigated land, about 45 million hectares, *i.e.*, 19.5%, are affected by salinity. Of the 1,500 million hectares under dry land agriculture, about 32 million hectares (2.1%) are salt-affected to varying degrees [14].

Mineral stress is another constraint in plant growth. Intensive agriculture practices in developed nations have lessened the availability of nutrients in the soil resulting in its low fertility and poor availability of nutrients required for plant growth. However, plants can initiate a number of cellular, molecular, and physiological changes in response to this stress. Plant reactions to abiotic stressors are complex and dynamic; they are both reversible and irreversible. Plants respond to stress in a number of ways depending on the tissue or organ affected by the stress, in addition to its intensity and duration. Some plants complete their life cycle during less stressful periods and hence escape the effects of stress. On the other hand, some plants have evolved stress tolerance, avoidance, or resistance mechanisms that isolate plant cells from stressful conditions [15]. In this chapter, we present an overview of various environmental factors that cause stress in plants and thereby affect their overall growth and yield.

ENVIRONMENTAL CONDITION THAT CAUSE STRESS

Plants are exposed to various types of abiotic stressors like drought, salinity, cold and heat, waterlogging, hypoxia, and anoxia, which have a negative influence on the survival, biomass production, and yields of staple food crops by up to 70%, hence, putting at risk food security worldwide [16 - 18]. Water deficiency stresses imparted by drought, salinity, and temperature severity are the most prevalent abiotic stresses that limit plant growth and productivity. The response of plants to these abiotic stressors is multigenic, with plants responding to these stressors at the molecular level by regulating the synthesis of various secondary metabolites, which act as physiological buffers to nullify the bad effects of abiotic stressors. To understand the functional dynamics of the stress signal perception and regulation of the associated molecular networks, the stress tolerance capacity and capability of plants are being un-ravelled through high throughput sequencing and functional genomics tools. Plant acclimatisation to abiotic stress is a complex and coordinated response involving hundreds of genes and various signal transduction pathways activated in response to the various environmental factors throughout the developmental period of the plant [19 - 21].

Hence, a thorough understanding of the molecular pathways functioning in response to the abiotic stress in plants is essential for targeting any improvement in plant biomass or yield. A better and thorough insight into the plant responsiveness to abiotic stress will help in both traditional and modern breeding applications towards improving stress tolerance in plants. Genetic regulatory mechanisms occurring at the level of transcriptional regulation, alternative splicing molecular mechanisms, and the rapid generation of signal transduction regulatory proteins *via* ubiquitination, sumoylation, phosphorylation, and chromatin remodeling tend to influence complex signal transduction networks that act in turn to regulate processes such as membrane transport, the ascent of sap to maintain cellular ion homeostasis and the synthesis of the secondary metabolites.

Microbiome inhabited by the plants also facilitates the reduction of environmental stress [22].

The most fundamental living system symbiotically closely associated with the plants right from seed on the earth is the microbial life. As soon as a seed comes into the soil to start its life cycle, the surrounding microbial life forms an integral part of plant growth and development. Microorganisms tend to form a symbiotic association with the plant parts like the roots, leaves, and stem. Plant microbiome provides fundamental support to plant life and helps in acquiring nutrients, resistance against diseases, and tolerance of abiotic stressors [23]. Intrinsic metabolic and genetic capabilities of the microbes make them suitable microorganisms to combat extreme conditions of the environment [24, 25]. The microbial interactions with the plants improve the metabolic capability of the plants to fight against abiotic stresses by evoking various kinds of local and systemic responses that are beneficial for the survival and adaptability of the plants to a particular environment [26].

PLANT RESPONSES TO STRESS

Plants are important components of earth. The ultimate source of oxygen on this planet, which makes it habitable for human survival and life, is due to plants. It is not easy for them also to survive on Earth without facing resistance from powers of Earth's energy. Plants encounter a variety of environmental insults in their lifetime in the form of various stresses. It may be biotic or abiotic stresses. Both biotic and abiotic stressors cause retardation in growth and fresh weight and retard food production from plants [27]. Abiotic stressors include conditions like drought, flooding, salinity, stress due to air, water, soil pollution, nutrient imbalance, radiation, stress related to heat such as high and low temperatures, heavy metal stress *etc.,* is due to unfavorable environmental conditions which include physical, chemical or mechanical factors [28, 29]. These conditions cause severe losses to agriculture and plant growth.

Stress at the cellular and molecular levels cause alterations of different physiological processes of plants, including inactivation and denaturation of enzymes, proteins, displacement / substitution of essential metal ions from bio-molecules, blocking of functional groups of metabolically important molecules, and conformational modifications and disruption of membrane integrity, which further lead to altered plant metabolism, inhibition of respiration, photosynthesis, and alerted activities of several important enzymes [30 - 32]. Abiotic stress like heavy metal pollution also stimulates the generation of free radicals and reactive oxygen species (ROS) such as hydrogen peroxide, singlet oxygen, superoxide radicals, and hydroxyl radicals and perhydroxyl radical, which disturb the redox homeostasis [31 - 33]. ROS are extremely reactive and rapidly interrupt normal cell metabolism. Membrane destabilization is generally attributed to lipid peroxidation, due to an increased production of active oxygen species. Many studies have been performed on the enhancement of plant tolerance to oxidative stress by modifying the plant antioxidant defense system [34]. In addition, it is believed that the alterations in antioxidant enzymes may be due to the synthesis of new isozymes or enhancement of pre-existing enzyme levels for the metabolism of ROS [35] (Fig. **1**).

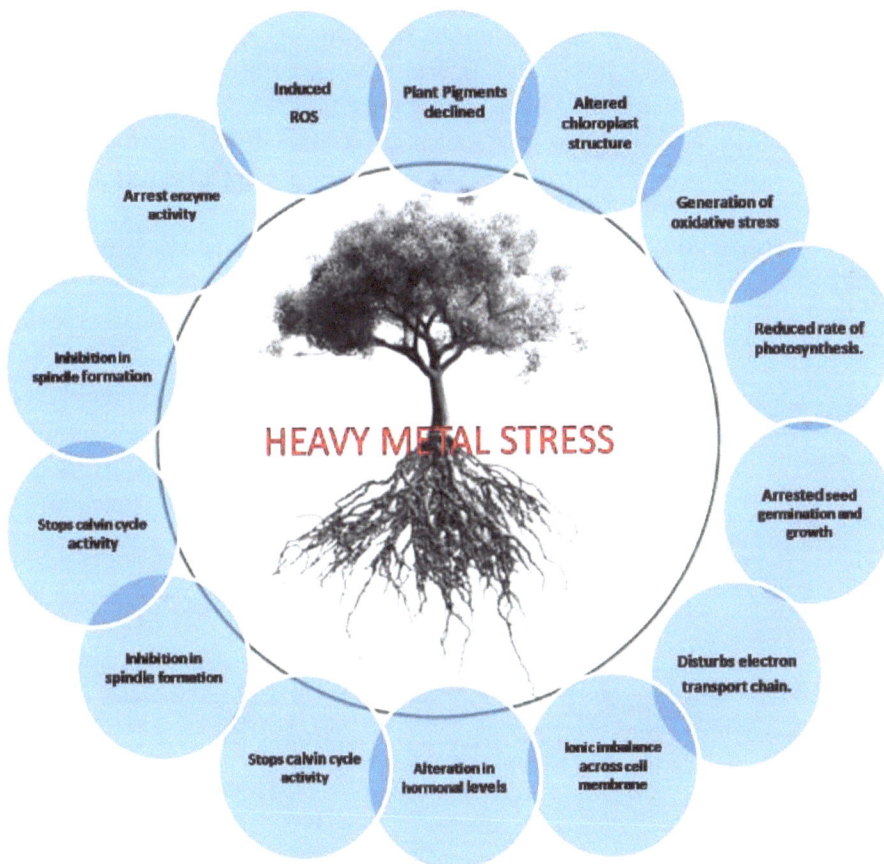

Fig. (1). Various response elicited plant systems during abiotic stress.

On the other hand, plants may develop an anti-oxidative system through enhancing anti-oxidative enzymes and antioxidants *e.g.* glutathione to combat the various abiotic stressors [36, 37]. The defence mechanism of plants towards stress includes enzymatic antioxidative machinery including catalase (CAT), ascorbate peroxidase (APX), glutathione peroxidase (GPX), superoxide dismutase (SOD), glutathione reductase (GR) while also increasing malondialdehyde (MDA) and H_2O_2 contents in plants [38, 39]. The Asada-Halliwell pathway is documented to be responsible for sequestration of H_2O_2 through various enzymes. Studies have reported that the antioxidative defence system of plants plays a significant role in sequestration of ROS by reducing the imbalance that arises due to stress [37, 40].

EFFECTS OF ABIOTIC STRESS

Overview of the Effect of Abiotic Stress on Productivity and Yield

Abiotic stress is a serious global problem limiting agricultural production and food security. Plants are exposed to different types of abiotic stressors such as salinity, drought, heavy metals, heat, cold, radiation *etc.* which poses harmful effects on the health and yield of crops thus affecting global food security [41 - 44]. The investigated effects of various abiotic stressors are given in the Table **1**. The detrimental effect of these stressors continue to increase due to growing urbanization, industrialization, pollution, climate change and leads to nutritional imbalances in the crops by reducing availability of water thus increasing toxicity and reducing products. Abiotic stress damages more than 50% crops worldwide due to their harmful effects [10, 45]. In plants, abiotic stresses disturb the cellular redox homeostasis resulting in the generation of reactive oxygen species (ROS) which leads to oxidative stress [46]. ROS generation can lead to cell damage in different organelles of the cell such as nucleus, mitochondria, chloroplast and peroxisome and ultimately plant cell death [47].

Table 1. Effect of various aboitic stresses on crop species.

Crop Species	Metal Treatment	Results	References
Calendula officinalis L	Salt stress	shoot dry matter, peroxidase activity (POD) and proline, protein and total chlorophyll declined	[48]
Fagophyrum tararicum	Cd metal	CONTENT of SOD, POD, CAT, APX enhanced in leaves and roots of tertary buckwheat under Cd stress.	[49]
Oudeneya Africana	drought	leaf area, photosynthesis rate and chlorophyll content reduced	[50]
Cynodon dactylon	Cold stress	SOD, POD, APX, GPX and GS activity increase	[51]
Zea mays L	As and Cd metal	Reduction in gas exchange attributes (photosynthetic rate, transpiration rate, stomatal conductance) and chlorophyll	[52]
Vigna mungo	As metal	The activities of antioxidative enzymes (SOD, POD, and APX) except CAT were increased.	[53]

(Table 1) cont.....

Crop Species	Metal Treatment	Results	References
Miscanthus sp.	Cd metal	Resulted in significant decrease in plant growth and dry biomass.	[54]
Oryza sativa	Pb metal	Decrease in shoot height, root length, biomass were observed	55
Glycine max	As metal	Changes in the expression of a key messenger (Phosphatidic acid) *via* phospholipase D and phospholipase C.	[56]

Salt Stress

Salinity stress being one of the major environmental threats causes both ionic toxicity and osmotic stress in plants [57, 58]. Salt stress severely affects the crop yield because of oxidative damage, nutritional disorders, and negative impact on ion imbalance and water relations and thus limits plant growth. Disturbance in cellular ion homeostasis negatively affects the uptake of soil nutrient resulting in nutrition deficiencies in plants under salt stress. It reduces the growth and development of crops by inhibiting several vital biochemical and physiological processes [59 - 61].

Siddiqui *et al.* [62] found that salt stress decreases the activities of enzymes of sulfate and nitrate assimilation pathway in plants, and thus increases the demand for sulfur and nitrogen. The accumulation of reactive oxygen species (ROS) in the plants under salt stress results in oxidative stress that leads to cellular damage, such as lipid peroxidation, DNA damage and enzyme inactivation [63, 64]. Moreover, salinity stress also causes membrane disintegration, ion imbalance, metabolic function loss, breakdown of DNA and following cell injury and cell necrosis [65].

Temperature Stress

Like other stresses, high (heat) and low (cold) temperature stress have been a major limiting factor to the growth and development of plants leading to reduced yield of crops. Heat stress leads to oxidative stress, enzyme inactivation, lipid peroxidation, membrane damage, degradation of proteins and ultimately DNA defragmentation in plants [66]. According to various reports, there is an increase in the expressions of various genes in plants under temperature stress [58]. Plants have evolved various molecular and physiological mechanisms to overcome heat stress. High temperature reduced germination potential of the seeds, photosynthetic rate and crop yield. Increased temperature have various harmful effects in the plants during the reproduction period such as loss of functional tapetal cells, formation of dysplastic anther, inhibition of the swelling pollen grains during flowering period, thus leads to the poor shedding of pollen grains as well as the indehiscence of anthers [67, 68]. Heat stress decreased the number of spikes and florets in rice plants as well as seed-set in sorghum [69].

High temperatures cause significant decrease in the growth and net assimilation rate in maize and sugarcane plants [70, 71] as well as reduction in the length of internodes, early leaf senescence and accumulation of biomass in sugarcane [72]. In plants, cold stress affects various biochemical and physiological processes and ROS-homeostasis [66, 73, 74]. High temperature reported to decrease productivity in peanut [75], tomato and rice [76] as well as reduce starch, oil and protein contents in cereals and oilseed crops [77]. High temperature was shown to damage various PSII components in barley and wheat [78] as well as reduce photosynthetic rate in two varities of rice *i.e.* IR64 and Huanghuazhan [76].

Heavy Metal Stress

Heavy metal stress poses harmful effects on the growth of the plants and induces many physiological changes in plants which results in the reduction of productivity, thus threaten horticultural cultivars. Cadmium toxicity causes changes in metabolic processes, mineral deficiency, osmotic imbalance, inhibited essential elements uptake as well as reduction in growth and photosynthetic efficiency in plants, which ultimately results in reduced productivity [79].

In pea plant, cadmium has been reported to decrease chlorophyll content, plant growth, photosynthetic and transpiration rates as well as damage the antioxidant defense system and cause nutrient imbalance [80]. The roots of alfalfa showed oxidative and glutathione depletion under cadmium and mercury stress [81]. Nickel stress results in the enhancement of O_2^{2-} and H_2O_2 contents in the roots of wheat seedlings [74]. Arsenic toxicity leads to the generation of reactive oxygen species (ROS) which induce oxidative stress and harmfully affected primary metabolism in *Pistia stratiotes* [82].

Droughts and Salinity

There are various morphological, physiological, biochemical and molecular changes that arise due to abiotic stress that adversely affect the growth and productivity of plants [83]. These gross changes in morphology are underpinned by improvements in leaf anatomy. Water deficient leaves usually have smaller cells and higher stomatal density [84]. In another study, comparable effects of high temperature and water deficiency on cell density were reported, but there is limited data available regarding changes in the leaf anatomy in response to high temperatures [85]. Due to hydraulic insufficiency and chemical signalling from the roots to the shoots through the xylem, shoot growth may be limited in drying soil [72]. Growth of plant is impaired by reducing water uptake into expanding cells under water deficit pressure and enzymatically altering the cell wall's rheological properties; for example, ROS activity on cell wall enzymes [86]. High temperature, salinity or drought stress, results in oxidative stress which can cause functional and structural proteins to be denatured [2]. In *Medicago sativa* (alfalfa), it was reported that length of hypocotyl, germination capacity and fresh and dry shoot and root weights were reduced by a deficiency of water induced by polyethylene glycol, while the length of root was increased [87].

Drought stress decreased grain yields in barley (*Hordeum vulgare*), by reducing the number of tillers, spikes and grains per plant and grain weight. Drought stress after anthesis was detrimental to grain yield regardless of the extent of the stress [88]. In maize, water stress decreased yield by delaying silking, thereby increasing the period between anthesis and silking. This trait was strongly associated with grain yield, in particular the number of ears and kernels per plant [89]. Turgor potential, relative water quality, stomatal conductance, transpiration, and water-use efficiency were reduced under drought stress in *Hibiscus rosa-sinensis* [90]. The degree of stomatal opening of K^+-treated plants initially suggested an instant decline in drought-treated sunflower. However, diffusive resistance in the leaves of K+-treated plants remained lower than those that received no K+ at similarly low soil water capacity [85]. A major effect of drought is decrease in photosynthesis resulting from decreased leaf expansion, impaired photosynthetic machinery, premature leaf senescence, and related decrease in food production [91].

Under drought conditions, a rapid decline in photosynthesis was followed by reduced maximum velocity of ribulose-1, Rubisco 5-bisphosphate carboxylation, ribulose-1, 5-bisphosphate regeneration, Rubisco and stromal fructose bis-phosphatase activities, and higher plant quantum efficiency of Photosystem II [92]. The shoot and root biomass, photosynthesis and root respiration was decreased by severe drought. Restricted root respiration and root biomass under extreme soil drying situations can enhance growth and physiological activity of drought-tolerant wheat which is advantageous over a dry cultivar in arid regions [93]. In another study, it was found that cotton crops are very susceptible to the stress of temperature, soil salinity, heat and drought [94]. Mild drought stress during the initial stage can increase root elongation but under long-term water stress conditions root morphological and physiological activities are seriously hampered [95]. Under conditions which are water challenging, plants perceive stress through different sensors involved in signaling response. These are transduced by different pathways where several signaling and transcriptional factors play essential and specific role [96].

The transport of water within a plant occurs under strain, as determined by the availability of soil water and the deficit in atmospheric vapor pressure, causing turgor pressure within cells. Under changing environmental conditions, physiological adjustments which maintain turgor pressure are crucial. Various elements, such as root anatomy, water quality, and soil salts, influence water transport in roots [97]. Specifically, vulnerable to heat stress are reproductive processes involving pollen and stigma viability, pollen tube growth, pollination anthesis and early embryo development [98].

It has also been shown that plant hormones are involved in root and shoot long distance signaling and hydraulic conductivity control. Abscisic acid (ABA) is the most crucial hormone involved in the regulation of abiotic stress tolerance, such as drought, salinity, cold, heat and wounding [99]. Plant production is strongly affected by water deficit. The shoot and root are the most affected on a morphological level, and both are main components of plant adaptation to drought. In response to drought stress, plants typically restrict the number and area of leaves merely to raising the water budget at the expense of yield loss. Since roots are the only source of soil water, root growth, density, proliferation, and size are key plant responses to drought stress [100]. Various environmental conditions that improve the rate of transpiration often increase the pH of the leaf sap, which can encourage abscisic acid accumulation and reduce the stomatal conductance. Also, increased concentration of cytokinin in the xylem sap directly promotes stomatal opening and affects stomata's sensitivity towards abscisic acid [101].

Root growth rates are widely used in cotton crop for estimations crop yield losses. Insufficient soil moisture restricts root growth and development and thus impairs the functioning of the aerial components. Water deficiency in the upper soil profile results in deeper root penetration for further exploration of moisture and nutrients, while excess water in the upper layer results in reduced root penetration [101]. Accelerated abscission of fruits and leaves from drought-stressed cotton crops may be associated with reduction of final yields [102]. Under severe drought conditions, decreased stomatal conductance and metabolic (non-stomatal) damage, such as limited carboxylation, become major photosynthesis limitations [103]. Water and heat stress are reported to degrade proteins, decrease electron transport, and release calcium and magnesium ions from their protein-binding partners [104]. Extended exposure to high temperatures often causes a reduction in chlorophyll content, disintegration of thylakoid grana, increased amylolytic activity, and disruption of transport of assimilates [105].

Floods

Most crops are vulnerable to short-lived flood events, resulting in growth and yield reductions. The leaf structure is influenced by higher temperatures that often cause thinner leaves with a higher leaf area to grow [106]. Another mechanism which enhances plant survival under flooded conditions is to prevent the build-up of potential phytotoxins. A particular type of haemoglobin, known as phytoglobin may play such a role through the detoxification of the nitric oxide produced during root tissue hypoxia. Conversely, phytoglobin can also regenerate NAD^+ and thus function as an alternative route to fermentation [107]. *Oryza sativa* commonly known as rice is highly tolerant to flooded conditions. It is not only capable of germinating without any oxygen, but green leaves and stems capable of elongation mediated by ethylene response [108]. Flooding may have a detrimental impact on physical and chemical properties of soil (structure, porosity, and pH) and microbial functional communities thus impacting soil production levels [109]. The loss of nitrogen (N) from flooded soil can be significant; that along with other deleterious effects on plants can result in lower crop productivity [110]. Soybean growth and grain yield are affected by the flooding from excessive rainfall or irrigation [111].

Flooding primarily prevents the diffusion of gas between the plant and its surrounding area due to physical consequences. Simple exchange of oxygen as well as CO_2 cannot easily take place through stomata and underwater cell walls. This results in a lack of oxygen within flooded parts of the plants, which primarily limits heterotrophic, mitochondrial energy production. In addition, poor availability of CO_2 in flooded leaves limits photosynthesis. The flooding therefore triggers an energy shortage within plant cells [112 - 114]. Plants growing in a flooded area with no ongoing photosynthesis will see the concentration of oxygen rapidly decrease, leading to hypoxic condition [115].

CONCLUSION

Plants and crops are exposed to various abiotic stressors like drought, salinity, cold and heat, water logging, hypoxia and anoxia which have a negative influence on its survival, biomass production and yields. However, plants and crops have antioxidative defence systems that enhance their tolerance to oxidative stress and ameliorate the harmful effects of these abiotic factors. Due to the severity of these abiotic factors, stress at the cellular and molecular levels, causes alterations of different physiological processes of plants, including inactivation and denaturation of defensive enzymes. However, to overcome the deterimental effects of these environmental factors certain effective strategies have been developed that helps to achieve the long-term goal of plant and crop improvement and also minimize the economic loss to the breeders.

CONSENT FOR PUBLICATION

Not applicable.

CONFLICT OF INTEREST

The author declares no conflict of interest, financial or otherwise.

ACKNOWLEDGEMENTS

We are grateful to DST-PURSE, Programme, Department of Science and Technology (DST), New Delhi (India) for supporting this work.

REFERENCES

[1]   Cramer GR, Urano K, Delrot S, Pezzotti M, Shinozaki K. Effects of abiotic stress on plants: a systems biology perspective. BMC Plant Biol 2011; 11(1): 163.
[http://dx.doi.org/10.1186/1471-2229-11-163] [PMID: 22094046]

[2]   Skirycz A, Inzé D. More from less: plant growth under limited water. Curr Opin Biotechnol 2010; 21(2): 197-203.
[http://dx.doi.org/10.1016/j.copbio.2010.03.002] [PMID: 20363612]

[3]   Cramer GR. Abiotic stress and plant responses from the whole vine to the genes. Aust J Grape Wine Res 2010; 16: 86-93.
[http://dx.doi.org/10.1111/j.1755-0238.2009.00058.x]

[4]   Koyro HW, Ahmad P, Geissler N. Abiotic stress responses in plants: an overview. Environmental adaptations and stress tolerance of plants in the era of climate change. New York: Springer 2012; pp. 1-28.
[http://dx.doi.org/10.1007/978-1-4614-0815-4_1]

[5]   Gilliham M, Able JA, Roy SJ. Translating knowledge about abiotic stress tolerance to breeding programmes. Plant J 2017; 90(5): 898-917.
[http://dx.doi.org/10.1111/tpj.13456] [PMID: 27987327]

[6]   Bray EA, Bailey-Serres J, Weretilnyk E. Biochemistry and molecular biology of plants, responses to abiotic stresses. American Society of Plant Biologists 2000; 1158-249.

[7]   Wang W, Vinocur B, Altman A. Plant responses to drought, salinity and extreme temperatures: towards genetic engineering for stress tolerance. Planta 2003; 218(1): 1-14.
[http://dx.doi.org/10.1007/s00425-003-1105-5] [PMID: 14513379]

[8]   Evans LT. Is crop improvement still needed? J Crop Improv 2005; 14(1-2): 1-7.
[http://dx.doi.org/10.1300/J411v14n01_01]

[9]   Vinocur B, Altman A. Cellular basis of salinity tolerance in plants. Environ Exp Bot 2005; 52: 113-22.

[10] Boyer JS. Plant productivity and environment. Science 1982; 218(4571): 443-8.
[http://dx.doi.org/10.1126/science.218.4571.443] [PMID: 17808529]

[11] Ericson JA, Freudenberger MS, Boege E. 14 population dynamics, migration, and the future of the calakmul biosphere reserve. Biological Diversity: Balancing Interests through Adaptive Collaborative Management 2001; 261.

[12] Lamaoui M, Jemo M, Datla R, Bekkaoui F. Heat and drought stresses in crops and approaches for their mitigation. Front Chem 2018; 6: 26.
[http://dx.doi.org/10.3389/fchem.2018.00026] [PMID: 29520357]

[13] Zhu JK. Plant salt tolerance. Trends Plant Sci 2001; 6(2): 66-71.
[http://dx.doi.org/10.1016/S1360-1385(00)01838-0] [PMID: 11173290]

[14] Parihar P, Singh S, Singh R, Singh VP, Prasad SM. Effect of salinity stress on plants and its tolerance strategies: a review. Environ Sci Pollut Res Int 2015; 22(6): 4056-75.
[http://dx.doi.org/10.1007/s11356-014-3739-1] [PMID: 25398215]

[15] Khan PS, Nagamallaiah GV, Rao MD, Sergeant K, Hausman JF. Abiotic stress tolerance in plants: insights from proteomics. In emerging technologies and management of crop stress tolerance. Academic Press 2014; pp. 23-68.
[http://dx.doi.org/10.1016/B978-0-12-800875-1.00002-8]

[16] Vorasoot N, Songsri P, Akkasaeng C, Jogloy S, Patanothai A. Effect of water stress on yield and agronomic characters of peanut. Songklanakarin J Sci Technol 2003; 25(3): 283-8.

[17] Kaur G, Kumar S, Nayyar H, Upadhyaya HD. Cold stress injury during the pod-filling phase in chickpea (*Cicer arietinum* L.): effects on quantitative and qualitative components of seeds. J Agron Crop Sci 194(6): 457-64.

[18] Thakur P, Kumar S, Malik JA, Berger JD, Nayyar H. Cold stress effects on reproductive development in grain crops: an overview. Environ Exp Bot 2010; 67(3): 429-43.
[http://dx.doi.org/10.1016/j.envexpbot.2009.09.004]

[19] Kumari S, Sabharwal VP, Kushwaha HR, Sopory SK, Singla-Pareek SL, Pareek A. Transcriptome map for seedling stage specific salinity stress response indicates a specific set of genes as candidate for saline tolerance in *Oryza sativa* L. Funct Integr Genomics 2009; 9(1): 109-23.
[http://dx.doi.org/10.1007/s10142-008-0088-5] [PMID: 18594887]

[20] Sharan A, Soni P, Nongpiur RC, Singla-Pareek SL, Pareek A. Mapping the 'Two-component system' network in rice. Sci Rep 2017; 7(1): 9287.
[http://dx.doi.org/10.1038/s41598-017-08076-w] [PMID: 28839155]

[21] Lakra N, Kaur C, Anwar K, Singla-Pareek SL, Pareek A. Proteomics of contrasting rice genotypes: identification of potential tar- gets for raising crops for saline environment. Plant Cell Environ 2017; 41(5): 947-69.
[http://dx.doi.org/10.1111/ pce.12946] [PMID: 28337760]

[22] Ngumbi E, Kloepper J. Bacterial-mediated drought tolerance: current and future prospects. Appl Soil Ecol 2014; 105: 109-25.
[http://dx.doi.org/10.1016/j.apsoil.2016.04.009]

[23] Turner TR, James EK, Poole PS. The plant microbiome. Genome Biol 2013; 14(6): 209.
[http://dx.doi.org/10.1186/gb-2013-14-6-209] [PMID: 23805896]

[24] Sessitsch A, Hardoim P, Döring J, *et al.* Functional characteristics of an endophyte community colonizing rice roots as revealed by metagenomic analysis. Mol Plant Microbe Interact 2012; 25(1): 28-36.
[http://dx.doi.org/10.1094/MPMI-08-11-0204] [PMID: 21970692]

[25] Singh R, Singh P, Sharma R. Microorganism as a tool of bioremediation technology for cleaning environment: A review. Proc Int Acad Ecol Environ Sci 2014; 4: 1-6.

[26] Nguyen D, Rieu I, Mariani C, van Dam NM. How plants handle multiple stresses: hormonal interactions underlying responses to abiotic stress and insect herbivory. Plant Mol Biol 2016; 91(6): 727-40.
[http://dx.doi.org/10.1007/s11103-016-0481-8] [PMID: 27095445]

[27] Di Salvatore M, Carafa AM, Carratù G. Assessment of heavy metals phytotoxicity using seed

germination and root elongation tests: a comparison of two growth substrates. Chemosphere 2008; 73(9): 1461-4.
[http://dx.doi.org/10.1016/j.chemosphere.2008.07.061] [PMID: 18768198]

[28]   Hopkins WG, Hüner N. Introduction to Plant Physiology. USA: Wiley 2004.

[29]   Agrios GN. Plant Pathology. Department of Plant Pathology. 5th eds., United States of America: University of Florida 2005.

[30]   Hossain MDB, Islam MS, Islam MR, Salam MDA, Yousuf MA. Synthesis and characterization of mixed ligand complexes of Co(II) and Fe(III) ions with maleic acid and heterocyclic amines. J Bangladesh Chem Soc 2012; 25: 139-45.
[http://dx.doi.org/10.3329/jbcs.v25i2.15066]

[31]   Hossain Z, López-Climent MF, Arbona V, Pérez-Clemente RM, Gómez-Cadenas A. Modulation of the antioxidant system in Citrus under waterlogging and subsequent drainage. J Plant Physiol 2009; 166(13): 1391-404.
[http://dx.doi.org/10.1016/j.jplph.2009.02.012] [PMID: 19362387]

[32]   Dubey D, Pandey A. Effect of nickel (Ni) on chlorophyll, lipid peroxidation and antioxidant enzymes activities in black gram (*Vigna mungo*) leaves. Int J Sci Nat 2011; 2(2): 395-401.

[33]   Anjum NA, Ahmad I, Mohmood I, *et al.* Modulation of glutathione and its related enzymes in plants' responses to toxic metals and metalloids-A review. Environ Exp Bot 2012; 75: 307-24.

[34]   Vitória AP, Lea PJ, Azevedo RA. Antioxidant enzymes responses to cadmium in radish tissues. Phytochemistry 2001; 57(5): 701-10.
[http://dx.doi.org/10.1016/S0031-9422(01)00130-3] [PMID: 11397437]

[35]   Kang KS, Lim CJ, Han TJ, Kim JC, Jin CD. Changes in the isozyme composition of antioxidant enzymes in response to amino triazole in leaves of *Arabidopsis thaliana*. J Plant Biol 1999; 42(3): 187-93.
[http://dx.doi.org/10.1007/BF03030477]

[36]   Delaunay A, Isnard AD, Toledano MB. H2O2 sensing through oxidation of the Yap1 transcription factor. EMBO J 2000; 19(19): 5157-66.
[http://dx.doi.org/10.1093/emboj/19.19.5157] [PMID: 11013218]

[37]   Mittler R. Oxidative stress, antioxidants and stress tolerance. Trends Plant Sci 2002; 7(9): 405-10.
[http://dx.doi.org/10.1016/S1360-1385(02)02312-9] [PMID: 12234732]

[38]   Gallego S, Benavides M, Tomaro M. Involvement of an antioxidant defence system in the adaptive response to heavy metal ions in *Helianthus annuus* L. cells. Plant Growth Regul 2002; 36(3): 267-73.
[http://dx.doi.org/10.1023/A:1016536319908]

[39]   Panda SK, Choudhury S. Chromium stress in plants. Braz J Plant Physiol 2005; 17: 95-102.
[http://dx.doi.org/10.1590/S1677-04202005000100008]

[40]   Skórzyńska-Polit E, Drążkiewicz M, Krupa Z. Lipid peroxidation and antioxidative response in *Arabidopsis thaliana* exposed to cadmium and copper. Acta Physiol Plant 2010; 32(1): 169-75.
[http://dx.doi.org/10.1007/s11738-009-0393-1]

[41]   Lobell DB, Field CB. Global scale climate–crop yield relationships and the impacts of recent warming. Environ Res Lett 2007; 2(7) 014002
[http://dx.doi.org/10.1088/1748-9326/2/1/014002]

[42]   Nagajyoti PC, Lee KD, Sreekanth TVM. Heavy metals, occurrence and toxicity for plants: a review. Environ Chem Lett 2010; 8: 199-216.
[http://dx.doi.org/10.1007/s10311-010-0297-8]

[43]   Reddy PP. Impacts of climate change on agriculture. Climate resilient agriculture for ensuring food security. India: Springer 2015; pp. 43-90.
[http://dx.doi.org/10.1007/978-81-322-2199-9_4]

[44]   Savvides A, Ali S, Tester M, Fotopoulos V. Chemical priming of plants against multiple abiotic stresses: mission possible? Trends Plant Sci 2016; 21(4): 329-40.
[http://dx.doi.org/10.1016/j.tplants.2015.11.003] [PMID: 26704665]

[45]   Poltronieri P, Taurino M, Bonsegna S, Domenico SD, Santino A. Nitric oxide: detection methods and possible roles during jasmonate regulated stress response. In: Khan M, Mobin M, Mohammad F, Corpas F, Eds. Nitric Oxide in Plants: Metabolism and Role in Stress Physiology. Cham: Springer 2014; pp. 127-38.
[http://dx.doi.org/10.1007/978-3-319-06710-0_7]

[46]   Asada K. Production and scavenging of reactive oxygen species in chloroplasts and their functions. Plant Physiol 2006; 141(2): 391-6.
[http://dx.doi.org/10.1104/pp.106.082040] [PMID: 16760493]

[47]   Mano J. Early events in environmental stresses in plants. Induction mechanisms of oxidative stress. In: Inze D, van Montagu M, Eds. Oxidative stress in plants. London: Taylor and Francis 2002; pp. 217-46.

[48]   Baniasadi F, Saffari VR, Moud AAM. Physiological and growth responses of *Calendula officinalis* L. plants to the interaction effects of polyamines and salt stress. Scientia Horticulturae 2018; 234: 312-7.

[49]   Lu Y, Wang Q-F, Li J, *et al.* Effects of exogenous sulfur on alleviating cadmium stress in tartary buckwheat. Scientific Reports 2019; 9(1).
[http://dx.doi.org/10.1038/s41598-019-43901-4]

[50]   Talbi S, Rojas JA, Sahrawy M, *et al.* Effect of drought on growth, photosynthesis and total antioxidant capacity of the saharan plant *Oudeneya africana*. Environmental and Experimental Botany 2020; 176: 104099.

[51]   Hu Z, Fan J, Xie Y, *et al.* Comparative photosynthetic and metabolic analyses reveal mechanism of improved cold stress tolerance in bermudagrass by exogenous melatonin. Plant Physiol Biochem 2016; 100: 94-104.
[http://dx.doi.org/10.1016/j.plaphy.2016.01.008] [PMID: 26807934]

[52]   Anjum SA, Tanveer M, Hussain S, Ashraf U, Khan I, Wang L. Alteration in growth, leaf gas exchange, and photosynthetic pigments of maize plants under combined cadmium and arsenic stress. Water Air Soil Pollut 2017; 228: 13.

[53]   Srivastava S, Sinha P, Sharma YK. Status of photosynthetic pigments, lipid peroxidation and anti-oxidative enzymes in *Vigna mungo* in presence of arsenic. J Plant Nutr 2016; 40(3): 298-306.
[http://dx.doi.org/10.1080/01904167.2016.1240189]

[54]   Guo H, Hong C, Chen X, Xu Y, Liu Y, Jiang D, *et al.* Different growth and physiological responses to cadmium of the three miscanthus species. Plos One 2016; 11(4).
[http://dx.doi.org/10.1371/journal.pone.0153475]

[55]   Thakur AK, Uphoff NT, Stoop WA. Scientific underpinnings of the system of rice intensification (SRI): what is known so far? Adv Agron 2016; 147-79.
[http://dx.doi.org/10.1016/bs.agron.2015.09.004]

[56]   Armendariz AL, Talano MA, Travaglia C, Reinoso H, Wevar Oller AL, Agostini E. Arsenic toxicity in soybean seedlings and their attenuation mechanisms. Plant Physiol Biochem 2016; 98: 119-27.
[http://dx.doi.org/10.1016/j.plaphy.2015.11.021] [PMID: 26686284]

[57]   Chen J, Wang W, Wu F, *et al.* Hydrogen sulfide enhances salt tolerance through nitric oxidemediated maintenance of ion homeostasis in barley seedling roots. Sci Rep 2015; 5: 1-19.

[58]   Corpas FJ, Palma JM, Leterrier M, del Río LA, Barroso JB. Nitric oxide and abiotic stress in higher plants. In: Hayat S, Mori M, Pichtel J, Ahmad A, Eds. Nitric Oxide in Plant Physiology. 2009; pp. 51-63.
[http://dx.doi.org/10.1002/9783527629138.ch4]

[59]   Munns R, Tester M. Mechanisms of salinity tolerance. Annu Rev Plant Biol 2008; 59: 651-81.

[http://dx.doi.org/10.1146/annurev.arplant.59.032607.092911] [PMID: 18444910]

[60] Simontacchi M, Galatro A, Ramos-Artuso F, Santa-María GE. Plant survival in a changing environment: the role of nitric oxide in plant responses to abiotic stress. Front Plant Sci 2015; 6: 977.
[http://dx.doi.org/10.3389/fpls.2015.00977] [PMID: 26617619]

[61] Zhu JK. Regulation of ion homeostasis under salt stress. Curr Opin Plant Biol 2003; 6(5): 441-5.
[http://dx.doi.org/10.1016/S1369-5266(03)00085-2] [PMID: 12972044]

[62] Siddiqui MH, Mohammad F, Khan MN, Naeem M, Khan MMA. Differential response of salt-sensitive and salt-tolerant *Brassica juncea* genotypes to N application: enhancement of N metabolism and anti-oxidative properties in the salt-tolerant type. Plant Stress 2009; 3: 55-63. b

[63] Fan HF, Du CX, Guo SR. Nitric oxide enhances salt tolerance in cucumber seedlings by regulating free polyamine content. Environ Exp Bot 2013; 86: 52-9.
[http://dx.doi.org/10.1016/j.envexpbot.2010.09.007]

[64] Siddiqui MH, Al-Whaibi MH, Basalah MO. Role of nitric oxide in tolerance of plants to abiotic stress. Protoplasma 2011; 248(3): 447-55.
[http://dx.doi.org/10.1007/s00709-010-0206-9] [PMID: 20827494]

[65] Hasegawa PM, Bressan RA, Zhu JK, Bohnert HJ. Plant cellular and molecular responses to high salinity. Annu Rev Plant Physiol Plant Mol Biol 2000; 51: 463-99.
[http://dx.doi.org/10.1146/annurev.arplant.51.1.463] [PMID: 15012199]

[66] Suzuki N, Mittler R. Reactive oxygen species and temperature stresses: a delicate balance between signaling and destruction. Physiol Plant 2006; 126: 45-51.
[http://dx.doi.org/10.1111/j.0031-9317.2005.00582.x]

[67] Oz MT, Eyidogan F, Yucel M, Öktem HA. Functional role of nitric oxide under abiotic stress conditions. In: Khan MN, Mobin M, Mohammad F, Corpas FJ, Eds. Nitric Oxide Action in Abiotic Stress Responses in Plants. Switzerland: Springer International Publishing 2005; pp. 21-41.

[68] Fahad S, Hussain S, Saud S, *et al.* A biochar application protects rice pollen from high-temperature stress. Plant Physiol Biochem 2015; 96: 281-7.
[http://dx.doi.org/10.1016/j.plaphy.2015.08.009] [PMID: 26318145]

[69] Kumar S, Kumari P, Kumar U, *et al.* Molecular approaches for designing heat tolerant wheat. J Plant Biochem Biotechnol 2013; 22: 359-71.
[http://dx.doi.org/10.1007/s13562-013-0229-3]

[70] Prasad PVV, Boote KJ, Allen LH. Adverse high temperature on pollen viability, seed-set, seed yield and harvest index of grain-sorghum [*Sorghum bicolor* (L.) Moench] are more severe at elevated carbon dioxide due to higher tissue temperature. Agric For Meteorol 2006; 139: 237-51.
[http://dx.doi.org/10.1016/j.agrformet.2006.07.003]

[71] Ashraf M, Hafeez M. Thermo tolerance of pearl millet and maize at early growth stages: growth and nutrient relations. Biol Plant 2004; 48: 81-6.
[http://dx.doi.org/10.1023/B:BIOP.0000024279.44013.61]

[72] Wahid A, Close TJ. Expression of dehydrins under heat stress and their relationship with water relations of sugarcane leaves. Biol Plant 2007; 51: 104-9.
[http://dx.doi.org/10.1007/s10535-007-0021-0]

[73] Ebrahim MK, Zingsheim O, El-Shourbagy MN, Moore PH, Komor E. Growth and sugar storage in sugarcane grown at temperature below and above optimum. J Plant Physiol 1998; 153: 593-602.
[http://dx.doi.org/10.1016/S0176-1617(98)80209-5]

[74] Xin Z, Browse J. Eskimo1 mutants of Arabidopsis are constitutively freezing-tolerant. Proc Natl Acad Sci USA 1998; 95(13): 7799-804.
[http://dx.doi.org/10.1073/pnas.95.13.7799] [PMID: 9636231]

[75] Zhao TJ, Sun S, Liu Y, *et al.* Regulating the drought-responsive element (DRE)-mediated signaling

pathway by synergic functions of trans-active and trans-inactive DRE binding factors in *Brassica napus*. J Biol Chem 2006; 281(16): 10752-9.
[http://dx.doi.org/10.1074/jbc.M510535200] [PMID: 16497677]

[76]   Fahad S, Hussain S, Saud S, *et al.* A combined application of biochar and phosphorus alleviates heat-induced adversities on physiological, agronomical and quality attributes of rice. Plant Physiol Biochem 2016; 103: 191-8.
[http://dx.doi.org/10.1016/j.plaphy.2016.03.001] [PMID: 26995314]

[77]   Maestri E, Klueva N, Perrotta C, Gulli M, Nguyen HT, Marmiroli N. Molecular genetics of heat tolerance and heat shock proteins in cereals. Plant Mol Biol 2002; 48(5-6): 667-81.
[http://dx.doi.org/10.1023/A:1014826730024] [PMID: 11999842]

[78]   Tóth SZ, Schansker G, Kissimon J, Kovács L, Garab G, Strasser RJ. Biophysical studies of photosystem II-related recovery processes after a heat pulse in barley seedlings (*Hordeum vulgare* L.). J Plant Physiol 2005; 162(2): 181-94.
[http://dx.doi.org/10.1016/j.jplph.2004.06.010] [PMID: 15779828]

[79]   Irfan M, Ahmad A, Hayat S. Effect of cadmium on the growth and antioxidant enzymes in two varieties of *Brassica juncea*. Saudi J Biol Sci 2014; 21(2): 125-31.
[http://dx.doi.org/10.1016/j.sjbs.2013.08.001] [PMID: 24600304]

[80]   Rodríguez-Serrano M, Romero-Puertas MC, Zabalza A, *et al.* Cadmium effect on oxidative metabolism of pea (*Pisum sativum* L.) roots. Imaging of reactive oxygen species and nitric oxide accumulation *in vivo*. Plant Cell Environ 2006; 29(8): 1532-44.
[http://dx.doi.org/10.1111/j.1365-3040.2006.01531.x] [PMID: 16898016]

[81]   Ortega-Villasante C, Rellán-Alvarez R, Del Campo FF, Carpena-Ruiz RO, Hernández LE, Hernandez LE. Cellular damage induced by cadmium and mercury in *Medicago sativa*. J Exp Bot 2005; 56(418): 2239-51.
[http://dx.doi.org/10.1093/jxb/eri223] [PMID: 15996984]

[82]   Hao F, Wang X, Chen J. Involvement of plasma-membrane NADPH oxidase in nickel-induced oxidative stress in roots of wheat seedlings. Plant Sci 2006; 170: 151-8.
[http://dx.doi.org/10.1016/j.plantsci.2005.08.014]

[83]   Farnese FS, Oliveira JA, Paiva EAS, *et al.* The involvement of nitric oxide in integration of plant physiological and ultrastructural adjustments in response to arsenic. Front Plant Sci 2017; 8: 516.
[http://dx.doi.org/10.3389/fpls.2017.00516] [PMID: 28469622]

[84]   Wang WX, Vinocur B, Shoseyov O, Altman A. Biotechnology of plant osmotic stress tolerance physiological and molecular considerations. In IV International Symposium on *In Vitro* Culture and Horticultural Breeding. 560: 285-92.

[85]   Shahinnia F, Le Roy J, Laborde B, *et al.* Genetic association of stomatal traits and yield in wheat grown in low rainfall environments. BMC Plant Biol 2016; 16(1): 150.
[http://dx.doi.org/10.1186/s12870-016-0838-9] [PMID: 27378125]

[86]   Hsiao TC, Xu LK. Sensitivity of growth of roots *versus* leaves to water stress: biophysical analysis and relation to water transport. J Exp Bot 2000; 51(350): 1595-616.
[http://dx.doi.org/10.1093/jexbot/51.350.1595] [PMID: 11006310]

[87]   Kumari P, Kumari R, Sharma B, Prasad SK, Singh RK. Abiotic stress response of field crops: recent approach. Int J Curr Microbiol Appl Sci 2019; 8(4): 1761-9.
[http://dx.doi.org/10.20546/ijcmas.2019.804.205]

[88]   Zeid IM, Shedeed ZA. Response of alfalfa to putrescine treatment under drought stress. Biol Plant 2006; 50(4): 635.
[http://dx.doi.org/10.1007/s10535-006-0099-9]

[89]   Samarah NH. Effects of drought stress on growth and yield of barley. Agron Sustain Dev 2005; 25: 145-9.

[http://dx.doi.org/10.1051/agro:2004064]

[90]　Cattivelli L, Rizza F, Badeck FW, *et al.* Drought tolerance improvement in crop plants: an integrated view from breeding to genomics. Field Crops Res 2008; 105(1-2): 1-14.
[http://dx.doi.org/10.1016/j.fcr.2007.07.004]

[91]　Egilla JN, Davies FT, Boutton TW. Drought stress influences leaf water content, photosynthesis, and water-use efficiency of Hibiscus rosa-sinensis at three potassium concentrations. Photosynthetica 2005; 43(1): 135-40.
[http://dx.doi.org/10.1007/s11099-005-5140-2]

[92]　Lindhauer MG. Influence of K nutrition and drought on water relations and growth of sunflower (*Helianthus annuus* L.). Zeitschrift für Pflanzenernährung und Bodenkunde 1985; 148(6): 654-69.
[http://dx.doi.org/10.1002/jpln.19851480608]

[93]　Wahid A, Rasul E. Photosynthesis in leaf, stem, flower, and fruit. In: Pessaraki M, Ed. Handbook of photosynthesis. 2005; pp. 479-97.

[94]　Zhou Y, Lam HM, Zhang J. Inhibition of photosynthesis and energy dissipation induced by water and high light stresses in rice. J Exp Bot 2007; 58(5): 1207-17.
[http://dx.doi.org/10.1093/jxb/erl291] [PMID: 17283375]

[95]　Liu HS, Li FM. Root respiration, photosynthesis and grain yield of two spring wheat in response to soil drying. Plant Growth Regul 2005; 46(3): 233-40.
[http://dx.doi.org/10.1007/s10725-005-8806-7]

[96]　Khan A, Tan DKY, Afridi MZ, *et al.* Nitrogen fertility and abiotic stresses management in cotton crop: a review. Environ Sci Pollut Res Int 2017; 24(17): 14551-66.
[http://dx.doi.org/10.1007/s11356-017-8920-x] [PMID: 28434155]

[97]　Luo HH, Zhang YL, Zhang WF. Effects of water stress and rewatering on photosynthesis, root activity, and yield of cotton with drip irrigation under mulch. Photosynthetica 2016; 54(1): 65-73.
[http://dx.doi.org/10.1007/s11099-015-0165-7]

[98]　Hirayama T, Shinozaki K. Research on plant abiotic stress responses in the post-genome era: past, present and future. Plant J 2010; 61(6): 1041-52.
[http://dx.doi.org/10.1111/j.1365-313X.2010.04124.x] [PMID: 20409277]

[99]　Boursiac Y, Chen S, Luu DT, Sorieul M, van den Dries N, Maurel C. Early effects of salinity on water transport in Arabidopsis roots. Molecular and cellular features of aquaporin expression. Plant Physiol 2005; 139(2): 790-805.
[http://dx.doi.org/10.1104/pp.105.065029] [PMID: 16183846]

[100]　Giorno F, Wolters-Arts M, Mariani C, Rieu I. Ensuring reproduction at high temperatures: the heat stress response during anther and pollen development. Plants (Basel) 2013; 2(3): 489-506.
[http://dx.doi.org/10.3390/plants2030489] [PMID: 27137389]

[101]　Lata C, Prasad M. Role of DREBs in regulation of abiotic stress responses in plants. J Exp Bot 2011; 62(14): 4731-48.
[http://dx.doi.org/10.1093/jxb/err210] [PMID: 21737415]

[102]　Kavar T, Maras M, Kidrič M, Šuštar-Vozlič J, Meglič V. Identification of genes involved in the response of leaves of *Phaseolus vulgaris* to drought stress. Mol Breed 2008; 21(2): 159-72.
[http://dx.doi.org/10.1007/s11032-007-9116-8]

[103]　Wilkinson S, Davies WJ. ABA-based chemical signalling: the co-ordination of responses to stress in plants. Plant Cell Environ 2002; 25(2): 195-210.
[http://dx.doi.org/10.1046/j.0016-8025.2001.00824.x] [PMID: 11841663]

[104]　Zhang H, Khan A, Tan DKY, Luo H. Rational water and nitrogen management improves root growth, increases yield and maintains water use efficiency of cotton under mulch drip irrigation. Front Plant Sci 2017; 8: 912.
[http://dx.doi.org/10.3389/fpls.2017.00912] [PMID: 28611817]

[105]   Pettigrew WT. Physiological consequences of moisture deficit stress in cotton. Crop Sci 2004; 44(4): 1265-72.
[http://dx.doi.org/10.2135/cropsci2004.1265]

[106]   Abid M, Tian Z, Ata-Ul-Karim ST, *et al.* Adaptation to and recovery from drought stress at vegetative stages in wheat (*Triticum aestivum*) cultivars. Funct Plant Biol 2016; 43(12): 1159-69.
[http://dx.doi.org/10.1071/FP16150] [PMID: 32480535]

[107]   Zandalinas SI, Sales C, Beltrán J, Gómez-Cadenas A, Arbona V. Activation of secondary metabolism in citrus plants is associated to sensitivity to combined drought and high temperatures. Front Plant Sci 2017; 7: 1954.
[http://dx.doi.org/10.3389/fpls.2016.01954] [PMID: 28119698]

[108]   Kozłowska M, Rybus-Zając M, Stachowiak J, Janowska B. Changes in carbohydrate contents of Zantedeschia leaves under gibberellin-stimulated flowering. Acta Physiol Plant 2007; 29(1): 27-32.
[http://dx.doi.org/10.1007/s11738-006-0004-3]

[109]   Poorter H, Niinemets U, Poorter L, Wright IJ, Villar R. Causes and consequences of variation in leaf mass per area (LMA): a meta-analysis. New Phytol 2009; 182(3): 565-88.
[http://dx.doi.org/10.1111/j.1469-8137.2009.02830.x] [PMID: 19434804]

[110]   Dordas C, Rivoal J, Hill RD. Plant haemoglobins, nitric oxide and hypoxic stress. Ann Bot 2003; 91(Spec No): 173-8.
[http://dx.doi.org/10.1093/aob/mcf115] [PMID: 12509338]

[111]   Voesenek LACJ, Bailey-Serres J. Flood adaptive traits and processes: an overview. New Phytol 2015; 206(1): 57-73.
[http://dx.doi.org/10.1111/nph.13209] [PMID: 25580769]

[112]   Vriezen WH, Zhou Z, Van Der Straeten D. Regulation of submergence-induced enhanced shoot elongation in *Oryza sativa* L. Ann Bot 2003; 91(Spec No): 263-70.
[http://dx.doi.org/10.1093/aob/mcf121] [PMID: 12509346]

[113]   Yang H, Sheng R, Zhang Z, Wang L, Wang Q, Wei W. Responses of nitrifying and denitrifying bacteria to flooding-drying cycles in flooded rice soil. Appl Soil Ecol 2016; 103: 101-9.
[http://dx.doi.org/10.1016/j.apsoil.2016.03.008]

[114]   Bange MP, Baker JT, Bauer PJ, *et al.* Climate change and cotton production in modern farming systems. CABI 2016.
[http://dx.doi.org/10.1079/9781780648903.0000]

[115]   Armstrong W, Woolhouse HWW. Adv Bot Res 1979.

CHAPTER 2

Drought and its Effects on Crop Physiology

Samiksha[1], **Sandeep Kaur**[2,3], **Drishtant Singh**[4], **Satwinderjeet Kaur**[2] and **Satwinder Kaur Sohal**[1,*]

[1] *Department of Zoology, Guru Nanak Dev University, Amritsar, Punjab 143005, India*

[2] *Department of Botanical and Environmental Sciences, Guru Nanak Dev University, Amritsar, Punjab 143005, India*

[3] *PG Department of Botany, Khalsa College, Amritsar, Punjab 143005, India*

[4] *Department of Molecular Biology and Biochemistry, Guru Nanak Dev University, Amritsar, Punjab 143005, India*

Abstract: During natural conditions, plants undergo a series of biogenic stressors that are devastating to plant growth. Among these, drought stress is the most common, which alters molecular and morphological parameters in plants and thus has detrimental effects due to environmental injury and physico-chemical disturbances. These have led to the development of technologies that increase the quality and sustainability of crops under deteriorating soil, declining natural resources, and environmental stress. Effective agronomic and genetic methods for crop protection provide best management practices to combat drought conditions. This chapter aims to contribute to the development of approaches for sustainable agricultural management practices suitable for crop production during drought stress.

Keywords: Crop production, Drought stress, Plant responses, Stress management.

INTRODUCTION

Climate change disruptions are the primary concern in many developing nations due to increased vulnerability and limited capacity to manage the adverse effects of climate change on efficient crop production [1]. For economic growth, countries are mainly focused on farming whilst the agricultural industry is entirely dependent on nature [2]. Drought stress is the most critical factor which restricts crop growth and quality [3, 4]. Demand for food, fiber, and various other commodities has strained our capacity to grow high-quality agricultural products. Drought stress further compounds this problem by adding detrimental effects on crop quality in addition to the broader impact of drought on crop production, pri-

* **Corresponding Author Satwinder Kaur Sohal:** Department of Zoology, Guru Nanak Dev University, Amritsar, Punjab 143005, India; Tel: +91-9914451077; E-mail address: satudhillon63@gmail.com

Tajinder Kaur & Saroj Arora (Eds.)

marily resting on soil moisture level, accessibility of nutrients [5], and crop varieties [6].

Various studies regarding drought as an extreme stress factor have been extensively reported. A previous study depicted that the biochemical and molecular approach to meet the environmental stress in plants is specific, which cannot be inferred directly [7]. Technologically, additional advanced solutions are still required to maintain current crop production and resolve the problems of food security, growing population pressure, and declining resource base in response to climate change. Being a rich source of various nutrients like minerals, vitamins and proteins, these crops can play a vital role in escaping the global nutrition production gap. Nonetheless, there must be an effective use of available germplasm and technologies to sustain production at an increased risk of drought stress. The present chapter deals with the impact of drought stress on crop growth, the assimilation of nutrients and water content. Crop enhancement and drought tolerance management solutions are also discussed.

PLANT RESPONSES TO DROUGHT STRESS

Drought stress causes negative consequences on crop development in different stages of crops, which rely heavily on the extent of stress and its duration, as well as the phase of plant growth. These effects are depicted by changes in plant morphology, physiology, biochemistry, and molecular processes.

Morphological and Yield Responses

During germination or initial seedling growth, some field plants are highly susceptible to drought. Water scarcity disrupts seedling growth by reducing the imbibition of water and seedling capacity [8]. Limited water supply leads to osmotic disruptions, excess reactive oxygen species (ROS) production, and impaired cellular metabolism, resulting in changes in the DNA and protein structure, membrane disruption, diminished production of ATPs, and respiration [9], which ultimately results in loss of seed germination. About a 25% decline in plant height during water-stressed conditions was confirmed in the study conducted on citrus seedlings by Wu *et al.* [10]. Adverse effects on *Abelmoschus esculentus* stem length have also been identified. Drought distress diminished leaf growth in a variety of species of plants [8, 11]. Multi-specific differences within two *Populus* species during drought stress have also been observed by Wullschleger *et al.* [12].

Drought stress also caused a decline in the plant height of various field crops [13]. Deficiency in soil water affects the root development but had less impact on aerial areas resulting in the incline of root to shoot ratio. Different effects of drought stress have been reported on root growth, for instance, the length of the root of *Catharanthus roseus* and sunflower increased [14]; however, the increase in wheat and maize root growth did not change [15]. A rise in the root length was due to increased abscisic acid levels in roots under drought conditions [16]. Under dry conditions and high humidity, the developed roots are generally fibrous and strong [17]. The development of seedlings can be affected by water shortage, which can cause a reduction of about 32% in wheat production. However, in dry conditions, the root size and root/shoot ratio increased in wheat. The loss of root to coleoptile water supply during drought stress prevents the development of coleoptiles [18].

Yield and Related Attributes

Drought stress causes a reduction in development and productivity, which ultimately decreases agronomic and crop yield attributes. In vegetables, water deficiency considerably lowered seed efficiency by restricting the growth of pods, increasing abortion of pods, and reducing seed size [19]. The number of branches and the overall seed yield considerably declined in soybean under drought stress [20]. Drought, in turn, has a detrimental effect on grain yields and crop yields. The decline in different characteristics of crop plants might be the result of stress-induced changes in biochemical and physiological practices and equivalent adverse consequences for plant reproductive organs with reduced productivity. In general, drought damages the plants; however, the seed development stage is perhaps more vulnerable than pre and post-reproductive phases to such stress factors.

Physiological and Biochemical Responses

Nutrient Relations

Besides diminishing plant growth and efficiency, plant nutrient uptake activity is also hampered during drought stress [19]. It restricts nutrient absorption and translocation in plants [21]. The low availability of soil moisture under drought stress diminished root growth and nutrient inflow [22]. Drought stress also restricts the nutrient assimilation function of enzymes. For example, during drought stress, the nitrate reductase activities in the Dhainicha nodules and *Phaseolus vulgaris* L. were considerably reduced [19]. To sum up, there is a major relationship between the acquisition of nutrients, soil moisture content, and soil temperature. Stress from dryness reduces plant nutrient supply, translocation, uptake, and metabolism.

Photosynthesis and Gaseous Exchange

Drought stress affects photosynthesis and other characteristics of gas exchange in crop plants like crop yields. It hampers the process of photosynthesis due to the closing of the stomata. In *Brassica* cultivars, management of moisture stress decreased photosynthesis, intercellular carbon dioxide, and chlorophyll content [23]. The weight of wheat grain, a component of the period of grain growth, was influenced by photosynthetic supplies [24]. The biochemical response of short-term water deficits was analyzed for water connections and the aggregation of metabolites in the CSH-14 sorghum hybrid at anthesis and grain filling stages. Arising from this, it was noticed that stomatal conductance declined considerably during stress, but that rewatering in part relieved the effect imparting a higher sensitivity to water stress [25]. Thylakoid membranes and photosynthetic pigments are destroyed by drought stress [26]. A decrease in chlorophyll concentration during water restrictions was also documented [27]. The concentration of chlorophyll 'a' was higher relative to chlorophyll 'b' in drought-resistant plants [28]. Decreased chlorophyll 'a' / 'b' proportion was recorded in *Brassica* species during droughts [29]. Restricted CO_2 concentration due to drought stress limits the photosynthetic activity, leading to the production of reactive oxygen species (ROS) that are deleterious to photosynthetic machinery [30]. Water scarcity enables the cell to contract due to changes in cellular volume as a result by which the cell content becomes more gelatinous, leading to protein denaturation [31]. The high concentration of cytoplasmic solutes can also lead to ion toxicity having a massive effect on the function of enzymes associated with photosynthesis as well as other plant activities [31]. Reduced phosphorylation and disrupted ATP synthesis have been identified as significant factors restricting photosynthesis during moderate drought [32]. Drought stress suppresses the photochemical productivity of the PS II photosystem by reducing electron transport, eliminating external proteins, and releasing calcium and magnesium ions [33]. Sales *et al.* [34] studied sugarcane and demonstrated that the photochemical activity, PSII energy pressure production, CO_2 assimilation, leaf gas exchange, transpiration, and electron transport rate were adversely affected by drought stress. Drought stress decreases the fixation rate of carbon by reducing the absorption of CO_2 into the leaf [19]. Indeed extreme drought conditions due to the reduced Rubisco activity limit photosynthesis [35]. The behavior of the photosynthetic electron transport chain is precisely balanced to the existence of CO_2 in chloroplasts and changes in photosystem II throughout drought conditions. Drought stress induced changes in the ratio of chlorophyll 'a' and 'b' and carotenoids [8]. Chlorophyll concentration was considered an important indicator in the screening of germplasm for drought tolerance [25].

Water Status and Osmotic Balance

The essential attributes defining the relationship between plant and water are leaf water potential, stomatal conductivity, transpiration rate, and relative water quantity and canopy temperature [19]. Considerable reductions in the leaf water capacity of sugarcane occur during drought stress [34]. Drought stress exposure has also been reported to drastically reduce the relative turgor pressure, leaf perspiration rate and water capacity [36]. Therefore we can say that drought stress induces disruption in the relationships of plant water. Water stress reduces the relative water content and high relative water content is a dry-resistant mechanism that is linked to osmotic control but the less tensile strength of the tissue cell wall [33]. Osmotic adaptation is the phenomenon of aggregation of soluble compounds in reaction to water stress by retaining the turgor pressure in the tissues. Another significant aspect of plant metabolic control is the quality of water usage, and that is the ratio of the stored dry matter to the water absorbed [37]. During drought stress wheat cultivars generally have higher water quality [38].

Osmotic change is the decline in osmotic capacity in reference to water deficiency conditions due to the cumulative accumulation of solutes [39]. By absorbing different osmotically activated molecules like soluble sugars, proline, potassium, and calcium, plants maintain water connections under osmotic stress. Osmoregulation enables plants to maintain photosynthetic rate, growth partitioning and assimilation rate in stressful environments [40]. Cotton cultivars with high concentrations of endogenous glycine betaine were better suited to the circumstances of drought than those that produced less of it [41].

Assimilate Partitioning

Drought disrupts the equilibrium of assimilates as so many of them are rooted in order to enhance the absorption of water [42]. Export of assimilates from origin to sink typically relies on the degree of photosynthesis and the amount of sucrose in leaves [43]. Water shortage hinders the mechanism of photosynthesis and limits the sucrose amount, which eventually decreases the export frequency through the source to sink [44]. Drought also restricts the capacity of the sink to use received assimilates adequately [44]. In addition, the operation of acid invertase is adversely affected which interferes with the loading and unloading of phloem.

Oxidative Status

Owing to lower carbon dioxide fixation and accelerated photorespiration, numerous studies have shown increased ROS formation and oxidative damage owing to drought stress [19]. The discrepancy between generation and detoxification of ROS leads to the metabolic condition known as oxidative stress

[45]. The main ROS produced in plants under unfavorable environmental conditions are singlet oxygen, hydrogen peroxide, hydroxyl radicals and superoxide radicals [46]. Lipid and protein peroxidation in pea increases 4 times during drought stress relative to standard conditions [47]. ROS is primarily generated in chloroplasty [48], but the response of oxygen to elements of the electron transport chain also leads to the production of ROS [49]. The pathways involved in the production of ROS can be enzymatic or non-enzymatic [46]. Wahid *et al*. [50] reported that ROS production also takes place during high temperature stress. Plants have developed excellent antioxidant defense mechanisms under adverse conditions such as drought stress to cope with oxidative damage. Increased levels of non-enzymatic and several antioxidants (SOD, POD, GR, APX, CAT, *etc.*) have been documented in plants to stabilize cell homeostasis and minimize oxidative harm [51]. Sustaining increased rates of anti-oxidants can thus be a successful strategy for plants to combat the adverse effects of ROS [52]. Phytohormones too are powerful protective molecules in plants that retain greater levels of antioxidants during stress. They help plants to acclimate to different situations *via* growth, nutrient distribution and source or sink transitions [53].

MANAGEMENT STRATEGIES

Drought stress has negative impacts on plant growth and productivity [19]. To deal with this stress, plants frequently develop a sequence of strategies *via* biological and molecular level alterations. Reports had demonstrated that these modifications can be monitored by cultivating tolerant plant genotypes, transgenic organisms, plant growth regulators, seed treatments, and plant mineral nutrient utilization.

Selection and Breeding Strategies

Quantifying the total yield of grains during drought stress environment is a quick and easy process for screening sensitive crops. Amid water deficit constraints, favorable tolerant genotypes can be detected by direct screening of grain yield [54]. The traditional breeding strategies involve the cultivation of crop plants in drought stress conditions for their yield and related attributes [55]. Generally, breeding techniques were used to exploit the genetic composition of crops to improve their drought tolerance [56]. Recurrent selection aided by markers requires the aggregation of more superior alleles responsible for drought tolerance in plants [57]. Kumar *et al*. [58] stated that several earlier studies on tolerance of drought stress did not inflict extreme stress, so real and precise drought tolerant lines were not chosen. While traditional breeding strategies have already been utilized for drought tolerance, advances in genomics and molecular breeding approaches would nonetheless have a considerable role to play in the production of drought tolerant cultivars [59]. In addition, the selection of crop cultivars may be useful for breeding programs to grow stress-tolerant plant types for their specific responses to drought stress.

Molecular and Genomics Approaches

The drought susceptibility response is attributed to the control of numerous gene expression caused by drought stress [60]. Drought-tolerant plants produced by various biotechnological approaches require a comprehensive understanding of the factual basis for drought tolerance [61]. Therefore, it is crucial to identify transcription factors involved in drought tolerance. In *Arabidopsis,* an active variant of DREB2 has been observed to activate stress-related genes that enhanced tolerance for drought stress [62]. Under water deficit conditions involving drought stress resistance, differential contributions of two contrasting barley genotypes were identified in low-molecular dehydrins [63]. Bruce *et al.* [64] explained the need for advancing the genomic related methods and tools to classify the primary genes associated in plant tolerance for drought stress control. Although, drought tolerance is a physiologically controlled concept, it is important to consider QTLs for membrane flexibility as well as other related phenologicalgenes [65].

Agronomic and Physiological Measures

Plants react to drought stress through a set of physiological and biochemical mechanisms. Physiological implementation of seed priming, compatible solutes, use of selenium, crop mineral nutrients, and growth hormones may be beneficial in mitigating the negative impacts of drought stress.

Application of Compatible Solutes

At higher amounts, appropriate solutes prevent plants from experiencing oxidative stress whilst not having damaging effects on enzymes, membranes or other macromolecules [66]. Compatible solutes include soluble sugars, sugar alcohols, and some organic acids. Aggregation of solutes, such as trehalose, betaines, glycine betaine, proline and sugar alcohol defends plants from drought stress [66]. Several plants were genetically altered to express enzymes that help to facilitate the formulation of numerous compatible solutes. Several approaches have expanded the tolerance of certain crops to abiotic stress. For numerous plant species in drought tolerance, foliar implementation of glycine betaine was found to be beneficial [19].

Application of Selenium

The administration of selenium has many positive benefits on the mechanism of growth and drought tolerance by enhancing the antioxidant activities of plants [66]. In recent decades, research has shown that the use of selenium not only enhances plant growth, but also increases plant drought tolerance and antioxidant capacity. The antioxidant and physiological properties of Se have piqued the attention of many biological agents and farmers over the past few decades. Although it never participates in any biochemical phase of plants, it can only minimize the harm to physiological stressors. Under drought environment, the reports on the position of Se are very limited. It can regulate the water status in plants and boost biomass production through activation of antioxidant devices in crops and water stressors.

Plant Growth Regulator Application

Many agricultural and metabolic methods are used to reduce the negative impacts of drought stress and to promote tolerance to drought stress in plants. The use of plant hormones is among the most promising and realistic approaches for increasing crop production during stressful conditions [67]. Research has shown that exogenous growth promoters can increase responsiveness in plants to specific abiotic stressors like drought, heavy metal stress and salt stress [9]. Plant growth promoters have such a complex function to play in raising plant potential to acclimate to abiotic stress circumstances. In response to abiotic stress, auxins (IAA), cytokinins (Cks), abscisic acid (ABA), ethylene, gibberellins (GA3), jasmonates (JA), brassinosteroids (BRs) and salicylic acid (SA) has been widely observed in plants. Plants usually synthesize different kinds of phytohormones through individual and concurrent abiotic stress based on the effectiveness of their defense mechanism [68].

During drought stress, ABA plays a crucial role in plant communication between root and shoot, resulting in a water-saving antiperspirant reaction, stomatal closure and restricted leaf spread [69]. During drought stress, ABA is engaged in root development as well as in anatomical changes [70]. Amid drought stress, enhanced storage of ABA reduces the Cks amount thus activating the ABA / Cks proportion in plants [71]. During water stress circumstances, plant growth regulators substantially improved chlorophyll concentration and increased soya water capacity [11]. Reguera *et al.* [72] reported that the tolerance of drought in rice was concordant with the production of Cks. Cks-induced dry tolerance has also been found in tobacco and wheat [19]. Kaya *et al.* [73] documented the use of GA3 in maize could also improve the tolerance for drought. Physiological GA3 implementation splits the rosette by rapidly spreading the segregated tissues during cold stress [74]. Salicylic acid is also known to be a powerful chemical compound for drought and cooling resistance in many plant species [8, 75].

Non-hormonal growth regulators, *e.g.* triazole, paclobutrazole, chlorocholin-chloride, meflidide, and unikonazole [76, 77] are used to improve drought and cooling tolerance in crop plants [77, 78]. Still and Pill [79] recorded that the foliar formulation of paclobutrazol appeared successful in enhancing the tolerance of drought in tomato seedlings. Such findings indicate that the exogenous implementation of numerous plant growth regulators is an effective tool for enhancing plant drought and cooling tolerance. Future work will, indeed, concentrate on the production of genetically modified or transgenic plants capable of producing different hormones using biotechnological and molecular approaches. Such transgenic plants should therefore be able to grow successfully in challenging conditions.

Use of Mineral Nutrients

Mineral nutrients play a significant role in enhancing plant resistance to drought stress [80]. In order to minimize the detrimental effects of exogenous stress, nitrogen (N) fertilization plays a major role [81]. Nitric oxide (NO) serves as a scavenger to ROS and prevents the plant from stress [78, 82]. Phosphorus (P) allows plants to retain leaf water capacity that in turn increases stomatological conductance and photosynthetic levels [19]. During drought, supplementation of potassium (K) and its absorption in plant parts may be effective for the achievement of good yields [83]. Pandey *et al.* [84] recorded K-induced rises in free sugar and proline in rice through water deficit circumstances. Kafkafi [85] stated that low-temperature injuries can be prevented incarnation plants by supplementing K in irrigation water.

Seed Priming

Crop priming is known to increase the resistance to drought stress in several plant species [86, 87]. Crop priming facilitates germination-related homeostatic mechanisms despite radical protrusion from crop [86, 88]. It is a good method for acclimatization toward stress circumstances [87] by attenuating physio-biochemical reactions in seedling growth [89]. Seed-priming with plant hormones and other substances can help in mitigating the negative impacts of drought stress at earlier stages of development. Crop priming with SA enhanced water and chlorophyll content, antioxidant activity, plant biomass and wheat yield while selenium priming regulates enzymatic and non-enzymatic antioxidant levels in rapeseed during drought stress [90].

Different methods comprise chemical priming, solid matrix priming, osmopriming, hormonal priming, hydropriming, redox priming, *etc*. Though priming enhances the rate and homogeneity of seedling development and growth, particularly during stress conditions [91], the efficacy of various priming factors differs with various strains with different crop species [92].

Hydropriming has been documented to be an easy, inexpensive and effective approach to increase seed potential for osmotic modification, to enhance seedling settlement and crop development during stressed conditions [93]. In this process, seeds are submerged in sterile water maintained at an acceptable temperature and the length of hydropriming is defined by the regulation of seed imbibition under germination [73]. Janmohammadi *et al*. [94] proposed hydropriming as an effective, inexpensive seed rejuvenation method for maize inbreeding lines, particularly where germination is disrupted by salinity and drought stress.

Osmoconditioning is the immersion of seeds in aerated, minimal-water systems. Osmopriming effectively introduces seeds to a lower environmental water capacity to limit imbibition rate and degree. The cycle of osmopriming is similar to extended fast imbibition of seeds which triggers a steady progression of numerous pre-germinative biochemical processes. It is therefore beneficial to use osmopriming as a method to analyze the transformation of seeds from either a dried or biologically calm to a hydrated and physiologic effective state [95]. Osmopriming is scientifically, practically and economically very accurate compared to hydropriming [96] since osmotic seed priming provides better and simpler outcomes and is much less costly than other water management approaches and gives farmers an extremely desirable option for increasing crop production and yield [97].

Nutrient priming is an innovative approach combining the beneficial properties of seed priming with enhanced nutrient availability [98]. For nutrient priming, seeds are primed with solutions addressing nutrient limitations rather than only immersed in water [99]. Of the mineral nutrients, K performs a special role in leading to the sustainability of crops during environmental stress [100]. The priming of seed in Zn^{2+} solution increases the production of lentil and wheat grain [101]. Ascorbic acid, an essential vitamin, is used for priming because of its antioxidant function. It was already demonstrated that a high amount of endogenous ascorbate is necessary to preserve the antioxidant ability that defends plants against oxidative stress [102]. Seed production of different crops could be enhanced through the use of plant growth promoters and hormones while priming and with other pre-seeding practices [103]. Abscisic acid (ABA) is widely engaged in the reaction to abiotic stressors like drought, cold temperatures and oxidative stress [104]. Expanded replication of hormone and vitamin priming in root tips has also been recorded [105]. Accelerated cell division within the anterior meristem of the seed root may be attributed to the introduction of development regulators to the priming solution, which induced a rise in plant growth. Pre-treatment of seeds and foliar spraying of seedlings at early phases in mustard and wheat with various thiol substances has been documented to enhance stress tolerance and, most notably, crop efficiency [106, 107].

Thiourea management of *Brassica juncea* seeds was observed to be beneficial in preserving the integrity and activity of mitochondria in seeds [108]. Plants may develop resistance to abiotic stress following treatment with many naturally derived substances like selenium, butenolide, CuSO4, KH2PO4, ZnSO4, ethanol, paclobutrazole, putrescine, and chitosan [92]. In *Zea mays*, zinc and copper sulphate seed priming improved the germination of caryopses and growth of the seed by 43 and 29 percent, respectively [97]. KH_2PO_4 seed priming showed strong ability to improve germination, development and production of wheat [109]. Priming treatment with 0.5 percent K_2HPO_4 and KNO_3 resulted in better performance in comparison to the high concentrations of such compounds [110].

Solid or semi-solid medium is often used in solid matrix priming (SMP) as an alternative to the liquid medium [111]. This approach is achieved by combining seeds with solid or semi-solid material and a defined volume of water. Throughout the SMP, the seeds are gradually supplied with water and therefore delayed or regulated imbibition proceeds, enabling the compensatory mechanisms to work. Frequently used solid matrices comprise moisturized vermiculite, extended calcined clay, bituminous soft carbon, Agro-lig, sodium polypropionate gel [112].

Priming of seeds has been used to enhance germination and the proliferation of many plants, and seed-priming approaches were discovered to be successful for improved germination and planting during controlled circumstances [113]. The drawbacks of seed priming would be smaller than those of the benefits in contrast. There are few records of the negative effects of seed priming.

Proteins as Molecular Factors to Regulate Drought Stress

Plants have developed a complex response to ecological stress over the long term of evolution, which includes multiple regulatory networks and interaction at different levels. When drought occurs, regulation of various proteins can be controlled by either dependent or autonomous means to promote plant growth to respond to drought conditions. Also with advancement of high-throughput sequencing and transcriptome analysis, protein interactions have been widely investigated in recent years. In the promoter of transcription factors that confer drought tolerance, such as ABRE and MYB active site in ZmWRKY40, drought-response elements commonly occur [114]. The promoter of different drought genes, the DRE / CRT factor is widely distributed and is the linking motif of DREB / CBF TFs, which is the target for ABRE in the transcription of the ABA dependent gene, demonstrated by the protein interactions between AREB / ABF and DREB1A / CBF3, DREB2A, and DREB2C [115, 116]. ZmWRKY58 has been shown to interact with ZmCaM2 [117], which is an essential Ca^{2+} sensitive protein and plays a major role in the response to abiotic stress [118, 119], indicating that there is cross-talk in drought tolerance of WRKY TF and Ca^{2+} signalling. In plant drought tolerance, indirect control between TFs through the modulation of the signalling transduction or metabolism pathway may also play an important role. ABI4 encodes a key ABA-responsive regulator, and the induced expression of ZmDREB2.3 / ZmABI4 was strongly correlated with maize tolerance to drought [120]. Some of the proteins/transcription factors involved in drought stress are the basic region/leucine zipper (bZIP) transcription factors, APETALA2/Ethylene Response Element Binding Factors, NAC (NAM, ATAF1/2, CUC2) transcription factors, WRKY, The nuclear factor Y, *etc.* [120].

CONCLUSION

Drought stress significantly hinders crop growth and production from germination to maturity. The decline in growth is due to decreased cell division and length due to reduced turgor. Germination is reduced due to lower water capacity and imbibition. In addition, specific growth indexes like plant height and number of leaves also decrease, leading to a reduction in photosynthesis and dry matter aggregation during drought suggesting a stomatological closure due to drought stress. Plants often experience physiological alterations when they are subjected to drought stress. Selection and recognition of the effects of drought on anatomical characteristics and biochemical modifications arising in exposure to drought can be beneficial in the choice and breeding of genotypes that are resistant to drought.

CONSENT FOR PUBLICATION

Not applicable.

CONFLICT OF INTEREST

The authors declare that they do not have any conflict of interest.

ACKNOWLEDGEMENTS

The grant received from UGC New Delhi under the scheme "University with Potential for Excellence (UPE)" and SAP programme for conducting the research work is gratefully acknowledged.

REFERENCES

[1] Ali F, Bano A, Fazal A. Recent methods of drought stress tolerance in plants. Plant Growth Regul 2017; 82(3): 363-75.
[http://dx.doi.org/10.1007/s10725-017-0267-2]

[2] Mendelsohn R. The impact of climate change on agriculture in Asia. J Integr Agric 2014; 13(4): 660-5.
[http://dx.doi.org/10.1016/S2095-3119(13)60701-7]

[3] Jahan MA, Hossain A, Jaime A, *et al.* Effect of naphthaleneacetic acid on root and plant growth and yield of ten irrigated wheat genotypes. Pak J Bot 2019; 51(2): 451-9.
[http://dx.doi.org/10.30848/PJB2019-2(11)]

[4] Sabagh EL, Hossain A, Islam A, *et al.* Salinity stress management for sustainable soybean production using foliar application of compatible antioxidants and soil application of organic fertilizers: a critical review. Aust J Crop Sci 2019; 13(02): 228-36.
[http://dx.doi.org/10.21475/ajcs.19.13.02.p1285]

[5] Gandah M, Bouma J, Brouwer J, Hiernaux P, Van Duivenbooden N. Strategies to optimize allocation of limited nutrients to sandy soils of the Sahel: a case study from Niger, West Africa. Agric Ecosyst Environ 2003; 94(3): 311-9.
[http://dx.doi.org/10.1016/S0167-8809(02)00035-X]

[6] Cooper PJ, Gregory PJ, Keatinge JD, Brown SC. Effects of fertilizer, variety and location on barley production under rainfed conditions in Northern Syria 2. Soil water dynamics and crop water use. Field Crops Res 1987; 16(1): 67-84.
[http://dx.doi.org/10.1016/0378-4290(87)90054-2]

[7] Mittler R. Abiotic stress, the field environment and stress combination. Trends Plant Sci 2006; 11(1): 15-9.
[http://dx.doi.org/10.1016/j.tplants.2005.11.002] [PMID: 16359910]

[8] Farooq M, Wahid A, Kobayashi N, Fujita D, Basra SMA. Plant drought stress: effects, mechanisms and management. Agron Sustain Dev 2009; 29: 185-212.
[http://dx.doi.org/10.1051/agro:2008021]

[9] Priestley DA. Seed Aging: Implications for Seed Storage and Persistence in the Soil. USA: Comstock Associates 1986; p. 304.

[10] Wu QS, Xia RX, Zou YN. Improved soil structure and citrus growth after inoculation with three arbuscular mycorrhizal fungi under drought stress. Eur J Soil Biol 2008; 44(1): 122-8.
[http://dx.doi.org/10.1016/j.ejsobi.2007.10.001]

[11] Zhang M, Duan L, Zhai Z, *et al.* Effects of plant growth regulators on water deficit-induced yield loss in soybean. In: Proceedings of the 4th international crop science congress. Brisbane, Australia. 2004; pp. 252-6.

[12] Wullschleger SD, Yin TM, DiFazio SP, *et al.* Phenotypic variation in growth and biomass distribution for two advanced-generation pedigrees of hybrid poplar. Can J For Res 2005; 35(8): 1779-89.
[http://dx.doi.org/10.1139/x05-101]

[13] Zheng M, Tao Y, Hussain S, *et al.* Seed priming in dry direct-seeded rice: consequences for emergence, seedling growth and associated metabolic events under drought stress. Plant Growth Regul 2016; 78(2): 167-78.
[http://dx.doi.org/10.1007/s10725-015-0083-5]

[14] Jaleel CA, Gopi R, Sankar B, Gomathinayagam M, Panneerselvam R. Differential responses in water use efficiency in two varieties of *Catharanthus roseus* under drought stress. C R Biol 2008; 331(1): 42-7.
[http://dx.doi.org/10.1016/j.crvi.2007.11.003] [PMID: 18187121]

[15]   Sacks MM, Silk WK, Burman P. Effect of water stress on cortical cell division rates within the apical meristem of primary roots of maize. Plant Physiol 1997; 114(2): 519-27.
[http://dx.doi.org/10.1104/pp.114.2.519] [PMID: 12223725]

[16]   Manivannan P, Jaleel CA, Sankar B, *et al*. Growth, biochemical modifications and proline metabolism in *Helianthus annuus* L. as induced by drought stress. Colloids Surf B Biointerfaces 2007; 59(2): 141-9.
[http://dx.doi.org/10.1016/j.colsurfb.2007.05.002] [PMID: 17560769]

[17]   Ashraf MY. Yield and yield components response of wheat (*Triticum aestivum* L.) genotypes grown under different soil water deficit conditions. Acta Agron Hung 1998; 46(1): 45-51.

[18]   El-Monayeri MO, Hegazi AM, Ezzat NH, Salem HM, Tahoun SM. Growth and yield of some wheat and barley varieties grown under different moisture stress levels. ASSJM 1984; 20(3): 231-43.

[19]   Hussain HA, Hussain S, Khaliq A, *et al*. Chilling and drought stresses in crop plants: implications, cross talk, and potential management opportunities. Front Plant Sci 2018; 9: 393.
[http://dx.doi.org/10.3389/fpls.2018.00393] [PMID: 29692787]

[20]   Frederick JR, Camp CR, Bauer PJ. Drought-stress effects on branch and mainstem seed yield and yield components of determinate soybean. Crop Sci 2001; 41(3): 759-63.
[http://dx.doi.org/10.2135/cropsci2001.413759x]

[21]   Hu Y, Schmidhalter U. Drought and salinity: a comparison of their effects on mineral nutrition of plants. J Plant Nutr Soil Sci 2005; 168: 541-9.
[http://dx.doi.org/10.1002/jpln.200420516]

[22]   Kuchenbuch R, Claassen N, Jungk A. Potassium availability in relation to soil moisture. Plant Soil 1986; 95(2): 233-43.
[http://dx.doi.org/10.1007/BF02375075]

[23]   Das R. Characterization of responses of *Brassica* cultivars to elevated CO_2 under moisture stress condition (Doctoral dissertation, Indian Agricultural Research Institute; New Delhi)

[24]   Li AG, Hou YS, Wall GW, Trent A, Kimball BA, Pinter PJ. Free-air CO_2 enrichment and drought stress effects on grain filling rate and duration in spring wheat. Crop Sci 2000; 40(5): 1263-70.
[http://dx.doi.org/10.2135/cropsci2000.4051263x]

[25]   Deka D, Singh AK, Singh AK. Effect of drought stress on crop plants with special reference to drought avoidance and tolerance mechanisms: a review. Int J Curr Microbiol Appl Sci 2018; 7(9): 2703-21.
[http://dx.doi.org/10.20546/ijcmas.2018.709.336]

[26]   Anjum SA, Wang LC, Farooq M, Hussain M, Xue LL, Zou CM. Brassinolide application improves the drought tolerance in maize through modulation of enzymatic antioxidants and leaf gas exchange. J Agron Crop Sci 2011; 197(3): 177-85.
[http://dx.doi.org/10.1111/j.1439-037X.2010.00459.x]

[27]   Din J, Khan SU, Ali I, Gurmani AR. Physiological and agronomic response of canola varieties to drought stress. J Anim Plant Sci 2011; 21(1): 78-82.

[28]   Jain M, Tiwary S, Gadre R. Sorbitol-induced changes in various growth and biochemici parameters in maize. Plant Soil Environ 2010; 56(6): 263-7.
[http://dx.doi.org/10.17221/233/2009-PSE]

[29]   Ashraf M, Mehmood S. Response of four *Brassica* species to drought stress. Environ Exp Bot 1990; 30(1): 93-100.
[http://dx.doi.org/10.1016/0098-8472(90)90013-T]

[30]   Basu S, Ramegowda V, Kumar A, Pereira A. Plant adaptation to drought stress. F1000 Res 2016; 5: 1554.

[31]   Hoekstra FA, Golovina EA, Buitink J. Mechanisms of plant desiccation tolerance. Trends Plant Sci

2001; 6(9): 431-8.
[http://dx.doi.org/10.1016/S1360-1385(01)02052-0] [PMID: 11544133]

[32]   Lawlor DW, Cornic G. Photosynthetic carbon assimilation and associated metabolism in relation to water deficits in higher plants. Plant Cell Environ 2002; 25(2): 275-94.
[http://dx.doi.org/10.1046/j.0016-8025.2001.00814.x] [PMID: 11841670]

[33]   Sourour A, Afef O, Mounir R, Mongi BY. A review: morphological, physiological, biochemical and molecular plant responses to water deficit stress. Int J Eng Sci 2017; 6: 1-4.
[http://dx.doi.org/10.9790/1813-0601010104]

[34]   Sales CR, Ribeiro RV, Silveira JA, Machado EC, Martins MO, Lagôa AM. Superoxide dismutase and ascorbate peroxidase improve the recovery of photosynthesis in sugarcane plants subjected to water deficit and low substrate temperature. Plant Physiol Biochem 2013; 73: 326-36.
[http://dx.doi.org/10.1016/j.plaphy.2013.10.012] [PMID: 24184453]

[35]   Bota J, Medrano H, Flexas J. Is photosynthesis limited by decreased Rubisco activity and RuBP content under progressive water stress? New Phytol 2004; 162(3): 671-81.
[http://dx.doi.org/10.1111/j.1469-8137.2004.01056.x]

[36]   de Campos MK, de Carvalho K, de Souza FS, *et al.* Drought tolerance and antioxidant enzymatic activity in transgenic 'Swingle'citrumelo plants over-accumulating proline. Environ Exp Bot 2011; 72(2): 242-50.
[http://dx.doi.org/10.1016/j.envexpbot.2011.03.009]

[37]   Monclus R, Dreyer E, Villar M, *et al.* Impact of drought on productivity and water use efficiency in 29 genotypes of Populus deltoides x Populus nigra. New Phytol 2006; 169(4): 765-77.
[http://dx.doi.org/10.1111/j.1469-8137.2005.01630.x] [PMID: 16441757]

[38]   Abbate PE, Dardanelli JL, Cantarero MG, Maturano M, Melchiori RJ, Suero EE. Climatic and water availability effects on water-use efficiency in wheat. Crop Sci 2004; 44(2): 474-83.
[http://dx.doi.org/10.2135/cropsci2004.4740]

[39]   Zhang J, Nguyen HT, Blum A. Genetic analysis of osmotic adjustment in crop plants. J Exp Bot 1999; 50(332): 291-302.
[http://dx.doi.org/10.1093/jxb/50.332.291]

[40]   Subbarao GV, Nam NH, Chauhan YS, Johansen C. Osmotic adjustment, water relations and carbohydrate remobilization in pigeonpea under water deficits. J Plant Physiol 2000; 157(6): 651-9.
[http://dx.doi.org/10.1016/S0176-1617(00)80008-5]

[41]   Naidu BP, Cameron DF, Konduri SV. Improving drought tolerance of cotton by glycinebetaine application and selection. In: Michalk DL, Pratley JE, Eds. Agronomy, Growing a Greener Future? Proceedings of the 9th Australian Agronomy Conference, Wagga Wagga, article 221. Australia. The Australian Society of Agronomy Inc 1998.

[42]   Leport L, Turner NC, French RJ, Barr MD, Duda R, Davies SL. Physiological responses of chickpea genotypes to terminal drought in a Mediterranean-type environment. Eur J Agron 2006; 11: 279-91.

[43]   Komor E. Source physiology and assimilate transport: the interaction of sucrose metabolism, starch storage and phloem export in source leaves and the effects on sugar status in phloem. Funct Plant Biol 2000; 27(6): 497-505.
[http://dx.doi.org/10.1071/PP99127]

[44]   Fahad S, Bajwa AA, Nazir U, *et al.* Crop production under drought and heat stress: plant responses and management options. Front Plant Sci 2017; 8: 1147.
[http://dx.doi.org/10.3389/fpls.2017.01147] [PMID: 28706531]

[45]   Baier M, Kandlbinder A, Golldack D, Dietz K. Oxidative stress and ozone: perception, signalling and response. Plant Cell Environ 2005; 28: 1012-20.
[http://dx.doi.org/10.1111/j.1365-3040.2005.01326.x]

[46]   Apel K, Hirt H. Reactive oxygen species: metabolism, oxidative stress, and signal transduction. Annu

Rev Plant Biol 2004; 55: 373-99.
[http://dx.doi.org/10.1146/annurev.arplant.55.031903.141701] [PMID: 15377225]

[47] Moran JF, Becana M, Iturbe-Ormaetxe I, Frechilla S, Klucas RV, Aparicio-Tejo P. Drought induces oxidative stress in pea plants. Planta 1994; 194(3): 346-52.
[http://dx.doi.org/10.1007/BF00197534]

[48] Ramachandra Reddy A, Chaitanya KV, Vivekanandan M. Drought-induced responses of photosynthesis and antioxidant metabolism in higher plants. J Plant Physiol 2004; 161(11): 1189-202.
[http://dx.doi.org/10.1016/j.jplph.2004.01.013] [PMID: 15602811]

[49] Moller IM. Plant mitochondria and oxidative stress: electron transport, NADPH turnover, and metabolism of reactive oxygen species. Ann Rev Plant Physiol 2001; 52: 561-91.

[50] Wahid A, Gelani S, Ashraf M, Foolad MR. Heat tolerance in plants: an overview. Environ Exp Bot 2007; 61: 199-223.
[http://dx.doi.org/10.1016/j.envexpbot.2007.05.011]

[51] Gill SS, Tuteja N. Reactive oxygen species and antioxidant machinery in abiotic stress tolerance in crop plants. Plant Physiol Biochem 2010; 48(12): 909-30.
[http://dx.doi.org/10.1016/j.plaphy.2010.08.016] [PMID: 20870416]

[52] Sharma P, Dubey RS. Drought induces oxidative stress and enhances the activities of antioxidant enzymes in growing rice seedlings. Plant Growth Regul 2005; 46(3): 209-21.
[http://dx.doi.org/10.1007/s10725-005-0002-2]

[53] Fahad S, Hussain S, Matloob A, *et al.* Phytohormones and plant responses to salinity stress: a review. Plant Growth Regul 2015; 75(2): 391-404.
[http://dx.doi.org/10.1007/s10725-014-0013-y]

[54] Verulkar SB, Mandal NP, Dwivedi JL, *et al.* Breeding resilient and productive genotypes adapted to drought-prone rainfed ecosystem of India. Field Crops Res 2010; 117: 197-208.
[http://dx.doi.org/10.1016/j.fcr.2010.03.005]

[55] Ahmad P, Wani MR, Azooz MM, Tran LS, Eds. Improvement of crops in the era of climatic changes. New York, NY: Springer 2014.
[http://dx.doi.org/10.1007/978-1-4614-8830-9]

[56] Maqbool MA, Aslam M, Ali H. Breeding for improved drought tolerance in Chickpea (CO$_2$ L.). Plant Breed 2017; 136(3): 300-18.
[http://dx.doi.org/10.1111/pbr.12477]

[57] Varshney RK, Kudapa H, Roorkiwal M, *et al.* Advances in genetics and molecular breeding of three legume crops of semi-arid tropics using next-generation sequencing and high-throughput genotyping technologies. J Biosci 2012; 37(5): 811-20.
[http://dx.doi.org/10.1007/s12038-012-9228-0] [PMID: 23107917]

[58] Kumar A, Bernier J, Verulkar S, Lafitte HR, Atlin GN. Breeding for drought tolerance: direct selection for yield, response to selection and use of drought-tolerant donors in upland and lowland-adapted populations. Field Crops Res 2008; 107(3): 221-31.
[http://dx.doi.org/10.1016/j.fcr.2008.02.007]

[59] Kumar M, Uniyal M, Kumar N, Kumar S, Gangwar R. Conventional and molecular breeding for development of drought tolerant maize cultivars. J Crop Sci Tech 2015; 4: 1-3.

[60] Lang NT, Buu BC. Fine mapping for drought tolerance in rice (*Oryza sativa* L.). Omonrice 2008; 16(1): 9-15.

[61] Xiong L, Wang RG, Mao G, Koczan JM. Identification of drought tolerance determinants by genetic analysis of root response to drought stress and abscisic Acid. Plant Physiol 2006; 142(3): 1065-74.
[http://dx.doi.org/10.1104/pp.106.084632] [PMID: 16963523]

[62] Todaka D, Shinozaki K, Yamaguchi-Shinozaki K. Recent advances in the dissection of drought-stress

regulatory networks and strategies for development of drought-tolerant transgenic rice plants. Front Plant Sci 2015; 6: 84.
[http://dx.doi.org/10.3389/fpls.2015.00084] [PMID: 25741357]

[63]   Skodaeek Z, Prasil IT. New possibilities for research of barley (*Hordeum vulgare* L.) drought resistance. Úroda 2011; 8: 24-9.

[64]   Bruce WB, Edmeades GO, Barker TC. Molecular and physiological approaches to maize improvement for drought tolerance. J Exp Bot 2002; 53(366): 13-25.
[http://dx.doi.org/10.1093/jexbot/53.366.13] [PMID: 11741036]

[65]   Fu BY, Xiong JH, Zhu LH, *et al.* Identification of functional candidate genes for drought tolerance in rice. Mol Genet Genomics 2007; 278(6): 599-609.
[http://dx.doi.org/10.1007/s00438-007-0276-3] [PMID: 17665216]

[66]   Kiani SP, Talia P, Maury P, *et al.* Genetic analysis of plant water status and osmotic adjustment in recombinant inbred lines of sunflower under two water treatments. Plant Sci 2007; 172(4): 773-87.
[http://dx.doi.org/10.1016/j.plantsci.2006.12.007]

[67]   Chen Z, Wang Z, Yang Y, Li M, Xu B. Abscisic acid and brassinolide combined application synergistically enhances drought tolerance and photosynthesis of tall fescue under water stress. Sci Hortic (Amsterdam) 2018; 228: 1-9.
[http://dx.doi.org/10.1016/j.scienta.2017.10.004]

[68]   Nadeem SM, Ahmad M, Zahir ZA, Kharal MA. Role of phytohormones in stress tolerance of plants. In: Hakeem KR, Akhtar MS, Eds. Plant, Soil and Microbes. Cham: Springer 2016; 2: pp. 385-421.

[69]   Wilkinson S, Kudoyarova GR, Veselov DS, Arkhipova TN, Davies WJ. Plant hormone interactions: innovative targets for crop breeding and management. J Exp Bot 2012; 63(9): 3499-509.
[http://dx.doi.org/10.1093/jxb/ers148] [PMID: 22641615]

[70]   Giuliani S, Sanguineti MC, Tuberosa R, Bellotti M, Salvi S, Landi P. Root-ABA1, a major constitutive QTL, affects maize root architecture and leaf ABA concentration at different water regimes. J Exp Bot 2005; 56(422): 3061-70.
[http://dx.doi.org/10.1093/jxb/eri303] [PMID: 16246858]

[71]   Wani SH, Kumar V, Shriram V, Sah SK. Phytohormones and their metabolic engineering for abiotic stress tolerance in crop plants. Crop J 2016; 4(3): 162-76.
[http://dx.doi.org/10.1016/j.cj.2016.01.010]

[72]   Reguera M, Peleg Z, Abdel-Tawab YM, Tumimbang EB, Delatorre CA, Blumwald E. Stress-induced cytokinin synthesis increases drought tolerance through the coordinated regulation of carbon and nitrogen assimilation in rice. Plant Physiol 2013; 163(4): 1609-22.
[http://dx.doi.org/10.1104/pp.113.227702] [PMID: 24101772]

[73]   Kaya MD, Okçu G, Atak M, Cıkılı Y, Kolsarıcı Ö. Seed treatments to overcome salt and drought stress during germination in sunflower (*Helianthus annuus* L.). Eur J Agron 2006; 24(4): 291-5.
[http://dx.doi.org/10.1016/j.eja.2005.08.001]

[74]   Khan TA, Fariduddin Q, Yusuf M. Low-temperature stress: is phytohormones application a remedy? Environ Sci Pollut Res Int 2017; 24(27): 21574-90.
[http://dx.doi.org/10.1007/s11356-017-9948-7] [PMID: 28831664]

[75]   Farooq M, Aziz T, Wahid A, Lee DJ, Siddique KH. Chilling tolerance in maize: agronomic and physiological approaches. Crop Pasture Sci 2009; 60(6): 501-16.
[http://dx.doi.org/10.1071/CP08427]

[76]   Lurie S, Ronen R, Lipsker Z, Aloni B. Effects of paclobutrazol and chilling temperatures on lipids, antioxidants and ATPase activity of plasma membrane isolated from green bell pepper fruits. Physiol Plant 1994; 91(4): 593-8.
[http://dx.doi.org/10.1111/j.1399-3054.1994.tb02993.x]

[77]   Feng Z, Guo A, Feng Z. Amelioration of chilling stress by triadimefon in cucumber seedlings. Plant

Growth Regul 2003; 39(3): 277-83.
[http://dx.doi.org/10.1023/A:1022881628305]

[78] Lukatkin AS, Brazaityte A, Bobinas C, Duchovskis P. Chilling injury in chilling-sensitive plants: a review. Agriculture 2012; 99(2): 111-24.

[79] Still JR, Pill WG. Germination, emergence, and seedling growth of tomato and impatiens in response to seed treatment with paclobutrazol. HortScience 2003; 38(6): 1201-4.
[http://dx.doi.org/10.21273/HORTSCI.38.6.1201]

[80] Waraich EA, Ahmad R, Halim A, Aziz T. Alleviation of temperature stress by nutrient management in crop plants: a review. J Soil Sci Plant Nutr 2012; 12(2): 221-44.
[http://dx.doi.org/10.4067/S0718-95162012000200003]

[81] Waraich EA, Ahmad R, Ashraf MY. Saifullah, Ahmad M. Improving agricultural water use efficiency by nutrient management in crop plants. Acta Agric Scand B Soil Plant Sci 2011; 61(4): 291-304.
[http://dx.doi.org/10.1080/09064710.2010.491954]

[82] Wendehenne D. David Wendehenne and Jörg Durner answer a few questions about this month's fast moving front in the field of plant & animal science. Trends Plant Sci 2001; 6(4): 177-83.
[http://dx.doi.org/10.1016/S1360-1385(01)01893-3] [PMID: 11286923]

[83] Valadabadi SA, Farahani HA. Studying the interactive effect of potassium application and individual field crops on root penetration under drought condition. J Agric Biotechnol Sustain Dev 2010; 2(5): 82.

[84] Pandey R, Agarwal RM, Jeevaratnam K, Sharma GL. Osmotic stress-induced alterations in rice (*Oryza sativa* L.) and recovery on stress release. Plant Growth Regul 2004; 42(1): 79-87.
[http://dx.doi.org/10.1023/B:GROW.0000014893.45112.55]

[85] Kafkafi U. The functions of plant K in overcoming environmental stress situations. In 22nd Colloquium, International Potash Institute. Switzerland. 1990; pp. 81-93.

[86] Hussain S, Khan F, Cao W, Wu L, Geng M. Seed priming alters the production and detoxification of reactive oxygen intermediates in rice seedlings grown under sub-optimal temperature and nutrient supply. Front Plant Sci 2016; 7: 439.
[http://dx.doi.org/10.3389/fpls.2016.00439] [PMID: 27092157]

[87] Samota MK, Sasi M, Awana M, *et al.* Elicitor-induced biochemical and molecular manifestations to improve drought tolerance in rice (*Oryza sativa* L.) through seed-priming. Front Plant Sci 2017; 8: 934.
[http://dx.doi.org/10.3389/fpls.2017.00934] [PMID: 28634483]

[88] Wang W, Chen Q, Hussain S, *et al.* Pre-sowing seed treatments in direct-seeded early rice: consequences for emergence, seedling growth and associated metabolic events under chilling stress. Sci Rep 2016; 6(1): 19637.
[http://dx.doi.org/10.1038/srep19637] [PMID: 26782108]

[89] Hasanuzzaman M, Fujita M. Selenium pretreatment upregulates the antioxidant defense and methylglyoxal detoxification system and confers enhanced tolerance to drought stress in rapeseed seedlings. Biol Trace Elem Res 2011; 143(3): 1758-76.
[http://dx.doi.org/10.1007/s12011-011-8998-9] [PMID: 21347652]

[90] Singh B, Usha K. Salicylic acid induced physiological and biochemical changes in wheat seedlings under water stress. Plant Growth Regul 2003; 39(2): 137-41.
[http://dx.doi.org/10.1023/A:1022556103536]

[91] Parera CA, Cantliffe DJ. Improved germination and modified imbibition of shrunken-2 sweet corn by seed disinfection and solid matrix priming. J Am Soc Hortic Sci 1991; 116(6): 942-5.
[http://dx.doi.org/10.21273/JASHS.116.6.942]

[92] Jisha KC, Vijayakumari K, Puthur JT. Seed priming for abiotic stress tolerance: an overview. Acta Physiol Plant 2013; 35(5): 1381-96.

[http://dx.doi.org/10.1007/s11738-012-1186-5]

[93]   Kaur S, Gupta AK, Kaur N. Effect of osmo-and hydropriming of chickpea seeds on seedling growth and carbohydrate metabolism under water deficit stress. Plant Growth Regul 2002; 37(1): 17-22.
[http://dx.doi.org/10.1023/A:1020310008830]

[94]   Janmohammadi M, Dezfuli PM, Sharifzadeh F. Seed invigoration techniques to improve germination and early growth of inbred line of maize under salinity and drought stress. Gen Appl Plant Physiol 2008; 34(3-4): 215-26.

[95]   Chen K, Arora R. Dynamics of the antioxidant system during seed osmopriming, post-priming germination, and seedling establishment in Spinach (*Spinacia oleracea*). Plant Sci 2011; 180(2): 212-20.

[96]   Moradi A, Younesi O. Effects of osmo-and hydro-priming on seed parameters of grain sorghum (*Sorghum bicolor* L.). Aust J Basic Appl Sci 2009; 3(3): 1696-700.

[97]   Foti R, Abureni K, Tigere A, Gotosa J, Gere J. The efficacy of different seed priming osmotica on the establishment of maize (*Zea mays* L.) caryopses. J Arid Environ 2008; 72(6): 1127-30.
[http://dx.doi.org/10.1016/j.jaridenv.2007.11.008]

[98]   Al-Mudaris MA, Jutzi SC. The influence of fertilizer-based seed priming treatments on emergence and seedling growth of *Sorghum bicolor* and Pennisetum glaucum in pot trials under greenhouse conditions. J Agron Crop Sci 1999; 182(2): 135-42.
[http://dx.doi.org/10.1046/j.1439-037x.1999.00293.x]

[99]   Arif M. Seed priming maize for improving emergence and seedling growth. SJA 2005; 21(4): 539-43.

[100]   Cakmak I. The role of potassium in alleviating detrimental effects of abiotic stresses in plants. J Plant Nutr Soil Sci 2005; 168(4): 521-30.
[http://dx.doi.org/10.1002/jpln.200420485]

[101]   Arif MU, Waqas MU, Nawab KH, Shahid MU. Effect of seed priming in Zn solutions on chickpea and wheat. In: African Crop Science Conference Proceeding. 8: 237-40.

[102]   Zhou ZS, Guo K, Elbaz AA, Yang ZM. Salicylic acid alleviates mercury toxicity by preventing oxidative stress in roots of *Medicago sativa*. Environ Exp Bot 2009; 65(1): 27-34.
[http://dx.doi.org/10.1016/j.envexpbot.2008.06.001]

[103]   Lee SS, Kim JH, Hong SB, Yun SH, Park EH. Priming effect of rice seeds on seedling establishment under adverse soil conditions. Hangug Jagmul Haghoeji 1998; 43(3): 194-8.

[104]   Fujita M, Fujita Y, Noutoshi Y, *et al.* Crosstalk between abiotic and biotic stress responses: a current view from the points of convergence in the stress signaling networks. Curr Opin Plant Biol 2006; 9(4): 436-42.
[http://dx.doi.org/10.1016/j.pbi.2006.05.014] [PMID: 16759898]

[105]   Shakirova FM, Sakhabutdinova AR, Bezrukova MV, Fatkhutdinova RA, Fatkhutdinova DR. Changes in the hormonal status of wheat seedlings induced by salicylic acid and salinity. Plant Sci 2003; 164(3): 317-22.
[http://dx.doi.org/10.1016/S0168-9452(02)00415-6]

[106]   Sahu MP, Singh D. Role of thiourea in improving productivity of wheat (*Triticum aestivum* L.). J Plant Growth Regul 1995; 14(4): 169-73.
[http://dx.doi.org/10.1007/BF00204908]

[107]   Sahu MP, Kumawat SM, D'souza SF, Ramaswamy NK, Singh G. Sulphydryl bioregulator technology for increasing mustard production. Research Bulletin RAU-BARC 2005; 1-52.

[108]   Srivastava AK, Ramaswamy NK, Mukopadhyaya R, Jincy MG, D'Souza SF. Thiourea modulates the expression and activity profile of mtATPase under salinity stress in seeds of *Brassica juncea*. Ann Bot 2009; 103(3): 403-10.
[http://dx.doi.org/10.1093/aob/mcn229] [PMID: 19033283]

[109] Korkmaz A, Pill WG. The effect of different priming treatments and storage conditions on germination performance of lettuce seeds. Eur J Hortic Sci 2003; 68(6): 260-5.

[110] Sarwar N, Yousaf S, Jamil FF. Induction of salt tolerance in chickpea by using simple and safe chemicals. Pak J Bot 2006; 38(2): 325.

[111] Copeland LO, McDonald MB. Seed Science and Technology. Wasington.

[112] Kubik KK, Eastin JA, Eastin JD, Eskridge KM. Solid matrix priming of tomato and pepper. Lancaster, Pa: In: Proc. Intl. Conf. Stand Establishment Hort. Crops 1988; pp. 86-96.

[113] Basra SM, Farooq M, Tabassam R, Ahmad N. Physiological and biochemical aspects of pre-sowing seed treatments in fine rice (*Oryza sativa* L.). Seed Sci Technol 2005; 33(3): 623-8.
[http://dx.doi.org/10.15258/sst.2005.33.3.09]

[114] Wang B, Li Z, Ran Q, Li P, Peng Z, Zhang J. ZmNF-YB16 overexpression improves drought resistance and yield by enhancing photosynthesis and the antioxidant capacity of maize plants. Front Plant Sci 2018; 9: 709.
[http://dx.doi.org/10.3389/fpls.2018.00709] [PMID: 29896208]

[115] Lee SJ, Kang JY, Park HJ, *et al.* DREB2C interacts with ABF2, a bZIP protein regulating abscisic acid-responsive gene expression, and its overexpression affects abscisic acid sensitivity. Plant Physiol 2010; 153(2): 716-27.
[http://dx.doi.org/10.1104/pp.110.154617] [PMID: 20395451]

[116] Singh D, Laxmi A. Transcriptional regulation of drought response: a tortuous network of transcriptional factors. Front Plant Sci 2015; 6: 895.
[http://dx.doi.org/10.3389/fpls.2015.00895] [PMID: 26579147]

[117] Cai R, Zhao Y, Wang Y, *et al.* Overexpression of a maize WRKY58 gene enhances drought and salt tolerance in transgenic rice. Plant Cell Tissue Organ Cult 2014; 119(3): 565-77.
[http://dx.doi.org/10.1007/s11240-014-0556-7]

[118] Liu HT, Li B, Shang ZL, *et al.* Calmodulin is involved in heat shock signal transduction in wheat. Plant Physiol 2003; 132(3): 1186-95.
[http://dx.doi.org/10.1104/pp.102.018564] [PMID: 12857801]

[119] Townley HE, Knight MR. Calmodulin as a potential negative regulator of Arabidopsis COR gene expression. Plant Physiol 2002; 128(4): 1169-72.
[http://dx.doi.org/10.1104/pp.010814] [PMID: 11950965]

[120] Leng P, Zhao J. Transcription factors as molecular switches to regulate drought adaptation in maize. Theor Appl Genet 2020; 133(5): 1455-65.
[http://dx.doi.org/10.1007/s00122-019-03494-y] [PMID: 31807836]

CHAPTER 3

Salinity and its Effect on Yield of Field/Horticultural Crops

Sandeep Kaur[1,2], Neha Sharma[1], Ajay Kumar[1], Samiksha[3], Shagun Verma[1], Satwinderjeet Kaur[1], Satwinder Kaur Sohal[3] and Sharad Thakur[4,*]

[1] *Department of Botanical and Environmental Sciences, Guru Nanak Dev University, Amritsar, Punjab 143005, India*

[2] *PG Department of Botany, Khalsa College, Amritsar, Punjab 143005, India*

[3] *Department of Zoology, Guru Nanak Dev University, Amritsar, Punjab 143005, India*

[4] *CSIR-Institute of Himalayan Bioresource Technology, Palampur, H.P. 176061, India*

Abstract: Salinity is considered a crucial environmental factor that limits the production of the crop in many parts of the world with marginal agricultural soils. It causes a reduction in agricultural productivity globally and renders an estimated one-third of irrigated land of the world unsuitable for the production of crops. A high concentration of salt can kill all the crops and plants. Salinity can affect the yield and growth of most crops, as the higher rate of salinity can cause both hyperosmotic and hyper ionic effects in plants, leading to an increase in the production of activated oxygen species, membrane disorganization, and metabolic toxicity. Its effects on the growth and development of plants include osmotic stress, ion toxicity, mineral deficiencies, biochemical and physiological perturbations, and combinations of these stressors. Salinity reduces Ca^{2+} availability that in turn decreases the mobility and transport of Ca^{2+} to growing regions of the plant when dominated by Na^+ salts and thereby affects the quality of both reproductive and vegetative organs. The horticultural crops are mostly glycophytes that evolved under the conditions of low salinity of the soil. Nutrient uptake is directly affected by salinity, such as Cl^- reducing NO^{3-} uptake or Na^+ reducing K^+ uptake. The performance of crops may be affected adversely by salinity-induced nutritional disorders. These disorders resulting from salinity may affect the availability of nutrients, transporter partitioning, and competitive uptake within the plant. This chapter will elucidate the deleterious effect of salinity on the growth and development of crop plants.

Keywords: Agricultural soils, Crop, Ion toxicity, Nutritional disorders, Salinity.

---

* **Corresponding Author Sharad Thakur:** CSIR-Institute of Himalayan Bioresource Technology, Palampur, H.P. 176061, India; Tel: +91-9877545749; E-mail: thakursharad23@gmail.com

INTRODUCTION

Salinity is considered a major abiotic factor that limits the productivity and growth of plants due to increased use of soil salinization and poor quality of water for irrigation in many regions of the world [1 - 3]. Salts are the natural and main components of water and soils. The ions responsible for causing salination include Na^+, Ca^{2+}, K^+, Cl^-, and Mg^{2+}. The absorption of important nutrients like Ca, K, N, and Mg is affected by the excessive concentrations of Na^+ and Cl^- and the soils can also become sodic. Sodic soils represent many challenges as they have a very poor quality of structure, which prevents or limits water drainage and infiltration. In the plant tissues, water stress like salt stress leads to osmotic imbalances [4]. In the case of salt stress, the excessive amount of salt, usually Na^+ and Cl^-, leads to direct toxic and nutritional effects [5]. The toxicity of a specific ion is caused by the accumulation of chloride, sodium, and boron in the cells and tissues of transpiring leaves, having a deleterious effect on the growth and yield of crops and plants. The accrual of harmful ions might inactivate enzymes, hinder protein synthesis and photosynthesis, and injure the chloroplasts and other related organelles [6].

Soil salinity is the amount of excessive salt content in the soil, and the process of enhancing the content of salt is known as salinization. The presence of excess salt not only disturbs the structure of the soil, but also has the ability to attract water and thereby block its absorption by the roots of the plant. As a result, the plants may show signs of drought even at the time of waterlogging or wetting of the soil. The salt damage may also cause water pooling on the surface of the soil without penetration. The weathering of minerals also increases the salinity concentration of soil [7]. Another yield-limiting factor is nutrient deficiency, and this issue is gradually aggravated due to intensive cultivation as well as the injudicious and imbalanced utilization of fertilizers. The deficiency of phosphorus and nitrogen is widespread, followed by the deficiency of Zn. Zinc is a crucial protein component and acts as a functional, structural, or regulatory cofactor of several enzymes and plays a key role in the metabolism of the plant [8].

Salinity adversely affects the growth, yield, and biomass of the plants as the leaves become thicker and smaller than those of normal plants. The excess amount of chloride (but not sulfate) enhances the palisade cells' elongation, resulting in increased succulence [9]. It is anticipated that in time to come, these ecological stressors will turn out to be more frequent and extreme. At the same time, in 2050, it is estimated that the population of the world will experience a grave food shortage as it is presumed to reach about 10 billion [10]. Overcoming the problem of salt stress would have a positive impact on the production of agricultural crops. Plants have developed certain biochemical and physiological mechanisms to regulate the stability of the intracellular environment through the accumulation of numerous solutes under saline conditions [11]. Proline, as an essential osmoprotectant, might contribute to protecting enzymes from oxidative damage and also provide osmotic adjustment during the condition of salinity [12]. The adjustment of osmotic level in plants maintains the turgor pressure of cells and the uptake of water by allowing the regularity of physiological metabolism [13].

Several attempts have been made to improve the ability of crops to tolerate salt by genetic transformations and traditional breeding programs. However, due to the complexity of the trait, commercial success has been very limited [14]. Salinity is one of the severe environmental and agrarian issues in semi-arid and arid areas that detrimentally affect the growth of crops and the productivity of agriculture [15, 16], thereby affecting 2.1% of dryland agriculture and 19.5% of irrigated land at a global scale [17]. Thus, the development and discovery of schemes or programs to ameliorate the adverse effects of environmental stressors have been receiving considerable attention. This chapter will highlight the effect of salinity on the growth and productivity of crop plants.

CAUSES OF SALINITY

Salinity can be defined as the concentration of dissolved mineral salts present in the soil. The mineral salts present in saline soil are comprised of electrolytes of cations and anions. The major cations found in saline soil are sodium, potassium, magnesium, and calcium. The major anions present are chloride, sulphate, carbonate, bicarbonate, and nitrate. Mineral salts soluble in water accumulate in the upper layer formed by the A and B zones of the soil profile and the lower layer of the soil comprising unconsolidated rock material affecting the crop production and environmental well-being. The natural causes of soil salinity are rain, wind, and weathering of rocks. Rain and weathering are major sources of salt. Mineral salts are major components of rocks and rainwater containing a small amount of salt which accumulates in the soil over a long period of time. Winds act as a major distributor of salts from weathered rocks and oceans to land, contributing to soil salinity. Natural catastrophes like tsunami play a vital role in introducing seawater to nearby lands. Tsunami waves contain huge amounts of ocean salts, and when they hit the ground, a generous amount gets deposited in the soil. Besides natural causes, anthropogenic activities are also responsible for soil salinity. Humans introduce salts in soil by following improper agricultural practices. Water used for irrigation and its poor drainage plays an important role in increasing soil salinity. Even if the water used for irrigation is of good quality, it contains a small amount of salt, which accumulates with time if water drainage systems are not efficient [18].

Types of Salinity

There are two types of salinity, natural and man-induced. Natural salinity is also known as primary and man-induced salinity is also called secondary salinity.

Natural Salinity

It is the accumulation of salts and minerals in soil over a very long period of time. This happens because of two natural processes, weathering of rocks and the deposition of salts from ocean to land. Due to the weathering of rocks, soluble salts containing chlorides of sodium, magnesium, calcium, etc., and carbonates and sulphates get deposited in the soil. The other major cause of salinity is the deposition of oceanic salts carried to land due to rain and wind. This type of soil is widely distributed in arid and semi-arid regions.

Man-induced Salinity

This type of salinity arises due to human activities such as cultivation practices, the poor drainage system, and use of salt-rich water for irrigation. These human activities drastically change the soil balance between water supplied to crops during irrigation and water used by crops. The sustainable cultivation practices play an important role in maintaining soil health. One of the major reasons responsible for soil salinity is the replacement of perennial vegetation by annual crops hence disturbing the hydrologic soil balance [19].

EFFECT OF SALINITY ON PLANTS

Plant Growth

Reduction in the plant growth rate is one of the initial effects of soil salinity. An excessive amount of salt in the root zone of the plant can hamper or inhibit the process of water withdrawal by plants from surrounding soil leading to a slow growth rate due to water deficiency. Large amounts of salts in soil hinder the capability of plants to withdraw water from the soil by interfering with the osmotic gradient between plant cells and surrounding soil. In some cases, even if a plant is able to withdraw this saline water from the soil into the plant it can cause damage to plant cells and tissues once it enters the transpiration stream consequently reducing the plant growth [19, 20].

Seed Germination

The process of seed germination is an initial and very crucial step in plant life and cannot be completed without water. Salinity adversely affects the uptake of water by seeds from soil. A study was carried out on three halophytes *viz. Glycine soja,*

Limonium sinense and *Sorghum sudanense* revealed that even though these plants are well adapted to saline habitat, salinity does affect the seed germination process. Seeds of these plants can germinate at higher salt concentration because these plants are physiologically adapted to survive that stress but they are able to germinate more rapidly and efficiently at low levels of salt concentration [21]. Zhang *et al.* [22] studied the effect of salinity on two halophytes namely *Haloxylon ammodendron, Suaeda physophora* and one xerophyte *Haloxylon persicum* and they found that salinity affects seed germination in all three plants but it merely interferes with the osmotic gradient in halophytes and is not detrimental but in the xerophytic plant, it caused ionic toxicity which can be fatal in extreme cases. Wu *et al.* [23] reported the slow rate of seed germination in sunflower seeds at higher sodium chloride concentration. The concentration of NaCl (100-200 mmol) reduced water uptake by seeds and delayed the process of seed germination. *Echinochloa crusgalli* an annual grass found in non-saline habitats has been extensively studied for the effects of various environmental factors on seed germination. It was reported that in this plant, the rate of seed germination was low at higher salt concentration. Some seeds were able to germinate in NaCl concentration up to 1.5% but seedlings were unable to survive the saline environment for a long period. It was noted that seed growth in salt stress is more sensitive than seed germination [24].

YIELD OF FIELD/HORTICULTURAL CROPS

Crop productivity and growth have been adversely affected over a few decades due to salinity. It was reported that 20% of the total land in the world that is used to produce one-third of the world's food is badly affected by high salt concentration. An increase in salinity level decreases the dry matter yield of root and shoots systems in wheat crops. Total crop yield is noticed to be drastically lowered in saline soil. Overall, total wheat productivity is reported to be lowered in soils containing high salt content because of poor water uptake, low amount of photosynthetic pigment, and ionic toxicity [25]. Shafi *et al.* [26], studied the effect of salinity on eleven genotypes of wheat and reported that genotype supplied with lower concentrations of Na^+ ions produced the greatest dry matter with the same results observed for grain yield. Soils with minimum salt concentration yielded more grain. Horticultural crops are also adversely affected by soil salinity. A study carried out to check the effect of salinity on watermelon fruits showed that total fruit yield and mean fruit mass was affected significantly by salinity levels [27].

Apart from other environmental abiotic stressors, soil salinity is one of the major abiotic stressors that limit agricultural productivity and food supply of the world [28, 29] by causing a major reduction in cultivated land area, crop productivity and quality [30, 31]. Earth's 10% of land is salt-affected and an estimated 10 million hectares of agricultural land is lost annually due to salinization and water logging [32]. However, the existing soil salinity problems in crop production will become worse due to rapidly growing human populations in many countries and the presence of limited water resources which are forcing growers to use poor quality water for irrigation [28]. Due to the occurrence of specific ionic toxicities, the osmotic stress results in the reduction of both crop growth and yield by inducing leaf damage and defoliation.

The onset and timing of foliar injury may determine the extent of damage to crop yields. The injury that occurs late in the season on annual crops may have little effect on yield. However, in tree crops developing fruit yields over 2 years period, foliar injury and leaf loss during both years will be more damaging than the injury that occurs only in the 2ⁿᵈ year. As this, continued salt stress will eventually damage the tree itself [33]. The response of salt stress on the crop yield is illustrated by taking an example of the plant *Miscanthus giganteus. Miscanthus giganteus* (hybrid of *Miscanthus sinensis* and *Miscanthus sacchariflorus*), a perennial rhizomatous grass which is used as a leading biomass crop, when irrigated with nine different NaCl concentrations (0, 2.86, 5.44, 7.96, 10.65, 14.68, 17.5, 19.97 and 22.4 dS m^{-1}), biomass yield was reduced by 50% at 10.65 dS m^{-1} NaCl. The accumulative effect of increasing soil salinity reduced stem height and elongation [34]. In another example, the effect of salinity on crop yield is studied by employing a geographical information system (GIS) and remote sensing technologies in the saline soils of Harran Plain. The results have shown that an increase in soil salinity above the threshold for cotton and wheat crops leads to linear decrease in crop yields [35].

OXIDATIVE STRESS DUE TO SALINITY

Oxidative stress is a complex chemical and physiological phenomenon which is due to the overproduction and accumulation of reactive oxygen species (ROS) leading to damage in cellular components and cause their dysfunction [36]. Soil salinity is one of the causes for the generation of reactive oxygen species including superoxide radical ($^{.}O_2^{-}$), hydrogen peroxide (H_2O_2), hydroxyl radical (OH^{-}) and singlet oxygen ($^1O_2$) resulting in oxidative stress [37 - 39]. In plant cells, high salinity induces the formation of ROS that cause oxidative damage to membrane lipids, proteins and nucleic acids when over accumulated [40, 41]. The level of ROS increase in the plant tissues is due to irregularities in the electron transport chain and accumulation of photoreducing powers. It was studied that with the rise in salinity the increase in the content of indices of oxidative stress *viz.* superoxide radical ($^{.}O_2^{-}$) and H_2O_2 in leaves of all the cultivars (cultivated varieties) were more noticeable and significant in salt-sensitive varieties and non-significant in resistant cultivars.

Salinity induced oxidative stress and antioxidant system in salt-tolerant and salt-sensitive cultivars of rice (*Oryza sativa* L.) was studied and it was observed that although the production of superoxide radicals($\cdot O_2^-$) was more in leaves of non-stressed plants of sensitive cultivars than that in resistant cultivars and there is increase in the amount of $\cdot O_2^-$ with the elevation in salinity stress in all the cultivars, but the level of increment in superoxide radical $\cdot O_2^-$ was much higher and significant in the sensitive varieties and non-significant in tolerant cultivars. Hydrogen peroxide (H_2O_2) followed a similar pattern that was exhibited by $\cdot O_2^-$. The basal values of O_2^- and H_2O_2 in the salt sensitive varieties were also about two-folds the amount of these ROS in salt-tolerant cultivars [42]. Two different rice leaves genotypes showed difference in their salt stress tolerance as detected by the isobaric tags for relative and absolute quantitation (iTRAQ)-based proteomic technique which is useful in identifying the differentially-expressed proteins. In both rice genotypes the identification of iTRAQ protein profiling demonstrated that the differentially-expressed proteins were found in the involvement of salt-stress responses regulation, in photosynthesis, oxidation-reduction responses and carbohydrate metabolism. In regard to their subcellular localization, they were exhibited to localize in chloroplasts and cytoplasm (67.2% of the total up-regulated proteins) [43].

Enhancing evidence reveals the crucial roles of a Salt Overly Sensitive (SOS) stress signalling pathway in salt tolerance and ion homeostasis [44]. The SOS signal transduction pathway is formed of three main proteins, SOS1, SOS2, and SOS3. SOS1, encodes a Na^+/H^+ antiporter of plasma membrane and at cellular level is crucial in Na^+ efflux regulation. It also facilitates root to shoot long distance Na^+ transport. In plants, this protein overexpression confers salt tolerance [45]. Rice nucleolin protein (OsNUC1), a multifunctional protein having two isoforms, OsNUC1-S and OsNUC1-L, involved in the tolerance of salt-stress. The OsNUC1-S gene overexpression leads to improvement in the productivity of rice under saline conditions, which is further associated with a gas exchange parameters (stomatal conductance, net photosynthesis, and transpiration rate) better behaviour and higher contents of carotenoids [46].

SALINITY AND NUTRITIONAL IMBALANCES

Plant's native soil environment is the source of mineral nutrients that are divided into macro and micro-groups on the basis of quantity required by plants. These nutrients are crucial for plant metabolism, biochemistry, growth, and development [43]. Most of the crop plants are glycophytes, which are adapted to grow in non-saline soil environment but their growth is inhibited or even completely prevented by NaCl concentrations of 100–200 mM, resulting in plant death [44]. These results indicate that nutrition imbalances occur in crop plants growing in saline soil conditions. Salt accumulation in the root zone results in the development of osmotic stress and disrupts cell ion homeostasis which adversely affects plant growth. This high salt accumulation may inhibit photosynthesis and protein synthesis, damage chloroplasts and other organelles & inactivate enzymes [45].

In saline soils, there is the presence of extreme ratios of Na^+/Ca^+, Na^+/K^+, Ca^{2+}, Mg^{2+}, and Cl^-/NO_3^- ions causing nutrient deficiencies or imbalances, due to the competition of Na^+ and Cl^- with nutrients such as K^+, Ca^{2+}, and NO_3^-. These interactions between the ions often lead to Na^+-induced Ca^{2+} and/or K^+ deficiencies and Ca^{2+}-induced Mg^{2+} deficiencies disrupting mineral nutrition acquisition by plants in two ways. Firstly, the ionic strength of the soil solution, regardless of its composition, can influence nutrient uptake and translocation. The second and more common mechanism by which salinity affects the mineral relations of plants is due to the reduction of nutrient availability by competing with major ions (Na^+ and Cl^-) in the soil solution [46]. Therefore, the uptake of nutrients and their use by the plants in an efficient way in salt-affected soils is reduced because of the effect of salinity on the availability of nutrients, competitive uptake of ions from the soil, transport within the plant or may be caused by physiological inactivation of a given nutrient resulting in an increase in the plants internal requirement for that essential element resulting in nutrition imbalances [47].

SALINITY AND IONIC TOXICITY

Despite years of study, among the most mysterious issues concerning plant salt stress remains the mechanism(s) by which Na^+ and Cl^- are rooted. Ion absorption can proceed through the symplastic and apoplastic pathways [47]. The apoplastic direction is a simple flow path from the exterior and the xylem. In most cases, the output of such a flow bypass is less than 1% of the volume flow by transpiration [48]. Salt can reach the root *via* non-selective ion channels (NSCCs) that deliver sodium through the plasma membrane [49]. NSCCs are controlled by various salt-induced signals like 3, 5-cyclic guanosine monophosphate, calcium, and ROS. Certain channels and transporters might also participate, although their specific role in plant sodium imports is being addressed. The occurrence of abundant soluble salts throughout the soil interferes with the absorption and metabolism of mineral nutrients that are important to plants. Acceptable ion proportions include a method for the physical response of crops in reference to its development and growth [50]. Though, enhanced salt absorption promotes particular ion toxicity such as high Na^+, Cl^- or sulfate (SO_4^{2-}) that reduces the absorption of vital nutrients such as potassium (K^+), phosphorus (P), calcium (Ca^{2+}) and nitrogen (N) [51].

Salinity raises the Na^+ content of *Vicia faba* while the Na^+/K^+ ratio has reduced thus indicating a negative correlation between Na^+ and K^+ [45]. In fact, some of the adverse effects of Na^+ tend to be linked to the mechanical and biological stability of the membranes [52]. Salinity stress induces a rise in the concentrations of Na^+ and Cl^- in *Atriplex griffithii* in root, stem and leaves, and the maximum ion aggregation was observed in leaves preceded by stem and root implying a strong relationship among Na^+ and Cl^- concentration. The amount of Ca^{2+} was diminished in the shoots and leaves of *A. Clawithii* plants cultivated at high salinity; furthermore, stable in roots and K^+ component was diminished with elevated salinity, especially in leaves. But on the other hand, the level of Mg^{2+} was not much influenced in stems and roots; however the decline in leaf has been more pronounced [53].

The decline in the concentration of Ca^{2+} and Mg^{2+} leaves due to salinity stress indicates enhanced membrane integrity and lowered chlorophyll content, respectively [54]. Regardless of the fact that almost all plants concentrate Na^+ and Cl^- ions at large concentrations within their shoot tissue if grown in saline soils, Cl^- toxicity is also a major cause of growth reduction. Tavakkoli *et al.* [10] investigated the degree to which the relevant ion toxicity of Na^+ and Cl^- minimizes the growth of four genotypes of barley cultivated in saline soils under varying salinity applications. Strong Na^+, Cl^-, and NaCl independently decreased barley production; moreover, drops in growth and photosynthesis became strongest during NaCl stress and were primarily proportional to the impact of Na^+ and Cl^- stress. They also stated that Na^+ and Cl^- elimination between barley genotypes were independent processes but that separate genotypes exerted various combinations of both the pathways. High levels of Na^+ diminished K^+ and Ca^{2+} absorption and decreased photosynthesis primarily by restricting stomatological conductivity, although elevated Cl^- concentration reduced photosynthetic potential due to non-stomatological consequences and chlorophyll depletion [55]. Nevertheless, there are several strategies that plants use to counter salt stress, sustain homeostasis and resolve ion toxicity [56 - 58]. Most of these processes involve limitations on the processes involved in salt absorption, regulation of long-range salt transport, compartmentalization of salt, extrusion of salt from the plant and prioritization of the preservation of the K^+ / Na^+ ratio in the cytosol.

OSMOTIC EFFECT OF SALINITY

Salinity poses a problem for plant species; elevated amounts of inorganic minerals in the atmosphere produce osmotic and water stress, however, at the same period produce some osmotic agents that reduce the osmotic ability of the cells and thus avoid water loss. Large amounts of salts in soil hinder the capability of plants to withdraw water from the soil by interfering with the osmotic gradient between plant cells and surrounding soil. Due to the occurrence of specific ionic toxicities, the osmotic stress results in the reduction of both crop growth and yield by inducing leaf damage and defoliation. Zhang *et al.* [59] studied the effect of salinity on two halophytes namely *Haloxylon ammodendron, Suaeda physophora* and one xerophyte *Haloxylon persicum* and they found that salinity affects seed germination in all three plants but it merely interferes with the osmotic gradient in halophytes and is not detrimental but in the xerophytic plant, it caused ionic toxicity which can be fatal in extreme cases.

Considering the physiological responses of plant to salinity, the influence of proteins regarding responses to both ionic and osmotic effects of salinity can be observed. The osmotic effect not only significantly causes osmotic, but also induced mechanical stress on the cells of plant, an increased biosynthesis of some proteins as well as organic compounds that are osmotically active showing osmoprotective functions like LEA proteins [49]. Salt accumulation in the root zone results in the development of osmotic stress and disrupts cell ion homeostasis which adversely affects plant growth. This high salt accumulation may inhibit photosynthesis and protein synthesis, damage chloroplasts and other organelles & inactivate enzymes [60].

PHOTOSYNTHETIC PIGMENTS AND PHOTOSYNTHESIS

Photosynthesis is among the most crucial biochemical pathway *via* which plants transform solar energy into chemical energy. The lowering in photosynthetic levels in salt stress crops is primarily due to the decline in the water potential. Photosynthesis is often hindered where high levels of Na^+ and/or Cl^- are produced in chloroplasts [61]. In numerous studies, declined chlorophyll concentration during salt stress is a widely recorded phenomenon. In a separate analysis, *O. sativa* subjected to 100 mM NaCl treatment, recorded a decline of 30, 36 and 45percent in chlorophyll a, carotenoid and chlorophyll b concentration relative to control [62]. The depletion of chlorophyll a and b concentration of leaves was noticed during NaCl application whereby chlorophyll b concentration of leaves has been more influenced than pigment level in *O. sativa* leaves.

Saha *et al.* [63] noted a linear decline in total chlorophyll, chlorophyll a, chlorophyll b, carotenoids and xanthophyll levels and strength of chlorophyll fluorescence in *Vigna radiata* during enhanced concentrations of NaCl application. The pigment amount reduced on aggregate by 31 percent for overall chlorophyll, 22 percent for chlorophyll a, 45 percent for chlorophyll b, 14 percent for carotene and 19 percent for xanthophylls in comparison to its regulation. From one of the cucumber tests, it was demonstrated that the overall chlorophyll amount of the leaf significantly decreased with an increase in NaCl concentrations. The reduction in cumulative chlorophyll concentration was 12, 21 and 30 percent at 2 and 3 and 5 dS m−1 of salt stress *versus* non-treated plants [64]. Related with a drop in pigment amounts, the chlorophyll fluorescence amplitude was also lost on an average by 16%. Generally, chlorophyll "a" predominates over chlorophyll "b" in plants, although their levels are similar to growing salinity. Decreased stress chlorophyll concentration is a widely documented phenomenon and can be attributed to different causes in numerous studies, one of which is linked to the degradation of the membrane [65].

Photosystem II (PS II) is a fairly vulnerable area of the salt stress photosynthetic system [66]. Kalaji *et al.* [67] documented that salinity distress impacts barley development by altering the chlorophyll fluorescence (PS II) and the oxygen function of the complex. In addition, Mittal *et al.* [68] demonstrated that salt stress influences the production of *Brassica juncea* by influencing photosynthetic (PS II) and electron transport levels and protein D1. Such factors that decrease photosynthetic levels during salt stress include dehydration of cell membranes that reduce carbon dioxide permeability, enhanced senescence, alterations enzyme activity due to structural modifications in cytoplasmic content, and negative feedback due to lower sink activity [69].

CONCLUSION

The growing population of human demands more food which can be achieve in a sustainable way by enhancing the yield of crop with the minimal addition of resources. Salinity stress is a key constraint to the plant growth, biomass production and yield in the field and horticultural crops. The production of crops needs a high input of water and fertilizers which in turn increases soil salinity. The management strategies involving irrigation and fertilization must keep in mind the influence of salinity on growth of crop, soil properties, crop salt tolerance, and effects on the use of water and soil salinity. Halophytes are physiologically well adapted to counteract this stress but there are some instances where their adaptation techniques fail. Non-saline plants can survive some amount of salt stress but most of the time plant growth and overall productivity is greatly affected. It is difficult to stop natural sources of salts in contributing to soil salinity but human-induced salinity can be stopped by following the proper agricultural practices. The use of biofertilizers also found effective in increasing the salt tolerance potential of field and horticultural crops by reducing salinization of soil.

CONSENT FOR PUBLICATION

Not applicable.

CONFLICT OF INTEREST

No potential conflict of interest was reported by all author(s).

ACKNOWLEDGEMENTS

We are grateful to the Indian Council of Medical Research and DST-FIST Programme, Department of Science and Technology (DST), New Delhi (India).

REFERENCES

[1]   Munns R, Tester M. Mechanisms of salinity tolerance. Annu Rev Plant Biol 2008; 59(59): 651-81.
        [http://dx.doi.org/10.1146/annurev.arplant.59.032607.092911] [PMID: 18444910]

[2]   Zhang JL, Shi H. Physiological and molecular mechanisms of plant salt tolerance. Photosynth Res 2013; 115(1): 1-22.
        [http://dx.doi.org/10.1007/s11120-013-9813-6]

[3]   Wu GQ, Jiao Q, Shui QZ. Effect of salinity on seed germination, seedling growth, and inorganic and organic solutes accumulation in sunflower (*Helianthus annuus* L.). PSE 2015; 61(5): 220-6.

[4]   Abdallah SB, Aung B, Amyot L, *et al.* Salt stress (NaCl) affects plant growth and branch pathways of carotenoid and flavonoid biosyntheses in *Solanum nigrum*. Acta Physiol Plant 2016; 38(3): 72.

[5]   Ashraf M. Biotechnological approach of improving plant salt tolerance using antioxidants as markers. Biotechnol Adv 2009; 27(1): 84-93.
        [http://dx.doi.org/10.1016/j.biotechadv.2008.09.003] [PMID: 18950697]

[6]   Taiz L, Zeiger E. Plant Physiology. 3ʳᵈ edn. Sunderland, UK: Sinauer 2002; p.]690.

[7]   Singh R. Effect of salinity on plant growth. IJRSI 2015; 2(1): 78-9.

[8]   Rani S, Sharma MK, Kumar N. Impact of salinity and zinc application on growth, physiological and yield traits in wheat. Curr Sci 2019; 116(8): 25.
        [http://dx.doi.org/10.18520/cs/v116/i8/1324-1330]

[9]   Strogonov BP. Physiological basis of salt tolerance of plants. In: Poljakoff-Mayber A, Meyer AM, Eds. Israel Program Sci. Jerusalem 1962; p. 279.

[10]   He L, Ban Y, Inoue H, Matsuda N, Liu J, Moriguchi T. Enhancement of spermidine content and antioxidant capacity in transgenic pear shoots overexpressing apple spermidine synthase in response to salinity and hyperosmosis. Phytochemistry 2008; 69(11): 2133-41.
        [http://dx.doi.org/10.1016/j.phytochem.2008.05.015] [PMID: 18586287]

[11]   Roy SJ, Negrão S, Tester M. Salt resistant crop plants. Curr Opin Biotechnol 2014; 26(26): 115-24.
        [http://dx.doi.org/10.1016/j.copbio.2013.12.004] [PMID: 24679267]

[12]   Gupta B, Huang B. Mechanism of salinity tolerance in plants: physiological, biochemical, and molecular characterization. Int J Genomics 2014; 2014: 701596.
        [http://dx.doi.org/10.1155/2014/701596] [PMID: 24804192]

[13]   Radić S, Štefanić PP, Lepeduš H, Roje V, Pevalek-Kozlina B. Salt tolerance of *Centaurea ragusina* L. is associated with efficient osmotic adjustment and increased antioxidative capacity. Environ Exp Bot 2013; 1(87): 39-48.
        [http://dx.doi.org/10.1016/j.envexpbot.2012.11.002]

[14]   Flowers TJ, Flowers SA. Why does salinity pose such a difficult problem for plant breeders? Agric Water Manag 2005; 78-24.(1-2): 15-24. 15

[15]   Jouyban Z. The effects of salt stress on plant growth. J Eng Appl Sci (Asian Res Publ Netw) 2012; 2(1): 7-10.

[16]   Muhammad Z, Hussain F, Rehmanullah M, Majeed A. Effect of halopriming on the induction of NaCl salt tolerance in different wheat genotypes. Pak J Bot 2015; 47(5): 1613-20.

[17]   FAO. Global network on integrated soil management for sustainable× use of salt-affected soils 2002. http://www.fao.org/ag/AGL/agll/spush/ intro. html

[18]   Manchanda G, Garg N. Salinity and its effects on the functional biology of legumes. Acta Physiol Plant 2008; 30(5): 595-618.
[http://dx.doi.org/10.1007/s11738-008-0173-3]

[19]   Parihar P, Singh S, Singh R, Singh VP, Prasad SM. Effect of salinity stress on plants and its tolerance strategies: a review. ESPR 2015; 22(6): 4056-75.
[http://dx.doi.org/10.1007/s11356-014-3739-1]

[20]   Li Y. Effect of salinity on seed germination and seedling growth in three salinity plants. Pak J Biol Sci 2008; 11(9): 1268-72.

[21]   Zhang S, Song J, Wang H, Feng G. Effect of salinity on seed germination, ion content and photosynthesis of cotyledons in halophytes or xerophyte growing in central Asia. Plant Ecol 2010; 3(4): 259-67.
[http://dx.doi.org/10.1093/jpe/rtq005]

[22]   Rahman M, Ungar IA. The effect of salinity on seed germination and seedling growth of *Echinochloa crusgalli*. OHIO J SCI 1990; 90(1): 13-5.

[23]   Shafi M, Bakhat J, Khan MJ, Khan MA, Anwar S. Effect of salinity on yield and ion accumulation of wheat genotypes. Pak J Bot 2010; 42(6): 4113-21.

[24]   Colla G, Roupahel Y, Cardarelli M, Rea E. Effect of salinity on yield, fruit quality, leaf gas exchange, and mineral composition of grafted watermelon plants. Hort Science 2006; 41(3): 622-7.
[http://dx.doi.org/10.21273/HORTSCI.41.3.622]

[25]   Mosisa B, Shanko D, Tamiru F, Kasirajan AK. Effect of NaCl salinity in genotypic variation of maize (*Zea mays* L.) during early growth stage in low land soil of Ethiopia. J Adv Botany Zool 2017; 5(4): 1-4.

[26]   Rengasamy P. World salinization with emphasis on Australia. J Exp Bot 2006; 57(5): 1017-23.

[27]   Yamaguchi T, Blumwald E. Developing salt-tolerant crop plants: challenges and opportunities. Trends Plant Sci 2005; 10(12): 615-20.
[http://dx.doi.org/10.1016/j.tplants.2005.10.002]

[28]   Shrivastava P, Kumar R. Soil salinity: a serious environmental issue and plant growth promoting bacteria as one of the tools for its alleviation. Saudi J Biol Sci 2015; 22(2): 123-31.

[29]   Amer KH. Corn crop response under managing different irrigation and salinity levels. Agric Water Manag 2010; 97(10): 1553-63.
[http://dx.doi.org/10.1016/j.agwat.2010.05.010]

[30]   Maas EV, Grattan SR. Crop yields as affected by salinity. Agricultural drainage 1999; 38: 55-108.
[http://dx.doi.org/10.2134/agronmonogr38.c3]

[31]   Stavridou E, Hastings A, Webster RJ, Robson PR. The impact of soil salinity on the yield, composition and physiology of the bioenergy grass Miscanthus× giganteus. Glob Change Biol Bioenergy 2017; 9(1): 92-104.
[http://dx.doi.org/10.1111/gcbb.12351]

[32]   Cullu MA. Estimation of the effect of soil salinity on crop yield using remote sensing and geographic

information system. Turk J Agric For 2003; 27(1): 23-8. 3

[33]   Demidchik V. Mechanisms of oxidative stress in plants: from classical chemistry to cell biology. Environ Exp Bot 2015; 1(109): 212-28.
[http://dx.doi.org/10.1016/j.envexpbot.2014.06.021]

[34]   AbdElgawad H, Zinta G, Hegab MM, Pandey R, Asard H, Abuelsoud W. High salinity induces different oxidative stress and antioxidant responses in maize seedlings organs. Front Plant Sc 2016; 7: 276. 8

[35]   Chaparzadeh N, D'Amico ML, Khavari-Nejad RA, Izzo R, Navari-Izzo F. Antioxidative responses of *Calendula officinalis* under salinity conditions. Plant Physiol Biochem 2004; 42(9): 695-701.
[http://dx.doi.org/10.1016/j.plaphy.2004.07.001]

[36]   Chawla S, Jain S, Jain V. Salinity induced oxidative stress and antioxidant system in salt-tolerant and salt-sensitive cultivars of rice (*Oryza sativa* L.). J Plant Biochem Biot 2013; 22(1): 27-34.

[37]   Pérez-López U, Robredo A, Lacuesta M, *et al.* The oxidative stress caused by salinity in two barley cultivars is mitigated by elevated CO_2. Physiol Plant 2009; 135(1): 29-42.
[http://dx.doi.org/10.1111/j.1399-3054.2008.01174.x] [PMID: 19121097]

[38]   Gill SS, Tuteja N. Reactive oxygen species and antioxidant machinery in abiotic stress tolerance in crop plants. Plant Physiol Biochem 2010; 48(12): 909-30.
[http://dx.doi.org/10.1016/j.plaphy.2010.08.016] [PMID: 20870416]

[39]   El-Ramady H, Alshaal T, Elhawat N, *et al.* Plant nutrients and their roles under saline soil conditions. In: Hasanuzzaman M, Fujita M, Oku H, Nahar K, Hawrylak-Nowak B, Eds. Plant Nutrients and Abiotic Stress Tolerance. Singapore : Springer 2018; pp. 297-324.
[http://dx.doi.org/10.1007/978-981-10-9044-8_13]

[40]   Machado RM, Serralheiro RP. Soil salinity: effect on vegetable crop growth. Management practices to prevent and mitigate soil salinization. Hortic 2017; 3(2): 30.
[http://dx.doi.org/10.3390/horticulturae3020030]

[41]   Niste M, Vidican R, Rotar I, Stoian V, Pop R, Miclea R. Plant nutrition affected by soil salinity and response of Rhizobium regarding the nutrients accumulation. Pro-Environment/ProMediu 2014; 7(18): 19.

[42]   Grattan SR, Grieve CM. Mineral nutrient acquisition and response by plants grown in saline environments.Handbook of plant and crop stress. 1999; 2: pp. 203-9.

[43]   Hussain S, Zhu C, Bai Z, *et al.* iTRAQ-based protein profiling and biochemical analysis of two contrasting rice genotypes revealed their differential responses to salt stress. Int J Mol Sci 2019; 20(3): 547.
[http://dx.doi.org/10.3390/ijms20030547] [PMID: 30696055]

[44]   Hasegawa PM, Bressan RA, Zhu J-K, Bohnert HJ. Plant cellular andmolecular responses to high salinity. Annu Rev Plant Physiol Plant Mol Biol 2000; 51: 463-99.
[http://dx.doi.org/10.1146/annurev.arplant.51.1.463] [PMID: 15012199]

[45]   Shi H, Quintero FJ, Pardo JM, Zhu J-K. The putative plasma membrane Na(+)/H(+) antiporter SOS1 controls long-distance Na(+) transport in plants. Plant Cell 2002; 14(2): 465-77.
[http://dx.doi.org/10.1105/tpc.010371] [PMID: 11884687]

[46]   Boonchai C, Udomchalothorn T, Sripinyowanich S, Comai L, Buaboocha T, Chadchawan S. Rice overexpressing OsNUC1-S reveals dierential gene expression leading to yield loss reduction after salt stress at the booting stage. Int J Mol Sci 2018; 19(12): 3936.
[http://dx.doi.org/10.3390/ijms19123936] [PMID: 30544581]

[47]   Maathuis FJ, Ahmad I, Patishtan J. Regulation of Na+ fluxes in plants. Front Plant Sc 2014; 16(5): 467.

[48]   Krishnamurthy P, Ranathunge K, Franke R, Prakash HS, Schreiber L, Mathew MK. The role of root

apoplastic transport barriers in salt tolerance of rice (*Oryza sativa* L.). Planta 2009; 230(1): 119-34.

[49] Joseph B, Jini D. Proteomic analysis of salinity stress-responsive proteins in plants. Asian J Plant Sci 2010; 9: 307-13.
[http://dx.doi.org/10.3923/ajps.2010.307.313]

[50] Demidchik V, Maathuis FJ. Physiological roles of nonselective cation channels in plants: from salt stress to signalling and development. New Phytol 2007; 175(3): 387-404.
[http://dx.doi.org/10.1111/j.1469-8137.2007.02128.x] [PMID: 17635215]

[51] Wang W, Vinocur B, Altman A. Plant responses to drought, salinity and extreme temperatures: towards genetic engineering for stress tolerance. Planta 2003; 218(1): 1-14.
[http://dx.doi.org/10.1007/s00425-003-1105-5] [PMID: 14513379]

[52] Zhu JK. Plant salt tolerance. Trends Plant Sci 2001; 6(2): 66-71.
[http://dx.doi.org/10.1016/S1360-1385(00)01838-0] [PMID: 11173290]

[53] Kurth E, Cramer GR, Läuchli A, Epstein E. Effects of NaCl and $CaCl_2$ on cell enlargement and cell production in cotton roots. Plant Physiol 1986; 82(4): 1102-6.
[http://dx.doi.org/10.1104/pp.82.4.1102] [PMID: 16665141]

[54] Khan MA, Ungar IA, Showalter AM. Effects of salinity on growth, water relations and ion accumulation of the subtropical perennial halophyte, *Atriplex griffithii* var. stocksii. Ann Bot 2000; 85(2): 225-32.
[http://dx.doi.org/10.1006/anbo.1999.1022]

[55] Parida AK, Das AB, Mittra B. Effects of salt on growth, ion accumulation, photosynthesis and leaf anatomy of the mangrove, *Bruguiera parviflora*. Trees (Berl) 2004; 18(2): 167-74.
[http://dx.doi.org/10.1007/s00468-003-0293-8]

[56] Tavakkoli E, Fatehi F, Coventry S, Rengasamy P, McDonald GK. Additive effects of Na^+ and Cl^- ions on barley growth under salinity stress. J Exp Bot 2011; 62(6): 2189-203.
[http://dx.doi.org/10.1093/jxb/erq422] [PMID: 21273334]

[57] Parida AK, Das AB. Salt tolerance and salinity effects on plants: a review. Ecotoxicol Environ Saf 2005; 60(3): 324-49.
[http://dx.doi.org/10.1016/j.ecoenv.2004.06.010] [PMID: 15590011]

[58] Zhang M, Qin Z, Liu X. Remote sensed spectral imagery to detect late blight in field tomatoes. Precis Agric 2005; 6(6): 489-508.
[http://dx.doi.org/10.1007/s11119-005-5640-x]

[59] Zhang S, Song J, Wang H, Feng G. Effect of salinity on seed germination, ion content and photosynthesis of cotyledons in halophytes or xerophyte growing in Central Asia. Plant Ecol 2010; 3(4): 259-67.
[http://dx.doi.org/10.1093/jpe/rtq005]

[60] Niste M, Vidican R, Rotar I, Stoian V, Pop R, Miclea R. Plant nutrition affected by soil salinity and response of Rhizobium regarding the nutrients accumulation. Pro-Environment/ProMediu 2014; 7(18): 19.

[61] Chutipaijit S, Cha-um S, Sompornpailin K. High contents of proline and anthocyanin increase protective response to salinity in'*Oryza sativa*'L. spp.'indica'. Aust J Crop Sci 2011; 5(10): 1191.

[62] Amirjani MR. Effect of salinity stress on growth, sugar content, pigments and enzyme activity of rice. Int J Bot 2011; 7: 73-81.
[http://dx.doi.org/10.3923/ijb.2011.73.81]

[63] Saha P, Chatterjee P, Biswas AK. NaCl pretreatment alleviates salt stress by enhancement of antioxidant defense system and osmolyte accumulation in mungbean (*Vigna radiata* L. Wilczek). Indian J Exp Biol 2010; 48(6): 593-600.

[64] Khan MM, Al-Mas'oudi RS, Al-Said F, Khan I. Salinity effects on growth, electrolyte leakage,

chlorophyll content and lipid peroxidation in cucumber (*Cucumis sativus* L.). International Conference on Food and Agricultural Sciences. Singapore: IACSIT Press 2013; pp. 28-32.

[65] Mane AV, Karadge BA, Samant JS. Salinity induced changes in photosynthetic pigments and polyphenols of *Cymbopogon nardus* (L.) Rendle. J chem pharm 2010; 2(3): 338-47.

[66] Allakhverdiev SI, Sakamoto A, Nishiyama Y, Inaba M, Murata N. Ionic and osmotic effects of NaCl-induced inactivation of photosystems I and II in *Synechococcus* sp. Plant Physiol 2000; 123(3): 1047-56.
[http://dx.doi.org/10.1104/pp.123.3.1047] [PMID: 10889254]

[67] Kalaji HM, Bosa K, Kościelniak J, Żuk-Gołaszewska K. Effects of salt stress on photosystem II efficiency and CO_2 assimilation of two Syrian barley landraces. Environ Exp Bot 2011; 73: 64-72.
[http://dx.doi.org/10.1016/j.envexpbot.2010.10.009]

[68] Mittal S, Kumari N, Sharma V. Differential response of salt stress on *Brassica juncea*: photosynthetic performance, pigment, proline, D1 and antioxidant enzymes. Plant Physiol Biochem 2012; 54: 17-26.
[http://dx.doi.org/10.1016/j.plaphy.2012.02.003] [PMID: 22369937]

[69] Iyengar ER, Reddy MP. Photosynthesis in highly salt-tolerant plants. Handbook of Photosynthesis. New York: Dekker 1996; pp. 897-909.

CHAPTER 4

Temperature Rising Patterns and Insights into the Impacts of Consequent Heat Stress on Edible Plants

Arpna Kumari[1,†], **Sneh Rajput**[1,†], **Saroj Arora**[1] and **Rajinder Kaur**[1,*]

[1] Department of Botanical and Environmental Sciences, Guru Nanak Dev University, Amritsar, Punjab 143005, India

Abstract: Rapid urbanization and land-use transition contribute to the rise in the thermal scale of cities as well as small towns and villages worldwide. The equilibrium between the incoming solar energy and the outgoing terrestrial energy regulates the temperature. Nevertheless, the temperature, as we know, varies from place to place, and it also affects the natural processes as well as surrounding flora and fauna. On the other hand, temperature beyond the physiologically optimal limit is called high temperature that adversely affects the growth and development of plants as it has significant impacts on both the vegetative and reproductive phases of the plant life cycle. The extremely high temperature is referred to as heat stress which is reported as one of the devastating abiotic stressors. In plants, heat stress triggers various morpho-physiological changes in plants that affect their growth and economic outcomes *via* accelerating reactive oxygen species generation, reduced carbon assimilation, degradation and denaturation of proteins, lipid peroxidation of membranes, *etc*. Several conventional and modern strategies have been employed to resolve heat stress-induced damages in plants. Therefore, the present work is an outcome of extensive literature focused on the factors responsible for temperature variations' patterns, morpho-physiological responses of crops, and impacts on the economic yields of edible plants.

Keywords: Heat stress, Germination, Growth and development, Physiological responses, Sustainable crop production.

INTRODUCTION

Since 1750 BC, the concentration of various greenhouse gases like carbon dioxide, methane, and nitrous oxide have increased significantly due to various anthropogenic activities [1]. The concentrations of carbon dioxide, methane, and nitrous oxide were 280 ppm, 715 ppb, and 270 ppm, respectively in 1750,

* **Corresponding Author Rajinder Kaur:** Department of Botanical and Environmental Sciences, Guru Nanak Dev University, Amritsar, Punjab 143005, India; Tel.: +91-9814860975; E-mail: rajinder.botenv@gndu.ac.in
† Equal Contribution

Tajinder Kaur & Saroj Arora (Eds.)

which were recorded to increase up to 407.4 ppm, 1874.7 ppb, and 329.9 ppm, respectively. An increase in greenhouse gases has resulted in an increase of 0.75 °C in global temperature from 1906 to 2005 [1]. The highest rate of warming has been recorded in the last decade. The Intergovernmental Panel on Climate Change (IPCC) has predicted that the temperature will increase from 1.8-4.0 °C by the end of this century [1]. For the Indian region, an increase of 1.56-5.44 °C is predicted by 2080 [2].

Agriculture is the backbone of the economy of many developing countries including India, but the rising temperature will affect the economic yield of edible plants *via* direct or indirect effects. Higher temperatures may affect the yield both quantitatively and qualitatively, alter growth rates, evapotranspiration and photosynthesis rate, moisture availability through changes in irrigation pattern, *etc* [3 - 14]. This can lead farmers to invest more in loss dipping inputs like pesticides rather than yield-enhancing inputs like fertilizers. The negative effect of high temperature on agriculture includes a decline in food production that can threaten food security, thus entail specific agricultural measures to combat this [14]. Developing nations will be severely hit by temperature variations as 50% of the population rely on agriculture, and 75% of the population lives in rural areas [15]. In emerging economies like India and China, farmers and plant breeders will face extreme challenges as there will be an upsurge in food requirements with increasing population and a decline in calorie accessibility, which will increase malnutrition in children by 20% [16, 17].

Earlier, it was believed that the rising temperature would bring a net benefit to agriculture. However, recent studies have shown a portentous effect of climate change on agriculture as crops can tolerate temperature variations only at threshold level. After that, there can be a sharp decline in productivity [18 - 21]. Studies have shown a non-linear pattern between temperature and crop productivity [22 - 24]. Within a range of 5.5-32°C, the temperature has a positive effect on crop production, whereas temperature higher or lower than that range will negatively affect crop production [20, 25]. Recent studies have shown that with an increase of 1°C temperature, India will lose 4 5 million tonnes in wheat production and cause a decline in production of soybean, potato, groundnut, and mustard by 3-7% [7, 14]. Evocative evidence shows that an increase in temperature increases the risk of pests' attack, more crop diseases, and weed [26].

Moreover, an increase in temperature may alter the precipitation pattern [27]. Higher precipitation rates can cause disease invasion in crops, whereas a low precipitation rate can have a detrimental effect on crops, especially during the growing stages [28]. Lower precipitation rates may cause mild to severe droughts in arid and semi-arid areas, which can reduce the quality and quantity of livestock and crop yield [14]. In India, there are two main cropping seasons, *i.e.*, Rabi and Kharif. Rabi season is mostly influenced by the south-west monsoon, whereas the Kharif season is influenced by the north-east monsoon. Reports have shown that the overall temperature rise is expected to be more during the Rabi season (winter) rather than the Kharif season (monsoon) [8]. Khan *et al.* (2009) reported that the average temperature in India would increase by 1.1-4.5°C in the Rabi season and 0.4-2.0 °C in the Kharif season [29]. The decline in crop yield can increase food prices in the state as well as the national level [30]. Thus, the temperature can play an important factor in affecting the agriculture and economy of India.

Kumar and Parikh (2001b) used the Ricardian approach and revealed that an increase of 2°C in temperature would result in 8.4% damage in farm-level net revenue [31]. Due to an increase in temperature, heat-induced biotic stress increases the attack of pests as well as diseases that prompt farmers to use pesticides in early growing seasons [32]. Jagnani *et al.* (2018) also postulated that a 10% increase in growing days also increases 10% usage of pesticides while reducing the usage of fertilizer by about 5% [32]. Rising temperature will further affect soil erosion, surface runoff, soil water content, biodiversity, organic nitrogen, and carbon content as well as cause an increase in salinization [33]. Therefore, adequate strategies are required to reduce the effect of temperature on soil fertility and improve crop growth and production. Elevated temperature induces physiological and morphological stress as well as molecular and biochemical alterations that affect the growth and yield of plants [15]. Heat stress is more protuberant during reproductive development as compared to the vegetative growth period in most crop species [34]. Elevated temperature induces various physiological changes in plants such as scorching of leaves and stem, root and shoot growth inhibition, leaf senescence, and abscission, or fruit damage that lead to dropping in food productivity [35]. In many cases, high temperature affects plant growth by disturbing the shoot net assimilation rate [36]. High temperature may also cause ion channel, water, and organic solute movement across the plant membrane, which can disrupt processes like photosynthesis and respiration. Moreover, a higher temperature may also cause electrolyte leakage from leaves [37].

TEMPERATURE RISING TRENDS IN INDIA AND WORLDWIDE

Monitoring and analysis of temperature on a regional as well as global scale have gained a lot of importance due to clear signs of global warming. According to the report of IPCC, the mean surface air temperature of the earth has been increased by 0.6°C [38]. Circadian irregularity in temperature trends has been reported from India, and it was found that warming over India has been exclusively contributed by an increase in temperature [39]. The mean annual temperature increased by 0.4°C per 100 years in India during 1901-1982 [40]. Kumar *et al.* (1994) reported that the circadian irregularity in temperature trends in India is quite different from other parts of the world, while the mean annual temperature trend is similar to the other parts of the world [41]. An increase in the mean temperature of India is mainly contributed by the changing patterns of maximum temperature while the trend of minimum temperature remains almost the same. India contributes 17.5% of the world population with only 4% of water resources and 2.4% of the land. A mild change even can decline agricultural productivity by 4.5-9% depending upon the extremities of the temperature [42]. Thus, it is extremely important to understand the rising temperature trends so that appropriate strategies and policies can be formulated concerning the agricultural production demand (Fig. **1a**).

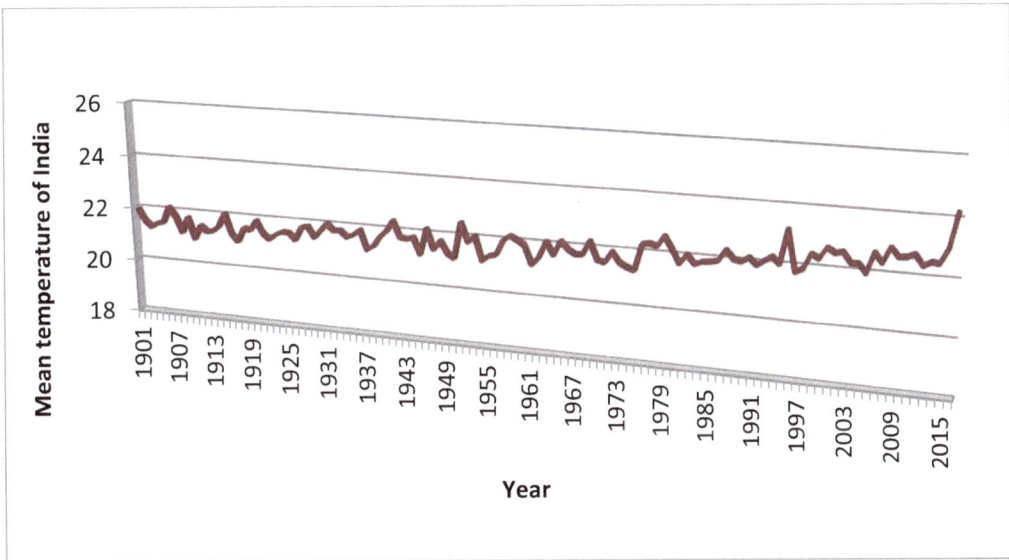

Fig. (1). (a) Temperature modulations trends of India from 1900 to 2020; Source: Open data platform India (Source: data.gov.in).

Since the industrial revolution, the world's mean temperature is increasing day by day. According to a survey conducted by the scientists of Goddard Institute for Space Studies of NASA, the global mean temperature has increased by more than 1°C since 1880. More than $2/3^{rd}$ of warming has increased since 1975 at a rate of 0.15-2°C/decade (Fig. **1b**).

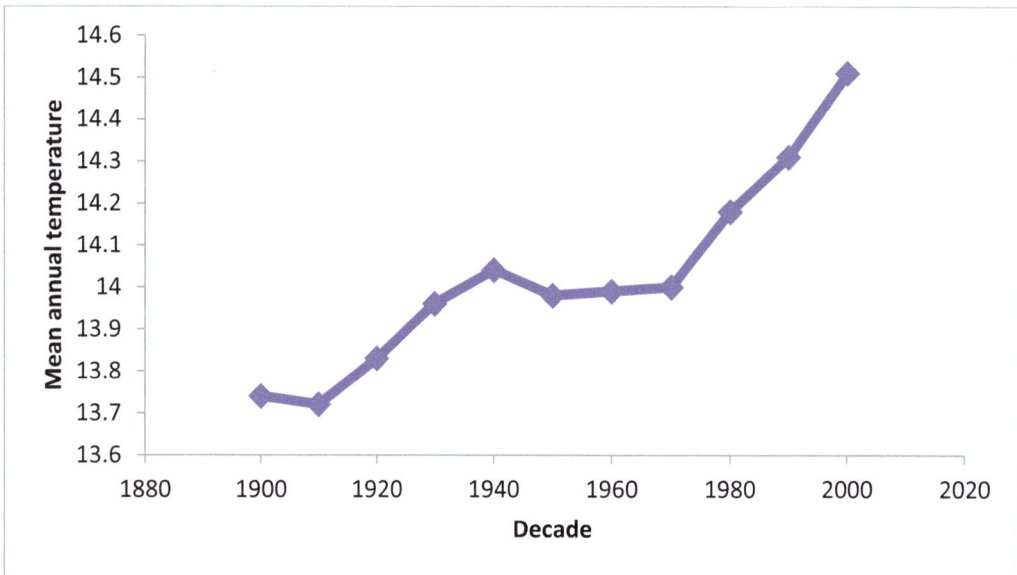

Fig. (1). (b): Temperature modulations trends of the world from 1900 to 2020; Source: https://www.currentresults.com/Environment-Facts/changes-in-earth-tem perature.php

The global mean temperature has not escalated at a steady rate during the decade. From 1880 to 1910, the mean temperature was soared around 13.6-13.7°C. Then the temperature rose about 0.1°C/decade from 1920-1940 whereas the global mean temperature remained around 14°C until 1980. Since 1980, the global mean temperature rose about at a rate of 0.2°C/decade. From 2000-2009 the global mean temperature was 0.61 °C, much higher than the average temperature from 1950-1980. If the temperature will keep on increasing at this rate, then the earth will be warmer by 2°C by the end of this century. According to the available data, the 1990s was the warmest decade recorded ever since the 1880s. According to the National Ocean and Atmospheric Administration (NOAA), NASA 2016 was the warmest year ever followed by 2019 with an average temperature of 0.95°C, just 0.04°C less than the average temperature of 2016.

Nowadays crop scientists are using growth simulation models to estimate the impact of climate change on crop growth and production [43 - 48]. It has been projected that climate change will adversely affect the areas situated at the low latitude and mild effects on mid and high latitude [44 - 48]. In summer 2018, Europe was badly affected by climatic extremities like temperature increase, heatwaves and drought which lead to extensive harvest fiascos and livestock fodder shortage in many countries [49]. Vogel and co-workers (2019) showed that temperature variations have a stronger effect on crop yield anomalies than precipitation extremes or droughts [50]. North America will be badly affected by temperature extremities in the case of maize, spring wheat, and soy production. In Asia, maize and rice production will be affected whereas in Europe spring wheat production will be affected [50]. Moreover, it has been a proven fact that increased warm day frequencies negatively impact crop yields. On the other hand, increased frequency of cold daily temperature also has a deleterious effect on all crop varieties except spring wheat [50]. Over the past three decades, global maize production has been declined to 3.8-12.5% [51, 52]. These effects may be more severe in the future as an increase in 1°C in temperature may cause 10-20% decrease in maize yield and an increase of 1-3°C will decline potato yield by 18-32% [53 - 55].

PLANTS RESPONSES TO THE RISING TEMPERATURE

Stress is a collective term used for both external abiotic and biotic constraints, which restrict plant growth and development by affecting photosynthesis and reducing plant capacity for carbon assimilation [56]. Today, people dealing with food production face most difficulties as crop outputs are not enough to meet food demands [57]. Heat stress is one of the major environmental constraints responsible for the low productivity of crops. Temperature variations sometimes lead to frequent heat waves that damage plants not only by changing the temperature during the day but also by the occurrence of heat stress episodes that impair normal homeostasis. One of the long-run consequences of increasing average surface temperature is its detrimental influence on agricultural productivity *via* its interference in metabolic processes of plants that results in significant morpho-physiological alterations. In the case of heat stress, it is observed to interfere with every cell process, resulting in macromolecules malfunction (*i.e.* protein denaturation and partial DNA and RNA strands melting) [58]. Furthermore, the survey of literature revealed that plants under heat stress showed adverse effects on germination, growth & development, biochemical, and physiological processes.

Germination, Growth, and Development

Germination is a crucial phase of the plant life cycle and this process starts with the absorption of water to set the metabolic events required to achieve seed germination [59]. Heat stress affects the entire plant life cycle as it exaggerates the impact of other abiotic stressors. It mediates changes in the physiological conditions of plants considerably during the germination phase which consequently affects the seedling stage.

Biochemical and Physiological Effects of Heat Stress

1. The modulation in the levels of photosynthetic and accessory pigment is

reported as a well-established index under stressed environments in the plants. The pigments play an important role in the regulation of photosynthesis in plants [60]. Photosynthesis is influenced by heat stress according to observations by several researchers. The decline in photosynthesis is directly linked to a decrease in crop yield. Plants undergo several complex reactions for the complete process of carbon assimilation. The photosynthetic apparatus consists of pigments, photosystems, light reactions, and dark reactions [61]. Thus, heat stress might affect any of these collectively or individually. The carbohydrates are the photosynthetic end products. The principal sources of sugar found to be actively involved in stress response approaches are starch, trehalose, raffinose, and fructans [62]. Carbohydrates provide osmoprotection to balance membrane structures and preserve cell turgidity.

2. Regarding proteins under heat stress, heat shock proteins (HSPs) are mainly observed to accumulate in plants that are also termed as molecular chaperones that play roles in folding and assembling proteins. The well-known families of heat stress proteins are the Hsp70 family, family Hsp90, family Hsp100, chaperonins (GroEL and Hsp60), and small Hsp protein. The criterion here used for this categorization of heat shock proteins is their molecular weight [63]. The study of Chandel *et al.* (2013) revealed that in rice, the induction of the different genes and HSPs that worked together in different cascades to combat the consequence of heat stress [64]. In the plants, heat shock factor/heat shock, *i.e.* HSF/HSP acts as a vital pathway that is involved in stress responses [65]. In this mechanism, HSF is reported to combine with heat shock elements, thus mediates the upregulation of heat stress-responsive genes [66]. Nevertheless, after the discovery of HSPs, it has been reported that these factors' roles are not limited to the heat stress management but is also allied with the other stressors management *viz.*, cold, salinity, light, oxidative stress, drought, and pathogenic infections [65].

3. The other plant stress biomarkers (malondialdehyde and hydrogen peroxide) and osmolytes (glycine betaine, proline, *etc.*) were observed to express under heat stress in different crop plants as per several previous studies. Different kinds of stressors, whether biotic or abiotic are reported to trigger a common stress response *i.e.* oxidative stress. The oxidative stress is mainly caused by the accelerated generation of reactive oxygen species (ROS) and produced *via* partial oxygen reduction. The major sites for ROS formation are chloroplasts, mitochondria, and peroxisomes in the cells as these organelles possess high metabolic activity. Thus, plants have a tiered defense system that relies on endogenous enzyme and non-enzyme antioxidants to deal with the increased ROS levels [61].

The recent reports on the effects of heat stress on germination, growth, biochemical, and physiological indices including the activities of antioxidative enzymes are given in the tabulated form (Table **1**).

Table 1. Effects of heat stress on germination, growth, biochemical and physiological indices of crop plants.

S. No.	Crop Plant (s)	Temperature (°C)	Responses	References
1.	*Vigna aconitifolia* (Jacq.) Marechal	42 °C	Total sugar and proline along with the increased activity of CAT, GPOX, and SOD was observed in most of the genotypes under heat stress.	[67]
2.	*Triticum aestivum* L.	22±1, 25±1, 30±1 and 35±1 °C	Sugars, free amino acids, and protein content correlated positively with each other during heat stress.	[68]
3.	*Glycine max* (L.) Merr.	20, 25, and 30 °C	Reduced germination potential of seeds mainly at 30 °C.	[69]
4.	*Solanum lycopersicum* L.	44 °C	Increased chlorophyll content, total soluble carbohydrate, antioxidative enzymes; decreased MDA content.	[70]
5.	*Lycopersicon esculentum* Mill.	38 °C	Formation of reactive oxygen species (superoxide and hydrogen peroxide) was recorded; enhanced proline, malondialdehyde contents, electrolyte leakage; declined growth.	[71]
6.	*Triticum aestivum* L.	35-40 °C	Decreased the relative water content, membrane stability index, chl a, chl b content.	[72]
7.	*Glycine max* (L.) Merr.	>32 °C	Declined grain number and weight through photosynthesis and redox state impairment.	[73]
8.	*Vicia faba* L.	38 °C	Declined growth attributes, total chlorophyll, and carbonic anhydrase, and Rubisco activities.	[74]
9.	*Triticum aestivum* L.	25/24 °C day and night	Decreased total chlorophyll content; enhanced in total carotenoid content, number of ears per plant, number of grains per ear.	[75]
10.	*Valerianella locusta* (L.) Betcke	35/22 °C day and night	Adversely affected the growth and oxidative status of the plant; reduced photosynthetic pigments concentration; increased amounts of H_2O_2.	[76]

(Table 1) cont.....

S. No.	Crop Plant (s)	Temperature (°C)	Responses	References
11.	*Triticum aestivum* L.	34/16 °C, day and night	Declined net CO_2 assimilation, starch, protein content in grain of wheat, seed set, kernel weight, and grain yield; increased proline content.	[77]
12.	*Brassica napus* L.	heat stress (30 °C day/25°C night)	Declined seed germination; reduced seed vigour.	[78]
13.	*Daucus carota* L.	24, 32.5, 35, 37.5, and 40 °C	Decreased percent germination, inhibition index, relative heat tolerance, and heat tolerance index.	[79]
14.	*Solanum lycopersicum* L.	42 °C	Damaged the cell membrane integrity; reduced chlorophyll content and leaf water potential.	[80]
15.	*Triticum aestivum* L.	36/30 °C day and night	Declined grain yield (53.05%); disintegration of membrane structure, chlorophyll and protein molecules were recorded.	[81]
16.	Maize hybrids	38/30 °C day and night	Caused oxidative stress by accelerating the generation of reactive oxygen species; enhanced malondialdehyde contents, reduced photosynthetic components, nutrients uptake, and yield attributes.	[82]
17.	Wheat genotypes	35/15 °C, average maximum/minimum temperature	Enhanced osmotic pressure, proline, glycine betaine, sugar contents; increased sodium and cell membrane stability; declined potassium, K/Na ratio, leaf area, chlorophyll a, chlorophyll b, total chlorophyll, and grain yield over control.	[83]

SUSTAINABLE AGRICULTURE, ECONOMIC YIELD, AND RISING TEMPERATURE

In poor countries, particularly rural areas, agriculture is still a very important source of income and employment [84]. For decades, the optimum yield of agricultural products has been adjusted to local temperatures. Several studies have shown the peaks of agricultural GDP at an average annual temperature of about 14 °C, whereas it is observed to decrease at higher temperatures [85, 86]. The study of Lobell *et al.* (2011) revealed the effect of weather on the development of main crops, finding that global warming has contributed to a decrease of about 3.8% and 5.5% of global corn and wheat development respectively [52]. The GDP of poor countries is reported to affect adversely by increasing temperature. For example, a temperature rise of 1 °C each year decreases the economic growth of poor countries by 1.3%, an impact caused mostly by a decrease in agricultural production and less by industrial output and political uncertainty [84]. In India, in the case of a severe climatic shock, the household incomes recorded to fall up to 60% and the incidence of poverty rises by 12 to 33% [87]. Drought and heat characteristics of these statements are prominently alleged; climate events destroy crop yields, exacerbate the debt burden of farmers, and provoke suicides in response [88]. Such arguments appear plausible with more than half of India's working population working under agriculture, one third below the global poverty line and nearly all experiencing increasing temperatures because of anthropogenic climate change. Previous reports suggested that most crop yield indices are influenced by heat stress, including seed yield per plant, seed per pod, and seed weight [89]. Some recent events of crop economic outcomes loss and suicidal case especially from India are discussed as below:

a. In the study of Chen *et al.* (2016), it was observed that spatio-temporal pattern of heat stress declined the yield of winter wheat significantly. They further revealed that heat and dry stress collectively declined the yield up to 1.28% yield per annum [90].

b. The high temperature observed to induce a significant negative impact on the anthesis stage of wheat genotypes. Furthermore, heat stress is observed to establish a direct association of photosynthesis with starch mobilization, pollen viability, and grain yield [91].

c. High ambient temperature is claimed to have adverse effects at the vegetative and reproductive growth of the crop plant which at the same time contributes to the yield reduction. The authors in this study observed the effects of short-term heat shock and long-term moderate heat on the 13 cultivars of tomato. A significant decrease in performance of all reproductive traits *viz.*, pollen viability, pollen number, female fertility, seeded-fruit set, and flower number per inflorescence was observed under long-term moderate heat (except on inflorescence number). On other hand, short-term heat shock was observed to affect negatively the vegetative traits [92].

d. As per the study of Lawas *et al.* (2018), the high ambient temperature reduces transpiration cooling thereby, drought and heat stress may occur together and may cause much more devasting yield-related perturbations. Thus, for minimizing the consequences induced by these stressors, molecular reactions to combined stress cannot be predicted from single stress. Moreover, heat shock proteins are observed to play an important role in adaptation to combined stress [93].

e. Heat stress is observed to short the development phases of wheat. It also caused a significant decline in total above-ground biomass (*i.e.* 41%). The reduction in total shoot biomass was recorded up to 77 and 58% at grain filling and maturity, respectively. The harvest index was significantly decreased (*i.e.*

36%) in heat-stressed plants. Thus, the study revealed that the plants using various stress management strategies adapt to a heat-stressed environment but also leads to a decline in biomass production [94].

f. Heat stress sensitivity responses of rice accessions during early seed development were studied. Among all rice accessions, the indica and tropical japonica accessions showed severe heat sensitivity, while tropical japonicas (three) and aus accession showed moderate heat sensitivity. The remaining rice accessions were recorded to be heat tolerant. Heat stress is observed to be negatively associated with the seed yield. The accessions displaying high heat sensitivity keep seed size at the cost of the number of fully developed mature seeds. Whereas, rice accessions demonstrating relative tolerance to the transient heat stress maintained the number of fully developed seeds but sacrificed on seed size, in particular seed length [95].

RECENT POSSIBLE APPROACHES TO OVERCOME TEMPERATURE INDUCED STRESS

The climate prediction models indicated that the rise of the global surface temperature by the end of the twenty-first century could reach 2 °C [96]. It is, thereby imperative to cultivate heat-tolerant plants to meet future food demands for the growing population worldwide [97]. The basic mechanisms which can be exploited for the development of heat-tolerant plants are shown in Fig. (2).

Hence, the demands have been raised for the production of heat-tolerant cultivars to respond to current and expected rises in heat stress throughout the world. Innumerous techniques have emerged with time. Recently, due to the rapid advances in genotyping and its falling cost have made widespread use of high-throughput genotyping, although accurate phenotyping has become the greatest constraint to unveil the genetic basis of essential and complex traits, slowing down the progress of breeding programs [99]. Therefore, continuous improvements are required to accelerate the heat-tolerant cultivars generation process *via* involving the recent techniques with the application of traditional approaches.

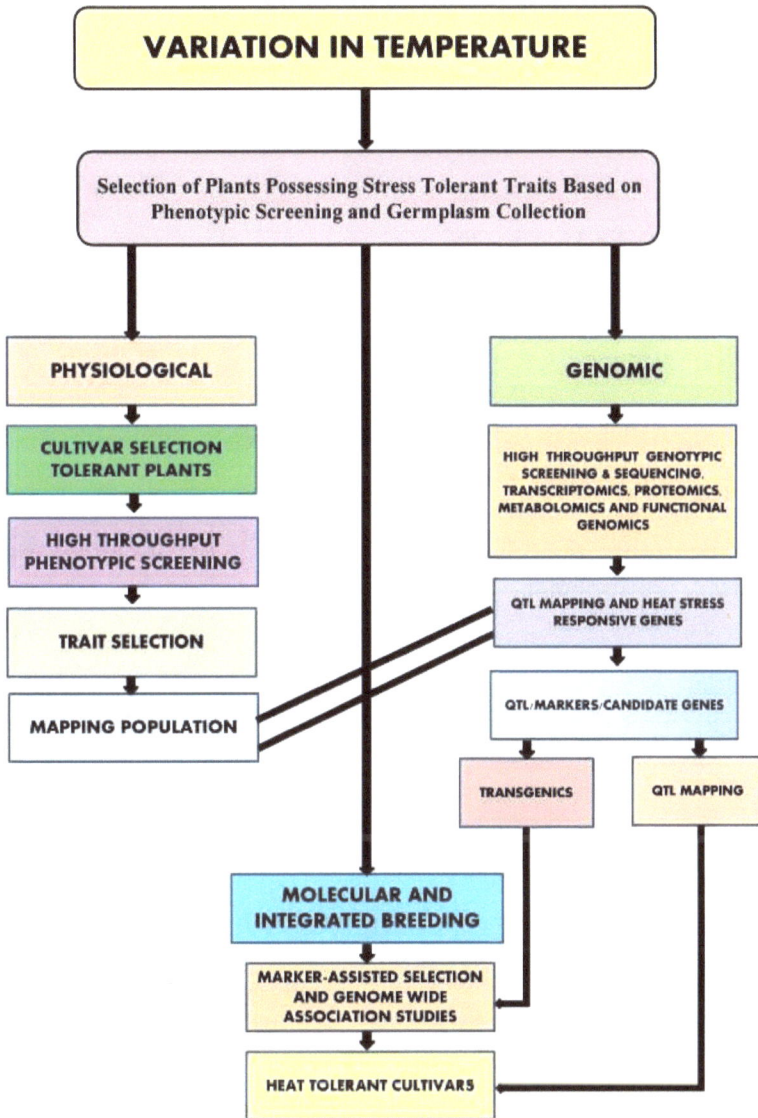

Fig. (2). The scheme of basic mechanism for the generation of heat-tolerant plants using physiological, molecular breeding, and genomics approaches (Adapted from Raza *et al.*, 2019 [98]).

CONCLUSION

Rising temperature particularly in terms of heat stress is one of the most extreme abiotic stressors that influence the growth and yield of crop plants. High-temperature influences all the functions of various enzymes including antioxidative enzymes and perturbs normal metabolism. Work over the last few years has led to a better understanding of the perception and response to heat stress, but there are still some gaps that need to be resolved urgently as per the need of the time. Under present conditions of climate change and population explosion, the production of heat resistant cultivars is necessary to satisfy day by day enhancing world food demands, and a better understanding of the molecular mechanisms involved in heat perception and reaction is imperative.

CONSENT FOR PUBLICATION

Not applicable.

CONFLICT OF INTEREST

The authors declare that they do not have any conflict of interest.

ACKNOWLEDGEMENTS

The authors are highly thankful to the Department of Botanical and Environmental Sciences, Guru Nanak Dev University, Amritsar for providing necessary facilities.

REFERENCES

[1]   Change IPOC. Climate change 2007: The physical science basis: Summary for policymakers 2007.

[2]   Parry M, Parry ML, Canziani O, Palutikof J, Van der Linden P, Hanson C, Eds. Climate change 2007-impacts, adaptation and vulnerability: Working group II contribution to the fourth assessment report of the IPCC. Cambridge University Press 2007; Vol. 4. b

[3]   Morison JI. Intercellular CO_2 concentration and stomatal response to CO_2. In: Zeiger E, Farquhar GD, Cowan IR, Eds. Stomatal Function. Stanford, CA: Stanford Univ. Press 1987; pp. 229-51.

[4]   McNaughton KG, Jarvis PG. Effects of spatial scale on stomatal control of transpiration. Agric For Meteorol 1991; 54(2-4): 279-302.
[http://dx.doi.org/10.1016/0168-1923(91)90010-N]

[5]   Peng S, Huang J, Sheehy JE, *et al*. Rice yields decline with higher night temperature from global warming. Proc Natl Acad Sci USA 2004; 101(27): 9971-5.
[http://dx.doi.org/10.1073/pnas.0403720101] [PMID: 15226500]

[6]   Porter JR, Semenov MA. Crop responses to climatic variation. Philos Trans R Soc Lond B Biol Sci 2005; 360(1463): 2021-35.
[http://dx.doi.org/10.1098/rstb.2005.1752] [PMID: 16433091]

[7]   Aggarwal PK. Global climate change and Indian agriculture: Impacts, adaptation and mitigation. Indian J Agric Res 2008; 78(11): 911.

[8]   Aggarwal PK, Baethegan WE, Cooper P, *et al*. Managing climatic risks to combat land degradation and enhance food security: Key information needs. Procedia Environ Sci 2010; 1: 305-12.
[http://dx.doi.org/10.1016/j.proenv.2010.09.019]

[9]   Douglas I. Climate change, flooding and food security in south Asia. Food Secur 2009; 1(2): 127-36.
[http://dx.doi.org/10.1007/s12571-009-0015-1]

[10]   Wang X, Cai J, Jiang D, Liu F, Dai T, Cao W. Pre-anthesis high-temperature acclimation alleviates damage to the flag leaf caused by post-anthesis heat stress in wheat. J Plant Physiol 2011; 168(6): 585-93.
[http://dx.doi.org/10.1016/j.jplph.2010.09.016] [PMID: 21247658]

[11]   Crawford AJ, McLachlan DH, Hetherington AM, Franklin KA. High temperature exposure increases plant cooling capacity. Curr Biol 2012; 22(10): R396-7.
[http://dx.doi.org/10.1016/j.cub.2012.03.044] [PMID: 22625853]

[12]   Rezaei EE, Webber H, Gaiser T, *et al.* Heat stress in cereals: Mechanisms and modelling. Eur J Agron 2015; 64: 98-113.
[http://dx.doi.org/10.1016/j.eja.2014.10.003]

[13]   Schauberger B, Archontoulis S, Arneth A, *et al.* Consistent negative response of US crops to high temperatures in observations and crop models. Nat Commun 2017; 8(1): 13931.
[http://dx.doi.org/10.1038/ncomms13931] [PMID: 28102202]

[14]   Kumar P, Tokas J, Kumar N, *et al.* Climate change consequences and its impact on agriculture and food security. Int J Chem Stud 2018; 6(6): 124-33.

[15]   Bita CE, Gerats T. Plant tolerance to high temperature in a changing environment: scientific fundamentals and production of heat stress-tolerant crops. Front Plant Sci 2013; 4: 273.
[http://dx.doi.org/10.3389/fpls.2013.00273] [PMID: 23914193]

[16]   Nelson MC, Story M. Food environments in university dorms: 20,000 calories per dorm room and counting. Am J Prev Med 2009; 36(6): 523-6.
[http://dx.doi.org/10.1016/j.amepre.2009.01.030] [PMID: 19356889]

[17]   Chhetri N, Chaudhary P. Green Revolution: Pathways to food security in an era of climate variability and change. J Disaster Res 2011; 6(5): 486-97.
[http://dx.doi.org/10.20965/jdr.2011.p0486]

[18]   Ackerman F, Stanton EA. Climate impacts on agriculture: a challenge to complacency? The Oxford Handbook of the Macroeconomics of Global Warming. Oxford, United Kingdom: Oxford University Press 2015.

[19]   Kang Y, Khan S, Ma X. Climate change impacts on crop yield, crop water productivity and food security-A review. Prog Nat Sci 2009; 19(12): 1665-74.
[http://dx.doi.org/10.1016/j.pnsc.2009.08.001]

[20]   Lobell DB, Baldos UL, Hertel TW. Climate adaptation as mitigation: The case of agricultural investments. Environ Res Lett 2013; 8(1): 015012.
[http://dx.doi.org/10.1088/1748-9326/8/1/015012]

[21]   Fezzi C, Bateman I. The impact of climate change on agriculture: Nonlinear effects and aggregation bias in Ricardian models of farmland values. J Assoc Environ Resour Econ 2015; 2(1): 57-92.
[http://dx.doi.org/10.1086/680257]

[22]   Schlenker W, Hanemann WM, Fisher AC. Water availability, degree days, and the potential impact of climate change on irrigated agriculture in California. Clim Change 2007; 81(1): 19-38.
[http://dx.doi.org/10.1007/s10584-005-9008-z]

[23]   Schlenker W, Roberts MJ. Nonlinear temperature effects indicate severe damages to U.S. crop yields under climate change. Proc Natl Acad Sci USA 2009; 106(37): 15594-8.
[http://dx.doi.org/10.1073/pnas.0906865106] [PMID: 19717432]

[24]   Bertone Oehninger E, Lin Lawell CY, Sanchirico J, *et al.* The effects of climate change on groundwater extraction for agriculture and land-use change 2016 Annual Meeting, July 31-August 2, Boston, Massachusetts 235724, Agricultural and Applied Economics Association.

[25]   Olen B, Wu J, Langpap C. Irrigation decisions for major west coast crops: Water scarcity and climatic determinants. Am J Agric Econ 2016; 98(1): 254-75.

[http://dx.doi.org/10.1093/ajae/aav036]

[26]   Barrett C, Jagnani M, Liu Y, You L. In the weeds: Effects of temperature on agricultural input decisions in moderate climates. 2018 Conference, July 28-August 2, 2018, Vancouver, British Columbia 277019, International Association of Agricultural Economists.

[27]   Neenu S, Biswas AK, Rao AS. Impact of climatic factors on crop production-A review. Agric Rev (Karnal) 2013; 34(2): 97-106.

[28]   Shah R, Srivastava R. Effect of global warming on Indian agriculture. Sustain Environ 2017; 2(4): 366.
       [http://dx.doi.org/10.22158/se.v2n4p366]

[29]   Khan SA, Kumar S, Hussain MZ, *et al.* Climate change, climate variability and Indian agriculture: Impacts vulnerability and adaptation strategies. Climate change and crops. Berlin, Heidelberg: Springer 2009; pp. 19-38.
       [http://dx.doi.org/10.1007/978-3-540-88246-6_2]

[30]   Udmale PD, Ichikawa YS. Drought impacts and adaptation strategies for agriculture and rural livelihood in the Maharashtra state of India. Open Agr J 2014; 8(1).

[31]   Kumar KK, Parikh J. Indian agriculture and climate sensitivity. Glob Environ Change 2001; 11(2): 147-54.
       [http://dx.doi.org/10.1016/S0959-3780(01)00004-8]

[32]   Jagnani M, Barrett CB, Liu Y, You L. Within-season producer response to warmer temperatures: Defensive investments by kenyan farmers. Economic J 2021; 131(633): 392-419.
       [http://dx.doi.org/10.1093/ej/ueaa063]

[33]   Mall RK, Gupta A, Sonkar G. Effect of climate change on agricultural crops. Current developments in biotechnology and bioengineering. Elsevier 2017; pp. 23-46.
       [http://dx.doi.org/10.1016/B978-0-444-63661-4.00002-5]

[34]   Zinn KE, Tunc-Ozdemir M, Harper JF. Temperature stress and plant sexual reproduction: uncovering the weakest links. J Exp Bot 2010; 61(7): 1959-68.
       [http://dx.doi.org/10.1093/jxb/erq053] [PMID: 20351019]

[35]   Vollenweider P, Günthardt-Goerg MS. Diagnosis of abiotic and biotic stress factors using the visible symptoms in foliage. Environ Pollut 2005; 137(3): 455-65.
       [http://dx.doi.org/10.1016/j.envpol.2005.01.032] [PMID: 16005758]

[36]   Wahid A. Physiological implications of metabolite biosynthesis for net assimilation and heat-stress tolerance of sugarcane (*Saccharum officinarum*) sprouts. J Plant Res 2007; 120(2): 219-28.
       [http://dx.doi.org/10.1007/s10265-006-0040-5] [PMID: 17024517]

[37]   Halford NG. New insights on the effects of heat stress on crops. J Exp Bot 2009; 60(15): 4215-6.
       [http://dx.doi.org/10.1093/jxb/erp311] [PMID: 19854798]

[38]   Houghton JT. Climate Change 2001: The Scientific Basis. Johnson, CA 2001.

[39]   Kothawale DR, Rupa Kumar K. On the recent changes in surface temperature trends over India. Geophys Res Lett 2005; 32(18).
       [http://dx.doi.org/10.1029/2005GL023528]

[40]   Hingane LS, Rupa Kumar K, Ramana Murty BV. Long-term trends of surface air temperature in India. J Climatol 1985; 5(5): 521-8.
       [http://dx.doi.org/10.1002/joc.3370050505]

[41]   Kumar KR, Kumar KK, Pant GB. Diurnal asymmetry of surface temperature trends over India. Geophys Res Lett 1994; 21(8): 677-80.
       [http://dx.doi.org/10.1029/94GL00007]

[42]   Ross RS, Krishnamurti TN, Pattnaik S, Pai DS. Decadal surface temperature trends in India based on a new high-resolution data set. Sci Rep 2018; 8(1): 7452.

[http://dx.doi.org/10.1038/s41598-018-25347-2] [PMID: 29748548]

[43]   Fader M, Rost S, Müller C, *et al.* Virtual water content of temperate cereals and maize: Present and potential future patterns. J Hydrol (Amst) 2010; 384(3-4): 218-31.
[http://dx.doi.org/10.1016/j.jhydrol.2009.12.011]

[44]   Liu J, Folberth C, Yang H, Röckström J, Abbaspour K, Zehnder AJ. A global and spatially explicit assessment of climate change impacts on crop production and consumptive water use. PLoS One 2013; 8(2): e57750.
[http://dx.doi.org/10.1371/journal.pone.0057750] [PMID: 23460901]

[45]   Balkovič J, van der Velde M, Skalský R, *et al.* Global wheat production potentials and management flexibility under the representative concentration pathways. Global Planet Change 2014; 122: 107-21.
[http://dx.doi.org/10.1016/j.gloplacha.2014.08.010]

[46]   Elliott J, Kelly D, Chryssanthacopoulos J, *et al.* The parallel system for integrating impact models and sectors (pSIMS). Environ Model Softw 2014; 62: 509-16.
[http://dx.doi.org/10.1016/j.envsoft.2014.04.008]

[47]   Folberth C, Yang H, Gaiser T, *et al.* Effects of ecological and conventional agricultural intensification practices on maize yields in sub-Saharan Africa under potential climate change. Environ Res Lett 2014; 9(4): 044004.
[http://dx.doi.org/10.1088/1748-9326/9/4/044004]

[48]   Rosenzweig C, Elliott J, Deryng D, *et al.* Assessing agricultural risks of climate change in the 21st century in a global gridded crop model intercomparison. Proc Natl Acad Sci USA 2014; 111(9): 3268-73.
[http://dx.doi.org/10.1073/pnas.1222463110] [PMID: 24344314]

[49]   Calls for farm support intensify as Europe struggles with heat wave, drought 2018. https://dw.com/en/calls-for- farmsupport -intensify- as- europe- struggles- with-heat wavedrought/a-44902321

[50]   Vogel E, Donat MG, Alexander LV, *et al.* The effects of climate extremes on global agricultural yields. Environ Res Lett 2019; 14(5): 054010.
[http://dx.doi.org/10.1088/1748-9326/ab154b]

[51]   Lobell DB, Field CB. Global scale climate-crop yield relationships and the impacts of recent warming. Environ Res Lett 2007; 2(1): 014002.
[http://dx.doi.org/10.1088/1748-9326/2/1/014002]

[52]   Lobell DB, Schlenker W, Costa-Roberts J. Climate trends and global crop production since 1980. Science 2011; 333(6042): 616-20.
[http://dx.doi.org/10.1126/science.1204531] [PMID: 21551030]

[53]   Challinor AJ, Watson J, Lobell DB, *et al.* Meta-analysis of crop yield under climate change and adaptation. Nat Clim Chang 2014; 4(4): 287-91.
[http://dx.doi.org/10.1038/nclimate2153]

[54]   Rose G, Osborne T, Greatrex H, *et al.* Impact of progressive global warming on the global-scale yield of maize and soybean. Clim Change 2016; 134(3): 417-28.
[http://dx.doi.org/10.1007/s10584-016-1601-9]

[55]   Hijmans RJ. The effect of climate change on global potato production. Am J Potato Res 2003; 80(4): 271-9.
[http://dx.doi.org/10.1007/BF02855363]

[56]   Grime JP. Evidence for the existence of three primary strategies in plants and its relevance to ecological and evolutionary theory. Am Nat 1977; 111(982): 1169-94.
[http://dx.doi.org/10.1086/283244]

[57]   Food and Agricultural Organization (FAO). Land and plant nutrition management service 2009. www.fao.org/wsfs/forum2050/

[58]   Sajid M, Rashid B, Ali Q, *et al.* Mechanisms of heat sensing and responses in plants. It is not all about Ca^{2+} ions. Biol Plant 2018; 62(3): 409-20.
[http://dx.doi.org/10.1007/s10535-018-0795-2]

[59]   Nonogaki H. Seed germination and reserve mobilization. eLS: John Wiley & Sons, Ltd 2008.

[60]   Bode S, Quentmeier CC, Liao PN, *et al.* On the regulation of photosynthesis by excitonic interactions between carotenoids and chlorophylls. Proc Natl Acad Sci USA 2009; 106(30): 12311-6.
[http://dx.doi.org/10.1073/pnas.0903536106] [PMID: 19617542]

[61]   Kumari A, Kaur R. Di-n-butyl phthalate-induced phytotoxicity in *Hordeum vulgare* seedlings and subsequent antioxidant defense response. Biol Plant 2020; 64: 110-8.
[http://dx.doi.org/10.32615/bp.2019.095]

[62]   Kumari A, Kaur R. Modulation of biochemical and physiological parameters in *Hordeum vulgare* L. seedlings under the influence of benzyl-butyl phthalate. Peer J 2019; 7: e6742.
[http://dx.doi.org/10.7717/peerj.6742] [PMID: 31041151]

[63]   Wang W, Vinocur B, Shoseyov O, Altman A. Role of plant heat-shock proteins and molecular chaperones in the abiotic stress response. Trends Plant Sci 2004; 9(5): 244-52.
[http://dx.doi.org/10.1016/j.tplants.2004.03.006] [PMID: 15130550]

[64]   Chandel G, Dubey M, Meena R. Differential expression of heat shock proteins and heat stress transcription factor genes in rice exposed to different levels of heat stress. J Plant Biochem Biotechnol 2013; 22(3): 277-85.
[http://dx.doi.org/10.1007/s13562-012-0156-8]

[65]   Jacob P, Hirt H, Bendahmane A. The heat-shock protein/chaperone network and multiple stress resistance. Plant Biotechnol J 2017; 15(4): 405-14.
[http://dx.doi.org/10.1111/pbi.12659] [PMID: 27860233]

[66]   Huang LJ, Cheng GX, Khan A, *et al.* CaHSP16.4, a small heat shock protein gene in pepper, is involved in heat and drought tolerance. Protoplasma 2019; 256(1): 39-51.
[http://dx.doi.org/10.1007/s00709-018-1280-7] [PMID: 29946904]

[67]   Harsh A, Sharma YK, Joshi U, *et al.* Effect of short-term heat stress on total sugars, proline and some antioxidant enzymes in moth bean (*Vigna aconitifolia*). Ann Agric Sci 2016; 61(1): 57-64.
[http://dx.doi.org/10.1016/j.aoas.2016.02.001]

[68]   Katakpara ZA, Gajera HP, Vaja KN, *et al.* Evaluation of heat tolerance indices in bread wheat (*Triticum aestivum* L.) genotypes based on physiological, biochemical and molecular markers. Indian J Plant Physiol 2016; 21(2): 197-207.
[http://dx.doi.org/10.1007/s40502-016-0222-7]

[69]   Lima JJ, de Freitas MN, Cantelmo NF, *et al.* Tolerance of soybean seeds to aluminium and heat stress during germination. IJAIR 2017; 5(6): 913-6.

[70]   Siddiqui M, Alamri SA, Mutahhar YY, *et al.* Nitric Oxide and calcium induced physiobiochemical changes in tomato (*Solanum lycopersicum*) plant under heat stress. Fresen Environ Bull 2017; 26: 663-72. a

[71]   Siddiqui MH, Alamri SA, Al-Khaishany MY, *et al.* Sodium nitroprusside and indole acetic acid improve the tolerance of tomato plants to heat stress by protecting against DNA damage. J Plant Interact 2017; 12(1): 177-86. b

[72]   Iqbal M, Raja NI, Mashwani ZU, *et al.* Assessment of AgNPs exposure on physiological and biochemical changes and antioxidative defence system in wheat (*Triticum aestivum* L) under heat stress. IET Nanobiotechnol 2019; 13(2): 230-6.
[http://dx.doi.org/10.1049/iet-nbt.2018.5041] [PMID: 31051456]

[73]   Ergo VV, Lascano R, Vega CR, *et al.* Heat and water stressed field-grown soybean: A multivariate study on the relationship between physiological-biochemical traits and yield. Environ Exp Bot 2018;

148: 1-1.
[http://dx.doi.org/10.1016/j.envexpbot.2017.12.023]

[74]   Siddiqui MH, Alamri SA, Al-Khaishany MYY, *et al.* Mitigation of adverse effects of heat stress on *Vicia faba* by exogenous application of magnesium. Saudi J Biol Sci 2018; 25(7): 1393-401.
[http://dx.doi.org/10.1016/j.sjbs.2016.09.022] [PMID: 30505187]

[75]   Ahmad MA, Prakash P. Impact of high night temperature stress on biochemical and yield parameters in Wheat (*Triticum aestivum* L.). Res J Agric Sci 2018; 9(5): 1137-9.

[76]   Hawrylak-Nowak B, Dresler S, Rubinowska K, Matraszek-Gawron R, Woch W, Hasanuzzaman M. Selenium biofortification enhances the growth and alters the physiological response of lamb's lettuce grown under high temperature stress. Plant Physiol Biochem 2018; 127: 446-56.
[http://dx.doi.org/10.1016/j.plaphy.2018.04.018] [PMID: 29689508]

[77]   Aiqing S, Somayanda I, Sebastian SV, *et al.* Heat stress during flowering affects time of day of flowering, seed set, and grain quality in spring wheat. Crop Sci 2018; 58(1): 380-92.
[http://dx.doi.org/10.2135/cropsci2017.04.0221]

[78]   Rashid M, Hampton JG, Rolston MP, *et al.* Heat stress during seed development affects forage brassica (*Brassica napus* L.) seed quality. J Agron Crop Sci 2018; 204(2): 147-54.
[http://dx.doi.org/10.1111/jac.12251]

[79]   Bolton A, Nijabat A, Mahmood-ur-Rehman M, *et al.* Variation for heat tolerance during seed germination in diverse carrot (*Daucus carota* L.) germplasm. HortScience 2019; 54(9): 1470-6.
[http://dx.doi.org/10.21273/HORTSCI14144-19]

[80]   Jahan MS, Wang Y, Shu S, *et al.* Exogenous salicylic acid increases the heat tolerance in tomato (*Solanum lycopersicum* L.) by enhancing photosynthesis efficiency and improving antioxidant defense system through scavenging of reactive oxygen species. Sci Hortic (Amsterdam) 2019; 247: 421-9.
[http://dx.doi.org/10.1016/j.scienta.2018.12.047]

[81]   Qaseem MF, Qureshi R, Shaheen H. Effects of Pre-anthesis drought, heat and their combination on the growth, yield and physiology of diverse wheat (*Triticum aestivum* L.) genotypes varying in sensitivity to heat and drought stress. Sci Rep 2019; 9(1): 6955.
[http://dx.doi.org/10.1038/s41598-019-43477-z] [PMID: 31061444]

[82]   Hussain HA, Men S, Hussain S, *et al.* Interactive effects of drought and heat stresses on morpho-physiological attributes, yield, nutrient uptake and oxidative status in maize hybrids. Sci Rep 2019; 9(1): 3890.
[http://dx.doi.org/10.1038/s41598-019-40362-7] [PMID: 30846745]

[83]   Hussain SA, Chachar QA, Keerio MI, *et al.* Physiological and biochemical response of wheat genotypes under temperature stress. Pak J Bot 2020; 52: 365-74.
[http://dx.doi.org/10.30848/PJB2020-2(34)]

[84]   Cattaneo C, Peri G. The migration response to increasing temperatures. J Dev Econ 2016; 122: 127-46.
[http://dx.doi.org/10.1016/j.jdeveco.2016.05.004]

[85]   Burke M, Hsiang SM, Miguel E. Global non-linear effect of temperature on economic production. Nature 2015; 527(7577): 235-9.
[http://dx.doi.org/10.1038/nature15725] [PMID: 26503051]

[86]   Dell M, Jones BF, Olken BA. Temperature shocks and economic growth: Evidence from the last half century. Am Econ J Macroecon 2012; 4(3): 66-95.
[http://dx.doi.org/10.1257/mac.4.3.66]

[87]   Thorat A, Vanneman R, Desai S, Dubey A. Escaping and falling into poverty in India today. World Dev 2017; 93: 413-26.
[http://dx.doi.org/10.1016/j.worlddev.2017.01.004] [PMID: 28966435]

[88]   Carleton TA. Crop-damaging temperatures increase suicide rates in India. Proc Natl Acad Sci USA

2017; 114(33): 8746-51.
[http://dx.doi.org/10.1073/pnas.1701354114] [PMID: 28760983]

[89]   Huang R, Liu Z, Xing M, *et al.* Heat stress suppresses *brassica napus* seed oil accumulation by inhibition of photosynthesis and BnWRI1 pathway. Plant Cell Physiol 2019; 60(7): 1457-70.
[http://dx.doi.org/10.1093/pcp/pcz052] [PMID: 30994920]

[90]   Chen Y, Zhang Z, Wang P, *et al.* Identifying the impact of multi-hazards on crop yield-a case for heat stress and dry stress on winter wheat yield in northern China. Eur J Agron 2016; 73: 55-63.
[http://dx.doi.org/10.1016/j.eja.2015.10.009]

[91]   Dwivedi SK, Basu S, Kumar S, *et al.* Heat stress induced impairment of starch mobilisation regulates pollen viability and grain yield in wheat: Study in Eastern Indo-Gangetic Plains. Field Crops Res 2017; 206: 106-14.
[http://dx.doi.org/10.1016/j.fcr.2017.03.006]

[92]   Xu J, Wolters-Arts M, Mariani C, *et al.* Heat stress affects vegetative and reproductive performance and trait correlations in tomato (*Solanum lycopersicum*). Euphytica 2017; 213(7): 156.
[http://dx.doi.org/10.1007/s10681-017-1949-6]

[93]   Lawas LMF, Zuther E, Jagadish SK, Hincha DK. Molecular mechanisms of combined heat and drought stress resilience in cereals. Curr Opin Plant Biol 2018; 45(Pt B): 212-7.
[http://dx.doi.org/10.1016/j.pbi.2018.04.002] [PMID: 29673612]

[94]   Hütsch BW, Jahn D, Schubert S. Grain yield of wheat (*Triticum aestivum* L.) under long-term heat stress is sink-limited with stronger inhibition of kernel setting than grain filling. J Agron Crop Sci 2019; 205(1): 22-32.
[http://dx.doi.org/10.1111/jac.12298]

[95]   Paul P, Dhatt BK, Sandhu J, *et al.* Divergent phenotypic response of rice accessions to transient heat stress during early seed development. Plant Direct 2020; 4(1): e00196.
[http://dx.doi.org/10.1002/pld3.196] [PMID: 31956854]

[96]   Houghton JT, Ding Y, Griggs DJ, *et al.* Intergovernmental panel on climate change (IPCC). Climate Change 2001: The Scientific Basis Contribution of Working Group I to the Third Assessment Report of the Intergovernmental Panel on Climate Change 2001; 881.

[97]   Katano K, Honda K, Suzuki N. Integration between ROS regulatory systems and other signals in the regulation of various types of heat responses in plants. Int J Mol Sci 2018; 19(11): 3370.
[http://dx.doi.org/10.3390/ijms19113370] [PMID: 30373292]

[98]   Raza A, Razzaq A, Mehmood SS, *et al.* Impact of climate change on crops adaptation and strategies to tackle its outcome: A review. Plants (Basel) 2019; 8(2): 34.
[http://dx.doi.org/10.3390/plants8020034] [PMID: 30704089]

CHAPTER 5

Physiological Role of Mineral Nutrients and their Uptake during Abiotic Stress

Ravinder Singh[1], Jyoti Mahajan[2,*], Saroj Arora[1], Renu Bhardwaj[1] and **Tajinder Kaur[3,*]**

[1] *Department of Botanical & Environmental Sciences, Guru Nanak Dev University, Amritsar, Punjab 143005, India*

[2] *PG Department of Agriculture, Khalsa College, Amritsar, Punjab 143005, India*

[3] *Department of Agriculture, Sri Guru Granth Sahib World University, Fatehgarh Sahib, Punjab 140407, India*

Abstract: Soil acts as a source of nutrient elements, and the availability of nutrients is determined by soil properties. The different elements are grouped according to their biochemical behavior and physiological functions. The nutrients that are needed in moderately large amounts are called macronutrients. The plant macronutrients comprise nitrogen, potassium, phosphorus, calcium, sulfur, and magnesium. The micronutrients are required for plant growth in much smaller quantities than macronutrients. These micronutrients contain boron, copper, iron, manganese, molybdenum, and zinc. All of these nutrients are absorbed through the roots. Water will transfers the nutrients from soil to the roots of the plant. In this chapter, we will discuss the physiological role of essential and non-essential elements and the effects of some general environmental stressors, such as salinity, drought, and metals, on nutrient uptake by plants.

Keywords: Deficiency symptoms, Drought, Essential elements, Heavy Metals, Macronutrients, Micronutrients, Salinity.

INTRODUCTION

Mineral nutrients are classified as macronutrients and micronutrients. Plants require these nutrients to complete their life cycle. Macronutrients are required in large quantities by plants such as carbon (C), hydrogen (H), oxygen (O), nitrogen (N), phosphorus (P), and potassium (K). Of these, C, H, and O are non-mineral nutrients as they are found in air and water. Micronutrients are required in smaller

---

* **Corresponding Author Jyoti Mahajan & Tajinder Kaur:** PG Department of Agriculture, Khalsa College, Amritsar, Punjab 143005, India; E-mail: mahajanjyoti9@gmail.com and Department of Agriculture, Sri Guru Granth Sahib World University, Fatehgarh Sahib, Punjab; Tel: +91-9646217854; E-mail: tsandhawalia15@yahoo.in

Tajinder Kaur & Saroj Arora (Eds.)

amounts than macronutrients. The micronutrients include boron (B), copper (Cu), chloride (Cl), iron (Fe), manganese (Mn), molybdenum (Mo), and zinc (Zn). Macronutrients are primarily chief structural components of plants, and micronutrients are essential for plant growth. Fig. (**1**) shows the classification of essential elements and their general function in plant growth. Mineral nutrients may also increase or decrease the tolerance of plants to various biotic stress factors, such as pathogens and pests, and abiotic environmental stress factors, such as drought, salinity, flooding, and chilling [1]. The physiological role of macro and micronutrients are listed in Table **1**.

The deficiency of mineral nutrients produces drastic effects on plant growth and development. The minimum supply or complete absence of any of the essential elements can exhibit typical symptoms that are specific to the particular elements in the plant functioning. This condition is known as nutrient deficiency, and its symptoms are called deficiency symptoms [2]. Although micronutrients are required in small traces, their deficiency in plants causes restricted growth, chlorosis, interveinal necrosis, and defoliation of leaves [3]. Table **2** highlights physiological and nutritional deficiency symptoms in plants. On the other hand, acute toxicity of micronutrients inhibits seed germination, photosynthesis, biosynthesis of chlorophyll, and plant growth, and under severe abiotic stress, the condition causes the death of the plant [4 - 8].

The non-essential elements, *e.g.*, cadmium (Cd) and chromium (Cr), are harmful to plant health [9]. Plants are also sensitive to the deficiency and excess concentrations of some metals and essential micronutrients like all living organisms. The metals such as Cd, Hg, As can be poisonous to plants at higher concentrations [10]. Various studies have been conducted to verify the effects of toxic metals on plants. Table **3** highlights the various studies conducted on the toxicity of essential and non-essential elements on plants. Cadmium and lead are reported to have an inhibitory effect on transpiration, photosynthesis, carbohydrate metabolism, and other metabolic activities [11]. Cadmium inhibits the chlorophyll biosynthesis, reduces total chlorophyll content and chlorophyll a/b ratio in *Brassica juncea*, while Pb toxicity causes toxicity of nucleoli, disturbance of mitosis, inhibition of root elongation, the appearance of chlorosis, inhibition of enzymatic activities, and decrease in photosynthesis [12]. Heavy metal contamination of agricultural soil has become a serious environmental concern due to its possible adverse ecological effects. The toxic metals are considered soil pollutants due to their extensive occurrence and acute and chronic toxic effect on plants grown in such soils [10].

Table 1. Physiological role of essential mineral nutrients in plants.

Mineral Nutrients			Primary Form of Uptake	Function	Deficiency Symptoms
Category	Nutrient	Symbol			
Macronutrients	Nitrogen	N	Nitrate, NO_3^-, ammonium, NH_4^+	Constituent of amino acids, proteins, nucleic acids, and some co-enzymes. Acts as donor atom in many enzymatically catalyzed reactions	Stunted growth, leaves turn yellow (chlorosis)
	Phosphorus	P	Phosphate, HPO_4^{2-}, H_2PO_4	Involved in photosynthesis, essential constituent of high energy intermediates like ATP, nucleic acids, and phospholipids in membranes	Purple or red anthocyanin pigmentation on leaves, premature leaf fall development of dead necrotic areas on the leaves, petioles, and fruits
	Potassium	K	Potassium ion, K^+	Regulation of stomatal opening and closing, essential for photosynthesis, maintains cytoplasmic pH	Development of dead necrotic areas at the tip and margin of leaves (scorching), greenback symptom in tomato is due to potassium deficiency
	Calcium	Ca	Calcium ion, Ca^{2+}	Important component of cell walls, control cell permeability, involved in cell signaling	Meristematic tissue of stem, leaves, and roots die, roots become short, brown, and stubby, younger leaves become distorted
	Magnesium	Mg	Magnesium ion, Mg^{2+}	Essential for chlorophyll (Magnesium porphyrin)	Interveinal chlorosis of the leaves, the appearance of anthocyanin pigment in leaves
	Sulphur	S	Sulfate, SO_4^{2-}	Essential for sulphur containing amino acids, *e.g.*, cysteine, cystine, and methionine	Chlorosis of young leaves

(Table 1) cont.....

Mineral Nutrients			Primary Form of Uptake	Function	Deficiency Symptoms
Category	**Nutrient**	**Symbol**			
Micronutrients	Boron	B	Boric acid, H_3BO_3	Facilitates the translocation of sugars, pollen germination, and pollen tube growth, regulation of stomatal opening	Death of shoot tip, leaves may have thick coppery texture, curl and becomes brittle, internal tissue disintegrates
	Chlorine	Cl	Chloride, Cl^-	Important for photosynthesis	-
	Copper	Cu	Cupric ion, Cu^{2-}	Involved in enzymatic reactions like Cu-Zn SOD, important for its oxidation-reduction properties	Necrosis of young leaves, defoliation, gummosis, dieback
	Iron	Fe	Ferrous ion, Fe^{2+}	Involved in cholorophyll biosynthesis, structural component of cytochromes and ferredoxin, important for its oxidation-reduction properties (Fe^{3+} to Fe^{2+})	Excessive chlorosis of leaves
	Molybdenum	Mo	Molybdate, MoO_4^{2-}	Important for its oxidation-reduction properties, co-factor of enzymes involved in nitrogen metabolism or nitrogen fixation.	Chlorotic interveinal mottling of lower leaves followed by marginal necrosis, infolding and wilting of leaves, inhibition of flower formation
	Manganese	Mn	Manganese ion, Mn^{2+}	Electron transfer in photolysis, essential in respiration and N metabolism	Appearance of chlorotic and necrotic spots in the interveinal areas of the leaves
	Nickel	Ni	Nickel ion, Ni^{2+}	Co-factor of enzyme urease in higher plants	-
	Zinc	Zn	Zinc ion, Zn^{2+}	Activator of some enzymes involved in synthesis of auxin promotes seed maturation and production	Interveinal chlorosis of older leaves, stunted growth due to small leaves and short internodes, clustering of leaves on branches (rosette)

Table 2. Physiological and nutritional deficiency symptoms.

Disorder	Symptoms	Deficiency
Chlorosis	The leaves become abnormally yellow or white in colour due to the reduction of chlorophyll content in the plant	Mg, N, P, S, Fe, Mn, Zn
Necrosis	Death of tissues	Mg, K, P, Ca, Cu
Bronzing	Development of bronze or copper colour in plant tissue	Cl
Die back	Growing tip collapse affecting the younger leaves and buds of the plant	K, Cu, Zn, B
Scorching	Burning of tissues accompanied by the light brown colour resulting from faulty spray, salt injury etc.	K, Fe, B
Lesions	A localized wound of leaf or stem tissue usually accompanied by loss of normal colour	Ca Mn & K (Necrotic lesions)
Rosette	Distortion in shape and appearance of leaves and clustering of leaves on short branches	Cu, Zn

Table 3. Effect of some essential and non-essential elements toxicity on the plants.

Element	Plant Species	Effects	References
Arsenic	*Nicotiana tabacum*	Impedes the photosynthesis, reduced essential nutrient content, ROS generation	[13]
	Oryza sativa	Hindered plant length and weight, oxidative stress enhanced	[14]
	Phaseolus aureus	Inhibited growth caused physiological disorders such as membrane damage, oxidative damage	[15]
	Nasturtium officinale	Decreased dry weight of roots and shoots and chlorophyll content of leaves, oxidative damage	[16]
Cadmium	*Nicotiana tabacum*	Increased H_2O_2 and superoxide production, plant growth reduced, and chlorophyll content declined	[17]
	Pisum sativum	Reduction in plant growth and photosynthetic pigment, cause oxidative injury by enhancing the production of ROS	[18]
	Vigna radiata	Severe oxidative stress, decreased plant height and root length, and reduced chlorophyll content	[19]
	Raphanus sativus	Oxidative stress and accelerated cell senescence in mesophyll area of leaf blade	[20]
Chromium	*Pisum sativum*	Reduce photosynthetic process and nutrient uptake by plant	[21]
	Vigna radiata	Increased lipid peroxidation and H_2O_2 generation	[22]

(Table 3) cont.....

Element	Plant Species	Effects	References
Copper	*Withania somnifera*	Inhibition of cell elongation and division, reduction of biomass, reducing membrane fluidity, and decreased concentration of carotenoid and photosynthetic pigment, ROS generation	[23]
	Brassica juncea	Reduced growth of plant, activities of nitrate reductase, and carbonic anhydrase. Decreased chlorophyll content and proline increased	[24]
	Brassica junica	Decreased shoot–root length, biomass, decreased chlorophyll content	[25]
Cobalt	*Vigna radiata*	Decreased root and shoot length of plant, Inhibition of cell division and elongation, Dry weight of root and shoot declined	[26]
	Lemna minor	decreases plant growth and biomass, chlorophyll content, starch accumulation, water potential, and transpiration rate	[27]
Mercury	*Cucumis sativus*	Decreased chlorophyll content and root, shoot, length induced oxidative stress	[28]
	Sesbania drummondii	Reduction in biomass, photosynthetic activity declined, and increased oxidative stress	[29]
	Triticum aestivum	Oxidative damage, reduced enzymatic activity	[30]
Nickel	*Cajanus cajan*	Reduction in seedling growth, decreased dry weight, increased lipid peroxidation, and elevated ROS generation	[31]
	Brassica napus	Stunted plant growth, brownish roots, chlorosis, and induced ROS generation	[32]
	Triticum aestivum	Growth reduction, decreased chlorophyll content, and increased ROS production	[33]
Lead	*Eichhornia crassipes*	Chlorophyll content decreased while increased MDA content in leaf and root tissues, reduction in biomass, Chlorosis and drying at edges in seedlings	[34]
	Vicia faba	Nutrient uptake was decreased	[35]
	Oryza sativa	Induced oxidative stress, alteration in antioxidative enzyme activity	[36]
Zinc	*Vetiveria zizanioides*	Total biomass decreased, proline content increased, total soluble protein decreased	[37]
	Lycopersicon esculentum	Decreased chlorophyll content, reduced growth and ROS generation	[38]
	Zea mays	Inhibited biomass production SOD, CAT, GPX, and POD, Decreased chlorophyll content Elevate the ROS production	[39]

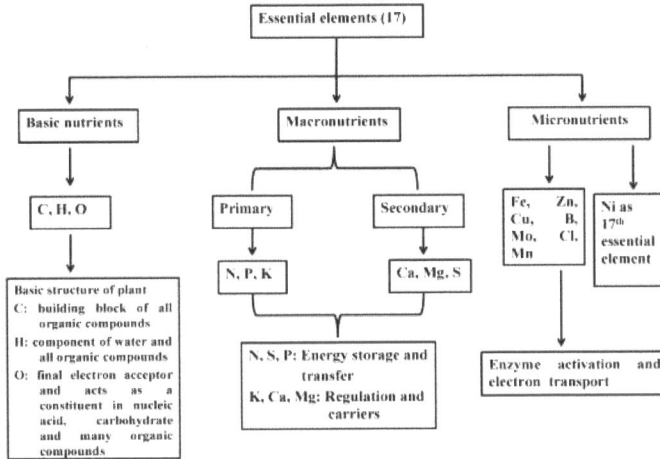

Fig. (1). Classification of essential elements.

ABIOTIC STRESS AND MINERAL UPTAKE

Abiotic stressors are considered a great limitation to production of crops and are estimated as the major cause of reduction in global crop yield [40, 41]. The imbalance between crop productivity and population growth has further worsened this condition. Hence, it is particularly very important to recognize the plants responses to such stress factors, mainly salinity, drought and heavy metals, so as to discover novel methods for improving crops both, quantitatively and qualitatively [42]. Plants obtain mineral nutrition from soil. Minerals are essential for proper growth and development of plants. Plants are unable to survive, unless, they obtain proper nutrition and water. The uptake and translocation of mineral nutrients depends on many environmental factors, such as salinity, water shortage, and the presence of xenobiotics. Therefore, understanding the relations among plant nutrients and abiotic stress factors is critical.

Salinity is one of the major widespread environmental stressors for crop plants in arid and coastal regions. Every year large areas of arable land are being removed from crop production due to increasing salt salinity. Increase in soil salinity is mainly caused by use of saline irrigation water and application of fertilizers [43]. Plants do not have the ability to tolerate high levels of salt in the cytoplasm. Therefore, they sequester surplus salts in vacuoles or compartmentalize the ions in different tissues [44]. The level of cationic nutrients (K, Ca, Mg) decreases with Na salt stress. Under saline conditions, soils have very high ratios of Na^+-Ca^{2+}, Na^+-K^+, Ca^{2+}-Mg^{2+} and Cl^--NO_3^- resulting in ionic imbalance and toxicity (Fig. **2**). Plants exposed to high salt stress have high concentrations of Na^+ and Cl^- and low concentration of K^+, Ca^{2+} and Mg^{2+}. High levels of sodium competes with K^+ uptake by plants, thus affecting the development, water balance and various enzymatic processes [44 - 46]. Salt stress also reduces N, P, K and Ca content in tissues [47]. High salinity also reduces N accumulation resulting in nitrogen deficiency [48].

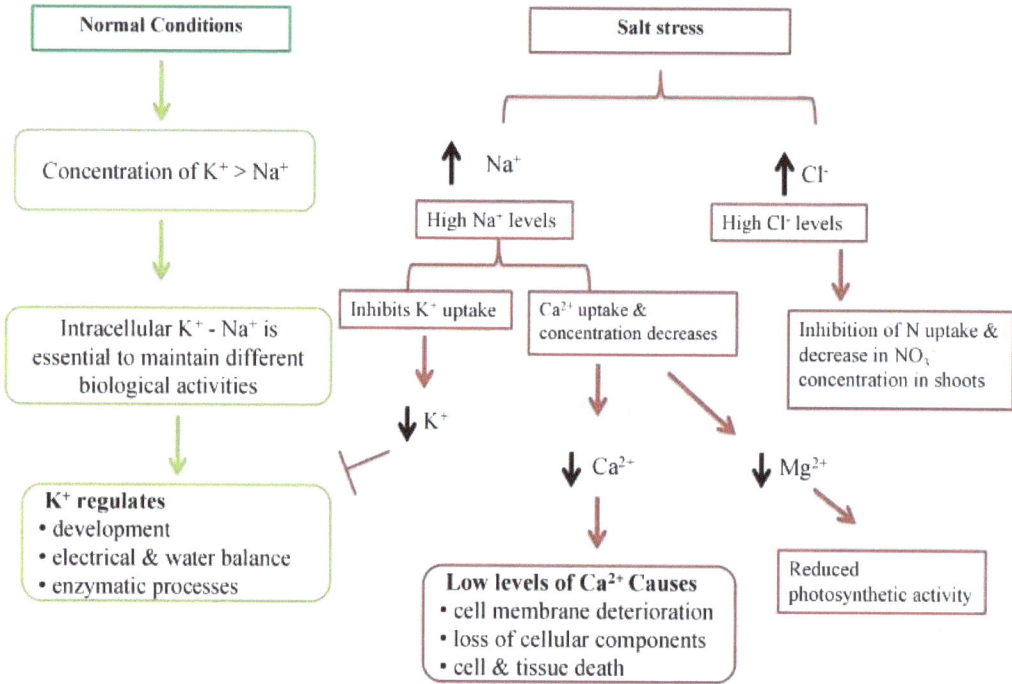

Fig. (2). Effect of salt stress on mineral uptake.

Drought is another most common environmental stress factor affecting plant growth, development, and reproduction. Mineral nutrients are generally taken as inorganic ions from soils essential for plant growth and development. Under drought conditions, there is reduced soil moisture leading to less nutrient uptake. The diffusion of mineral nutrients from the soil to the root surface is reduced which in turn decreases the translocation speed to the leaves. Drought also reduces the rate of transpiration affecting the movement of nutrients from the roots to shoots. In general, drought affects the movement of mineral nutrients from soil to plant tissues due to reduced soil moisture [49]. Calcium is involved in plant drought resistance by conserving water [50]. Potassium is responsible for osmo-regulation in plant cells during drought stress [51]. Drought stress in plants also leads to increased accumulation of magnesium in leaves and help in maintaining water content. Deficiency of magnesium affects the photosynthetic activity of plants since it is an essential component of chloroplast. In general, increase in concentration of magnesium under drought conditions helps in maintaining the process of photosynthesis [52]. Drought can also have a great impact on plant nutrient relations. It decreases the concentration of nitrogen (N) and phosphorus (P) in plant tissue [53]. It can also affect the nutrient uptake from soil due to reduced nutrient supply through mineralization, reduced diffusion and mass flow in the soil.

Another important environmental factor affecting crop productivity and quality is heavy metal contamination of soil and water. Heavy metal contamination of agricultural soil has become a serious environmental concern due to their possible adverse ecological effects. Although, some heavy metals play important role in plant growth and development at low levels, but at high concentrations they produce severe toxicity symptoms in plants. However, heavy metals like Cd, Cr, Pb, Hg, etc., are extremely toxic even at very low concentrations. Essential and non-essential heavy metals normally produce common toxic effects on plants, for example, chlorosis, low biomass accumulation, inhibition of growth and photosynthesis, altered water balance and nutrient assimilation, and senescence, which ultimately cause plant death [42]. Heavy metals such as Cd also interfere with nutrient uptake by changing plasma membrane permeability and may vary the conformation of proteins, including enzymes, transporters, or regulator proteins [54, 55]. Arsenic may compete directly with nutrients or change metabolic processes in plants. As(V) can act as a phosphate analogue and be transported into the cell where it can interfere with essential cellular processes, such as oxidative phosphorylation and ATP synthesis [56 - 58]. Chromium (Cr) is a non-essential element resulting in severe toxicity in plants. It also interferes with uptake of mineral nutrients such as B, Ca, Cu, Fe, K, Mg, P, and S [59]. Copper (Cu) accumulation also has a marked affect on uptake of certain essential and non-essential minerals such Fe, Zn, and K, as their content decreases with increasing levels of Cu [60]. High levels of lead (Pb) compete with transport of micronutrients, such as Mn and Cu into roots. Previous study reported that as the concentration of Pb increased, the concentration of Cu, Fe, Mn, P and S, decreased in different parts of the cabbage plant [61]. Mercury (Hg) can replace certain nutritional elements, such as Mn, Mg and Zn thereby altering the uptake and translocation of mineral nutrients [62]. Nickel (Ni) is an essemtial element for plants at low concentrations, however, at high concentrations it shows phytotoxic effects by affecting nutrient absorption by roots, decreasing accumulation of macronutrients (K, Ca, and Mg) and micronutrients (Zn, Mn, Fe, and Cu) [63, 64].

CONCLUSION

The enormous nutrient deficiencies and toxicities cause loss in productivity and the development of the various crops that may result into the appearance of abnormal and undefined visual symptoms. Moreover, understanding the role of each essential nutrient in the plant can help to determine its deficiency or toxicity symptoms. The various general deficiency symptoms include stunted growth, chlorosis, inter-veinal chlorosis, purple or red discoloration as well as necrosis. Excessive amount of nutrients will inhibit the uptake of other nutrients, thus potentially causing the deficiency symptoms to occur as well. Various abiotic stress factors such as salinity, drought and heavy metals also interferes with mineral nutrient uptake thus, affecting the growth and development of plants. However, studies conducted on effect of abiotic factors on mineral uptake and translocaton is limited. More research is required on the function of macronutrients and micronutrients as well as beneficial elements under various environmental stressors.

CONSENT FOR PUBLICATION

Not applicable.

CONFLICT OF INTEREST

The authors declare that they do not have any conflict of interest.

ACKNOWLEDGEMENTS

Authors would like to thank Guru Nanak Dev University, Amritsar, Punjab and Sri Guru Granth Sahib World University, Fatehgarh Sahib, Punjab for providing library and internet facilities for survey of literature during preparation of this chapter.

REFERENCES

[1]   Hajiboland R. Effect of micronutrient deficiencies on plants stress responses. In Abiotic stress responses in plants. New York, NY: Springer 2012; pp. 283-329.

[2]   Bradley L, Hosier S. Guide to symptoms of plant nutrient deficiencies

[3]   Noreen S, Fatima Z, Ahmad S, Ashraf M. Foliar application of micronutrients in mitigating abiotic stress in crop plants. In: Hasanuzzaman M, Fujita M, Oku H, Nahar K, Hawrylak-Nowak B, Eds. Plant Nutrients and Abiotic Stress Tolerance. Singapore: Springer 2018; pp. 95-117. [http://dx.doi.org/10.1007/978-981-10-9044-8_3]

[4]   Ouzounidou G, Eleftheriou EP, Karataglis S. Ecophysical and ultrastructural effects of copper in *Thlaspi ochroleucum* (Cruciferae). Can J Bot 1992; 70(5): 947-57. [http://dx.doi.org/10.1139/b92-119]

[5]   Nussbaum S, Schmutz D, Brunold C. Regulation of assimilatory sulfate reduction by cadmium in *Zea mays* L. Plant Physiol 1988; 88(4): 1407-10. [http://dx.doi.org/10.1104/pp.88.4.1407] [PMID: 16666474]

[6]   Munzuroglu O, Geckil H. Effects of metals on seed germination, root elongation, and coleoptile and hypocotyl growth in *Triticum aestivum* and *Cucumis sativus*. Arch Environ Contam Toxicol 2002; 43(2): 203-13.

[http://dx.doi.org/10.1007/s00244-002-1116-4] [PMID: 12115046]

[7]   Rehman SA, Iqbal MZ. Seed germination and seedling growth of trees in soil extracts from Korangi and Landhi industrial areas of Karachi, Pakistan. J New Seeds 2007; 8(4): 33-45.
[http://dx.doi.org/10.1300/J153v08n04_03]

[8]   Sett R. Tolerance of plants in response to abiotic stress factors. Recent Adv Petrochem Sci 2017; 1(5): 475-83. RAPSCI.MS.ID.555573.
[http://dx.doi.org/10.19080/RAPSCI.2017.01.555573]

[9]   Jadhav AH, Dalal SR, Shinde RD, Deshmukh RP. Effect of micronutrients on growth and flower production of gerbera under polyhouse conditions. Advances in Plant Sci 2005; 18(2): 755.

[10]   Nagajyoti PC, Lee KD, Sreekanth TV. Heavy metals, occurrence and toxicity for plants: a review. Environ Chem Lett 2010; 8(3): 199-216.
[http://dx.doi.org/10.1007/s10311-010-0297-8]

[11]   Küpper H, Küpper F, Spiller M. Environmental relevance of heavy metal-substituted chlorophylls using the example of water plants. J Exp Bot 1996; 47(2): 259-66.
[http://dx.doi.org/10.1093/jxb/47.2.259]

[12]   Stobart AK, Griffiths WT, Ameen-Bukhari I, Sherwood RP. The effect of Cd^{2+} on the biosynthesis of chlorophyll in leaves of barley. Physiol Plant 1985; 63(3): 293-8.
[http://dx.doi.org/10.1111/j.1399-3054.1985.tb04268.x]

[13]   Han D, Xiong S, Tu S, Liu J, Chen C. Interactive effects of selenium and arsenic on growth, antioxidant system, arsenic and selenium species of *Nicotiana tabacum* L. Environ Exp Bot 2015; 117: 12-9.
[http://dx.doi.org/10.1016/j.envexpbot.2015.04.008]

[14]   Dixit G, Singh AP, Kumar A, *et al.* Reduced arsenic accumulation in rice (*Oryza sativa* L.) shoot involves sulfur mediated improved thiol metabolism, antioxidant system and altered arsenic transporters. Plant Physiol Biochem 2016; 99: 86-96.
[http://dx.doi.org/10.1016/j.plaphy.2015.11.005] [PMID: 26741538]

[15]   Malik JA, Goel S, Kaur N, Sharma S, Singh I, Nayyar H. Selenium antagonises the toxic effects of arsenic on mungbean (*Phaseolus aureus* Roxb.) plants by restricting its uptake and enhancing the antioxidative and detoxification mechanisms. Environ Exp Bot 2012; 77: 242-8.
[http://dx.doi.org/10.1016/j.envexpbot.2011.12.001]

[16]   Namdjoyan S, Kermanian H. Exogenous nitric oxide (as sodium nitroprusside) ameliorates arsenic-induced oxidative stress in watercress (*Nasturtium officinale* R. Br.) plants. Sci Hortic (Amsterdam) 2013; 161: 350-6.
[http://dx.doi.org/10.1016/j.scienta.2013.07.035]

[17]   Iannone MF, Groppa MD, Benavides MP. Cadmium induces different biochemical responses in wild type and catalase-deficient tobacco plants. Environ Exp Bot 2015; 109: 201-11.
[http://dx.doi.org/10.1016/j.envexpbot.2014.07.008]

[18]   Agrawal SB, Mishra S. Effects of supplemental ultraviolet-B and cadmium on growth, antioxidants and yield of *Pisum sativum* L. Ecotoxicol Environ Saf 2009; 72(2): 610-8.
[http://dx.doi.org/10.1016/j.ecoenv.2007.10.007] [PMID: 18061671]

[19]   Nahar K, Hasanuzzaman M, Alam MM, Rahman A, Suzuki T, Fujita M. Polyamine and nitric oxide crosstalk: Antagonistic effects on cadmium toxicity in mung bean plants through upregulating the metal detoxification, antioxidant defense and methylglyoxal detoxification systems. Ecotoxicol Environ Saf 2016; 126: 245-55.
[http://dx.doi.org/10.1016/j.ecoenv.2015.12.026] [PMID: 26773834]

[20]   Vitória AP, Lea PJ, Azevedo RA. Antioxidant enzymes responses to cadmium in radish tissues. Phytochemistry 2001; 57(5): 701-10.
[http://dx.doi.org/10.1016/S0031-9422(01)00130-3] [PMID: 11397437]

[21]   Tripathi DK, Singh VP, Prasad SM, Chauhan DK, Dubey NK. Silicon nanoparticles (SiNp) alleviate chromium (VI) phytotoxicity in *Pisum sativum* (L.) seedlings. Plant Physiol Biochem 2015; 96: 189-98.
[http://dx.doi.org/10.1016/j.plaphy.2015.07.026] [PMID: 26298805]

[22]   Shanker AK, Djanaguiraman M, Sudhagar R, Chandrashekar CN, Pathmanabhan G. Differential antioxidative response of ascorbate glutathione pathway enzymes and metabolites to chromium speciation stress in green gram (*Vigna radiata* (L.) R. Wilczek. cv CO4) roots. Plant Sci 2004; 166(4): 1035-43.
[http://dx.doi.org/10.1016/j.plantsci.2003.12.015]

[23]   Khatun S, Ali MB, Hahn EJ, Paek KY. Copper toxicity in *Withania somnifera*: growth and antioxidant enzymes responses of *in vitro* grown plants. Environ Exp Bot 2008; 64(3): 279-85.
[http://dx.doi.org/10.1016/j.envexpbot.2008.02.004]

[24]   Fariduddin Q, Yusuf M, Hayat S, Ahmad A. Effect of 28-homobrassinolide on antioxidant capacity and photosynthesis in *Brassica juncea* plants exposed to different levels of copper. Environ Exp Bot 2009; 66(3): 418-24.
[http://dx.doi.org/10.1016/j.envexpbot.2009.05.001]

[25]   Yusuf M, Khan TA, Fariduddin Q. Interaction of epibrassinolide and selenium ameliorates the excess copper in *Brassica juncea* through altered proline metabolism and antioxidants. Ecotoxicol Environ Saf 2016; 129: 25-34.
[http://dx.doi.org/10.1016/j.ecoenv.2016.03.001] [PMID: 26974871]

[26]   Jaleel CA, Jayakumar K, Zhao C, Azooz MM. Antioxidant potentials protect *Vigna radiata* (L.) Wilczek plants from soil cobalt stress and improve growth and pigment composition. Plant Omics 2009; 2(3): 120-6.

[27]   Begović L, Mlinarić S, Antunović Dunić J, *et al.* Response of *Lemna minor* L. to short-term cobalt exposure: The effect on photosynthetic electron transport chain and induction of oxidative damage. Aquat Toxicol 2016; 175: 117-26.
[http://dx.doi.org/10.1016/j.aquatox.2016.03.009] [PMID: 27015565]

[28]   Cargnelutti D, Tabaldi LA, Spanevello RM, *et al.* Mercury toxicity induces oxidative stress in growing cucumber seedlings. Chemosphere 2006; 65(6): 999-1006.
[http://dx.doi.org/10.1016/j.chemosphere.2006.03.037] [PMID: 16674986]

[29]   Israr M, Sahi S, Datta R, Sarkar D. Bioaccumulation and physiological effects of mercury in *Sesbania drummondii*. Chemosphere 2006; 65(4): 591-8.
[http://dx.doi.org/10.1016/j.chemosphere.2006.02.016] [PMID: 16564071]

[30]   Li X, Yang Y, Jia L, Chen H, Wei X. Zinc-induced oxidative damage, antioxidant enzyme response and proline metabolism in roots and leaves of wheat plants. Ecotoxicol Environ Saf 2013; 89: 150-7.
[http://dx.doi.org/10.1016/j.ecoenv.2012.11.025] [PMID: 23260180]

[31]   Sresty TV, Sresty TV, Madhava Rao KV. Antioxidative parameters in the seedlings of pigeonpea (*Cajanus cajan* (L.) Millspaugh) in response to Zn and Ni stresses. Plant Sci 2000; 157(1): 113-28.
[http://dx.doi.org/10.1016/S0168-9452(00)00273-9] [PMID: 10940475]

[32]   Kazemi N, Khavari-Nejad RA, Fahimi H, Saadatmand S, Nejad-Sattari T. Effects of exogenous salicylic acid and nitric oxide on lipid peroxidation and antioxidant enzyme activities in leaves of *Brassica napus* L. under nickel stress. Sci Hortic (Amsterdam) 2010; 126(3): 402-7.
[http://dx.doi.org/10.1016/j.scienta.2010.07.037]

[33]   Parlak KU. Effect of nickel on growth and biochemical characteristics of wheat (*Triticum aestivum* L.) seedlings. NJAS Wagening J Life Sci 2016; 76: 1-5.
[http://dx.doi.org/10.1016/j.njas.2012.07.001]

[34]   Malar S, Shivendra Vikram S, Jc Favas P, Perumal V. Lead heavy metal toxicity induced changes on growth and antioxidative enzymes level in water hyacinths [*Eichhornia crassipes* (Mart.)]. Bot Stud

(Taipei, Taiwan) 2016; 55(1): 54.
[http://dx.doi.org/10.1186/s40529-014-0054-6] [PMID: 28597420]

[35]   Shahid M, Pinelli E, Pourrut B, Dumat C. Effect of organic ligands on lead-induced oxidative damage and enhanced antioxidant defense in the leaves of *Vicia faba* plants. J Geochem Explor 2014; 144: 282-9.
[http://dx.doi.org/10.1016/j.gexplo.2014.01.008]

[36]   Verma S, Dubey RS. Lead toxicity induces lipid peroxidation and alters the activities of antioxidant enzymes in growing rice plants. Plant Sci 2003; 164(4): 645-55.
[http://dx.doi.org/10.1016/S0168-9452(03)00022-0]

[37]   Xu W, Li W, He J, Balwant S, Xiong Z. Effects of insoluble Zn, Cd, and EDTA on the growth, activities of antioxidant enzymes and uptake of Zn and Cd in *Vetiveria zizanioides*. J Environ Sci (China) 2009; 21(2): 186-92.
[http://dx.doi.org/10.1016/S1001-0742(08)62249-4] [PMID: 19402420]

[38]   Sbartai H, Djebar MR, Sbartai I, Berrabbah H. Bioaccumulation du Cd et du Zn chez les plants de tomates (*Lycopersicon esculentum* L.). C R Biol 2012; 335(9): 585-93.
[http://dx.doi.org/10.1016/j.crvi.2012.08.001] [PMID: 23026089]

[39]   Islam F, Yasmeen T, Riaz M, Arif MS, Ali S, Raza SH. Proteus mirabilis alleviates zinc toxicity by preventing oxidative stress in maize (*Zea mays*) plants. Ecotoxicol Environ Saf 2014; 110: 143-52.
[http://dx.doi.org/10.1016/j.ecoenv.2014.08.020] [PMID: 25240234]

[40]   Acquaah G. Principles of Plant Genetics and Breeding. UK: John Wiley & Sons, Ltd 2009.

[41]   Jewell MC, Campbell BC, Godwin ID. Transgenic plants for abiotic stress resistance. In: Kole C, Michler CH, Abbott AG, Hall TC, Eds. Transgenic Crop Plants. Berlin, Heidelberg: Springer 2010; pp. 67-132.

[42]   Singh S, Parihar P, Singh R, Singh VP, Prasad SM. Heavy metal tolerance in plants: role of transcriptomics, proteomics, metabolomics, and ionomics. Front Plant Sci 2016; 6: 1143.
[http://dx.doi.org/10.3389/fpls.2015.01143] [PMID: 26904030]

[43]   Epstein E, Bloom AJ. Mineral nutrition of plants: principles and perspectives. 2nd ed., Sunderland, UK: Sinauer Assoc. Inc. 2005.

[44]   Zhu JK. Regulation of ion homeostasis under salt stress. Curr Opin Plant Biol 2003; 6(5): 441-5.
[http://dx.doi.org/10.1016/S1369-5266(03)00085-2] [PMID: 12972044]

[45]   Aziz I, Khan MA. Experimental assessment of salinity tolerance of Ceriops tagal seedlings and saplings from the Indus delta, Pakistan. Aquat Bot 2001; 70(3): 259-68.
[http://dx.doi.org/10.1016/S0304-3770(01)00160-7]

[46]   Duman F. Uptake of mineral elements during abiotic stress. In: Ahmad P, Prasad M, Eds. Abiotic Stress Responses in Plants. New York, NY: Springer 2012; pp. 267-81.
[http://dx.doi.org/10.1007/978-1-4614-0634-1_15]

[47]   Patel PR, Kajal SS, Patel VR, Patel VJ, Khristi SM. Impact of saline water stress on nutrient uptake and growth of cowpea. Braz J Plant Physiol 2010; 22(1): 43-8.
[http://dx.doi.org/10.1590/S1677-04202010000100005]

[48]   Siddiqui MH, Mohammad F, Khan MN, Al-Whaibi MH, Bahkali AH. Nitrogen in relation to photosynthetic capacity and accumulation of osmoprotectant and nutrients in *Brassica* genotypes grown under salt stress. Agric Sci China 2010; 9(5): 671-80.
[http://dx.doi.org/10.1016/S1671-2927(09)60142-5]

[49]   da Silva EC, Nogueira RJ, da Silva MA, de Albuquerque MB. Drought stress and plant nutrition. Plant stress 2011; 5(Special Issue 1): 32-41.

[50]   Shao HB, Chu LY, Jaleel CA, Zhao CX. Water-deficit stress-induced anatomical changes in higher plants. C R Biol 2008; 331(3): 215-25.

[http://dx.doi.org/10.1016/j.crvi.2008.01.002] [PMID: 18280987]

[51]   Roberts SK. Regulation of K+ channels in maize roots by water stress and abscisic acid. Plant Physiol 1998; 116(1): 145-53.
[http://dx.doi.org/10.1104/pp.116.1.145]

[52]   Mahouachi J. Changes in nutrient concentrations and leaf gas exchange parameters in banana plantlets under gradual soil moisture depletion. Sci Hortic (Amsterdam) 2009; 120: 460-6.
[http://dx.doi.org/10.1016/j.scienta.2008.12.002]

[53]   He M, Dijkstra FA. Drought effect on plant nitrogen and phosphorus: a meta-analysis. New Phytol 2014; 204(4): 924-31.
[http://dx.doi.org/10.1111/nph.12952] [PMID: 25130263]

[54]   Van Assche F, Clijsters H. Effects of metals on enzyme activity in plants. Plant Cell Environ 1990; 13(3): 195-206.
[http://dx.doi.org/10.1111/j.1365-3040.1990.tb01304.x]

[55]   Gonçalves JF, Antes FG, Maldaner J, *et al.* Cadmium and mineral nutrient accumulation in potato plantlets grown under cadmium stress in two different experimental culture conditions. Plant Physiol Biochem 2009; 47(9): 814-21.
[http://dx.doi.org/10.1016/j.plaphy.2009.04.002] [PMID: 19419882]

[56]   Meharg AA, Macnair MR. An altered phosphate uptake system in arsenate-tolerant Holcus lanatus L. New Phytol 1990; 116(1): 29-35.
[http://dx.doi.org/10.1111/j.1469-8137.1990.tb00507.x]

[57]   Meharg AA, Hartley-Whitaker J. Arsenic uptake and metabolism in arsenic resistant and nonresistant plant species. New Phytol 2002; 154(1): 29-43.
[http://dx.doi.org/10.1046/j.1469-8137.2002.00363.x]

[58]   Tripathi RD, Srivastava S, Mishra S, *et al.* Arsenic hazards: strategies for tolerance and remediation by plants. Trends Biotechnol 2007; 25(4): 158-65.
[http://dx.doi.org/10.1016/j.tibtech.2007.02.003] [PMID: 17306392]

[59]   Shanker AK, Cervantes C, Loza-Tavera H, Avudainayagam S. Chromium toxicity in plants. Environ Int 2005; 31(5): 739-53.
[http://dx.doi.org/10.1016/j.envint.2005.02.003] [PMID: 15878200]

[60]   Bouazizi H, Jouili H, Geitmann A, El Ferjani E. Copper toxicity in expanding leaves of *Phaseolus vulgaris* L.: antioxidant enzyme response and nutrient element uptake. Ecotoxicol Environ Saf 2010; 73(6): 1304-8.
[http://dx.doi.org/10.1016/j.ecoenv.2010.05.014] [PMID: 20561682]

[61]   Sinha P, Dube BK, Srivastava P, Chatterjee C. Alteration in uptake and translocation of essential nutrients in cabbage by excess lead. Chemosphere 2006; 65(4): 651-6.
[http://dx.doi.org/10.1016/j.chemosphere.2006.01.068] [PMID: 16545426]

[62]   Patra M, Bhowmik N, Bandopadhyay B, Sharma A. Comparison of mercury, lead and arsenic with respect to genotoxic effects on plant systems and the development of genetic tolerance. Environ Exp Bot 2004; 52(3): 199-223.
[http://dx.doi.org/10.1016/j.envexpbot.2004.02.009]

[63]   Rahman H, Sabreen S, Alam S, Kawai S. Effects of nickel on growth and composition of metal micronutrients in barley plants grown in nutrient solution. J Plant Nutr 2005; 28(3): 393-404.
[http://dx.doi.org/10.1081/PLN-200049149]

[64]   Ahmad MS, Ashraf M, Hussain M. Phytotoxic effects of nickel on yield and concentration of macro- and micro-nutrients in sunflower (*Helianthus annuus* L.) achenes. J Hazard Mater 2011; 185(2-3): 1295-303.
[http://dx.doi.org/10.1016/j.jhazmat.2010.10.045] [PMID: 21074317]

CHAPTER 6

Nitric Oxide Mediated Modulation of Antioxidative Responses under Salinity Stress in Different Plant Species: A Review

Jaskaran Kaur[1], Varinder Kaur[2] and Jatinder Kaur Katnoria[1,*]

[1] *Department of Botanical and Environmental Sciences, Guru Nanak Dev University, Amritsar, Punjab 143005, India*

[2] *Centre for Advanced Studies-Department of Chemistry, Guru Nanak Dev University, Amritsar, Punjab 143005, India*

Abstract: Nitric oxide (NO) is a small-sized, short-lived, highly diffusible, gaseous, and bioactive molecule that regulates various physiological and pathological processes in plants. It also plays a role in development, germination, flowering, senescence as well as response to abiotic stress in plants. In recent years, the role of NO in tolerance of abiotic stress posed by salinity, drought, heat, cold, UV-B, and heavy metals in plants has been identified and gained importance in the field of plant research. Salinity stress triggers the production of reactive oxygen species (ROS) that leads to oxidative stress in plants, resulting in cellular destruction *via* lipid peroxidation, enzyme inactivation, and DNA damage. To combat this stress and to minimize its harmful effects, certain plants activate various ROS-scavenging enzyme activities. The role of exogenous NO, which acts as an indicator in triggering ROS scavenging enzyme activities and regulates antioxidant activities in plants to alleviate the toxic effects of salt stress, has been well established. NO has been considered to play a key role in reducing the excessive production of ROS in cells, improving secondary defense mechanisms, alleviating osmotic damage, and maintaining cell permeability. Thus, understanding the mechanisms of action of NO that help in effectively utilizing the crop cultivation under saline conditions and support better growth of the plants is the need of an hour. Considering this, the present review focuses upon the recent knowledge of the crucial role of NO in providing tolerance to plants under salt stress.

Keywords: Antioxidant enzymes, Abiotic stress, Hydrogen peroxide, Nitric oxide, Oxidative stress, Reactive oxygen species, Salinity, Sodium nitroprusside.

* **Corresponding Author Jatinder Kaur Katnoria:** Department of Botanical and Environmental Sciences, Guru Nanak Dev University, Amritsar, Punjab 143005, India; Tel: +91-9501012458; E-mail: jkat08@yahoo.com

Tajinder Kaur & Saroj Arora (Eds.)

INTRODUCTION

Salinity is the accretion of excessive salt contents in the soil and is one of the major stressors, especially in arid and semi-arid regions, that progresses the inhibition of crop growth, ultimately leading to crop death [1, 2]. Stress induced by salt is considered an alarming condition for the agricultural productivity of soil because it can severely limit food crop production and result in decreased crop yields [3]. In nations where irrigation is a necessary utility for crop production, salinity is considered to be an ever-present threat [4]. Soil salinity has been found to be raised all over the world in past decades due to various anthropogenic reasons, including agricultural practices as well as changes in climatic conditions [5]. High salinity in agricultural soil in some regions has been documented to cause losses to 20% of the crop production annually [6]. Although abiotic stress like the one induced by salinity causes harmful effects in the plant system, yet certain plants have the capability to survive under such stress by producing their defense mechanisms. As roots are the prime organs to sense the high salinity in regions under crop cultivation, modulation of the root system architecture in response to salt stress has been reported by Wang *et al*. [7]. Similarly, in order to relieve the oxidative damage, certain mangrove plants have been reported to develop an effective antioxidant system with low-molecular weight antioxidants and antioxidant enzymes [8, 9]. The antioxidative enzyme activity has been correlated to redox state under salt stress by certain authors [10, 11]. Although plants have the potential to combat oxidative stress to a certain extent, the phenomenon of free radical scavenging can further be improved by the application of exogenous ROS scavengers like nitric oxide (NO) donors [12, 13]. Various antioxidative compounds like putrecine [11], sodium nitroprusside (SNP) [14, 15], 2-phenyl-4,4,5,5-tetremethy-limidazolinone1-oxyl 3-oxide [16], diethy-lenetriamine (DETA) [17] and sodium ferrocyanide ($Na_4Fe(CN)_6$) [18] have been used for production of nitric oxide. The proceeding part of the review focuses on the effects of salinity stress, the role of the antioxidative defense system in combating salinity stress, and the use of exogenous nitric oxide donors in combating the oxidative stress induced by salinity in different plants species.

EFFECTS OF SALINITY

Salinity, like other abiotic stressors, is shown to disturb intracellular ion homeostasis of plants [19] and results in membrane dysfunction [20]. Egbichi *et al*. [17] documented decreased dry weight of root and shoot as well as nodule weight and the number of nodules in *Glycine max* L. under salinity-induced stress. Salinity induced premature leaf senescence and loss of photosynthetic efficiency, leading to reduced carbon assimilation and ultimately reduced crop yield, which has also been reported [21]. Effects of treatment of sodium chloride (NaCl) on plant height, stem thickness, fresh weight, dry weight, and accumulation of polyamines in leaves of *Cucumis sativus* L. (Cucumber) was reported [22]. Hyper accumulation of Na^+ (sodium ion) in the cytoplasmic matrix has shown to pose direct harmful effects on plasma membrane (PM), causing electrolytes leakage and effects on various metabolic processes in the cytosol [23].

High salinity in the soil was shown to elicit two primary effects, such as osmotic stress and ionic stress in plants, which further reduced the ability of plants to take up water and minerals like potassium and calcium [24, 25]. Salinity has been reported to cause attenuation of metabolic activity [8, 26, 27], alteration in the functioning of tonoplast vesicles [4, 28], reduction in plant water potential [20], and alteration in nutrient uptake [29] in many plant species. Apart from all these, it has been shown to cause oxidative stress *via* enhanced reactive oxygen species (ROS) like superoxide radicals, hydrogen peroxide, and hydroxyl radicals [11, 30]. As the physiological changes have been associated with various harmful effects in plants, the antioxidant defense mechanism plays an important role in suppressing abiotic stress and increasing plant growth. The role of antioxidative enzymes in the alleviation of salinity stress has been discussed in the following section.

ROLE OF ANTIOXIDANT ENZYMES IN ALLEVIATION OF SALT STRESS

The reactive oxygen species can cause cellular injury through oxidation of lipids, proteins as well as nucleic acids, ultimately leading to cell death [23, 31]. However, to combat this oxidative stress, many plant systems have evolved antioxidative enzymes like superoxide dismutase (SOD), ascorbate peroxidase (APX), catalase (CAT), peroxidase (POX), glutathione reductase (GR), and dehydro ascorbate reductase (DHAR) [11]. Meneguzzo *et al.* [10] co-related the sodium chloride (NaCl) salt stress-induced alterations in antioxidative enzyme activities with tolerance to salinity in wheat crops. Alleviation of salt stress by decreasing ion leakage and malondialdehyde (MDA), carbonyl and hydrogen peroxide (H_2O_2) content in *Hordeum vulgare* L. (Barley) was observed by Li *et al.* [32].

High contents of certain antioxidants such as ascorbic acid, carotenoids, and polyphenols have been reported to contribute to resisting salt-induced oxidative damage by Rajaravindran and Natarajan [33] and Zhu *et al.* [34]. Superoxide dismutase (SOD), guaiacol peroxidase (GPX), and catalase (CAT) could also effectively remove free radicals in a study conducted by Chen *et al.* [35]. As the equilibrium between free radical generation and its sequestration is the determining factor for the survival of plant systems under the stress environment, therefore, exogenous application of free radical scavenger can help in the sequestration of various reactive oxygen species. As nitric oxide has been found to be effective in the modulation of antioxidative responses in various plant species, the emphasis has been given in the proceeding section for the same.

NITRIC OXIDE MEDIATED MODULATION OF ANTIOXIDATIVE RESPONSES

For maintaining the normal cell functioning under abiotic stress, ionic adjustment has been identified as one of the mechanisms that help the plant system either by reduction of toxic ions uptake into the cytoplasmic matrix or by the sequestration of these ions into the apoplast [21, 36, 37]. In past studies, nitric oxide (NO) has been recognized for its involvement in the mobilization of the intracellular cationic channel for calcium ions [38]. In recent years, NO has been utilized for the enhancement of tolerance of plants towards abiotic stress posed by salinity, drought, heat, cold, UV-B, and heavy metals [39 - 44]. The cytoprotective role of nitric oxide has been tested with DNA, lipids, proteins, and chlorophyll in plants [35].

Nitric oxide has been well established as a main signaling molecule in facilitation of several biotic and abiotic stress-induced physiological and pathological responses in plants. The treatment of maize seedlings with NO increased vacuolar H^+-PPase and H^+-ATPase activities, resulting in increased Na^+/H^+ exchange and H^+-translocation [4]. Nitric oxide (NO) alleviated the oxidative damage in leaves of *Aegiceras corniculatum* L. (Black mangrove) by reducing hydrogen peroxide (H_2O_2) content, polyphenol oxidase (PPO) activity and lipid peroxidation, while increasing guaiacol peroxidase (GPX) activity and contents of reduced glutathione as well as polyphenols to cope with salt stress [14]. Application of sodium nitroprusside (SNP), a NO donor, to *Atnoa1* (mutant variety of *Arabidopsis thaliana*) resulted in higher ratio of Na^+ to K^+ in shoots as compared to wild-type plants because of the increased Na^+ accumulation and decreased K^+ accumulation when exposed to NaCl. Upon treatment with NOS inhibitor (300 mM of N^G-nitr--LArg) and NO scavenger (400 mM of 2-phenyl-4,4,5,5-tetreme-hy-limidazolinone1-oxyl 3-oxide), wild-type plants of *Arabidopsis thaliana* showed decreased endogenous NO levels as well as enhanced NaCl-induced increase in Na^+ to K^+ ratio [16].

SNP could significantly delay leaf senescence by increasing photosynthetic rate, chlorophyll content and expression of LHCB, SOS1 and NHX1 genes and reduced the expression of abscisic acid (ABA) biosynthesis genes, NCED2, NCED9 genes and ABA content [45, 46]. Exogenous application of both SNP (a NO donor) and putrescine on the plants of *Cicer arietinum* L. showed positive effects on antioxidant enzymes under NaCl stress. Putrescine has been more effective in scavenging superoxide radical by increasing the superoxide dismutase (SOD) activity under NaCl stress whereas SNP was effective in hydrolyzing H_2O_2 by enhancing the activities of catalase (CAT), ascorbate peroxidase (APX) and peroxidase (POX) under salt stress. SNP and putrescine treatments showed no significant change on the activity of glutathione reductase (GR) and dehydroascorbate reductase (DHAR) enzymes [11].

Ahmad *et al*. [15] reported that exogenous application of nitric oxide in the form of its donor SNAP (S-nitroso-N-acetylpenicillamine) protected chickpea plants against salt stress-induced oxidative damage by enhancing the growth parameters, leaf relative water content (LRWC), chlorophyll content and accumulation of different osmolytes such as glycine betaine (GB), proline, soluble sugars and soluble proteins. External supply of NO mitigated the deleterious effects of NaCl stress by reducing electrolyte leakage, hydrogen peroxide (H_2O_2) and malondialdehyde (MDA) contents. NO also increased the activities of antioxidant enzymes such as APX (ascorbate peroxidase), CAT (catalase), GR (glutathione reductase) and SOD (superoxide dismutase) which were found to be correlated with the up-regulation of APX, CAT and SOD. NO application enhanced the salt tolerance of soybean plants by reducing accumulation of hydrogen peroxide and increasing ascorbate peroxidase (APX) activity [17].

Shi *et al*. [18] reported that NO significantly reduced the accumulation of TBARS, thus decreased lipid peroxidation of mitochondria in the seedlings of *Cucumis sativus* L. Sodium nitroprusside increased the inhibition of H^+-ATPase and H^+-PPase in plasma memebrane and/or tonoplast under salt stress in cucumber seedlings. Combined application of sodium nitroprusside (SNP), a NO donor and $CaCl_2$ was observed to be more effective in *Brassica juncea* L. (Mustard) than their individual application in reducing NaCl-induced stress by increasing the activities of superoxide dismutase (SOD), ascorbate peroxidase (APX), peroxidase (POX), glutathione reductase (GR), catalase (CAT), carbonic anhydrase (CA) and nitrate reductase (NR) enzymes, leaf chlorophyll content, leaf relative water content (LRWC), accumulation of osmolytes, accumulation of Pro and GB, potaasium (K^+) and calcium (Ca^+) concentration of leaf while decreasing sodium (Na^+) concentration of leaf, electrolyte leakage, TABRS (thiobarbituric acid relative substances) and H_2O_2 content in a study conducted by Khan *et al*. [21]. Li *et al*. [32] reported that SNP increased the activities of antioxidant enzymes like catalases (CAT), superoxide dismutases (SOD) and ascorbate peroxidases (APX) and ferritin accumulation at the protein level to protect barley seedlings from NaCl-stress damage in Barley seedlings. The application of SNP on the other hand significantly increased the expression of SOS1 and NHX1 genes and reduced the expression of ABA biosynthesis genes, NCED2, NCED9 genes and ABA content in cotton leaves and roots. The SNP aslo significantly increased the expression of cytokinin biosynthesis gene and IPT1 gene and ZR and iPA content of leaf and root under salt stress [45].

The exogenous application of nitric oxide alleviated the damage of salt stress in Jatropha seedlings by enhancing ascorbate (AsA), glutathione (GSH), catalase (CAT) and glutathione reductase (GR) enzyme activities in both endosperm and embryo axis. NO also decreased Na^+ and Cl^- accumulation, which potentially reduced oxidative damage and ROS accumulation. NO up-regulated the JcCAT1, JcCAT2, JcGR1 and JcGR2 gene expression in embryo Axis and also decreased H_2O_2 (hydrogen peroxide) content and membrane damage [47]. Application of SNP increased dry weight of both roots and leaves of *Kosteletzkya virginica* (Mallowes) activities of peroxidase (POD) in roots and leaves, superoxide dismutase (SOD) in leaves and catalase (CA) in roots and leaves [48]. In *Lycopersicon esculentum* L. (Tomato), SNP application completely neutralized the detrimental effects generated by the 50 mM NaCl concentration at 60 DAS (seedlings of 60 days) and not by higher concentrations of NaCl. The growth parameters (fresh and dry weight of both shoots and roots) relative water content, photosynthetic attributes (SPAD chlorophyll and photosynthetic rate) and activity of carbonic anhydrase enzyme was increased by SNP application following the NaCl treatment [49].

The pretreatment of rice seedlings with a low level of sodium nitroprusside (SNP) and Hydrogen peroxide (H_2O_2) increased quantum yield for photosystem II which was reduced under salt stress. Northern blot analysis showed that treatment with H_2O_2 and SNP increased the inducible expression of stress-related genes (SPS, P5CS, and HSP26), thus increased abiotic stress tolerance in rice seedlings [50]. Zhao *et al.* [51] documented that the exogenous application of 0.2 mm SNP to both DR and SR callus of *Phragmities comunis* decreased Na percentage and increased the percentage of Ca as well as K to Na ratio. SNP also reduced membrane permeability (MPs) in both callus which was earlier increased by 45% in the DR callus and by 185% in the SR callus under the NaCl-induced stress. In the DR callus, the activity of NOS was decreased to 60% upon treatment with SNP alone as well as SNP in combination with NaCl, while in SR callus, the NOS activity was reduced to 20% with the treatment of only SNP (0.2 mm) as well as SNP and NaCl. In *Populus euphratica* callus, SNP (0.3 mM) treatment decreased Na percentage by 38% while increased K percentage and K/Na ratio by 155% and 405%, respectively. DPI reversed NO and H_2O_2 effect by increasing Na percentage and decreasing K percentage and K/Na ratio, while PTIO and NMMA only reversed NO effect. NMMA and DPI decreased the activity of plasma membrane (PM) H^+-ATPase which was earlier increased under salt stress.

PTIO and NMMA reversed NO effect of increase of PM H^+-ATPase activity only but not of H_2O_2 during their study [52].

Nitric oxide treatment increased proline and ascorbate content while decreased H_2O_2 content in tomato plants. Application of SNP (100 μM and 300 μM) increased leaf dry mass and root length of tomato plants which were reduced under salt stress [53]. The application of NO enhanced fresh and dry weights of the edible parts of spinach as well as growth and photosynthetic rate of the plants. It also decreased contents of hydrogen peroxide and malondialdehyde in leaves. Application of NO enhanced the activities of antioxidant enzymes such as CAT (catalase), SOD (superoxide dismutase) and POX (peroxidase). Further, NO significantly increased the levels of ascorbate, total phenolics, glutathione, flavonoids and proline as well as the DPPH scavenging activity in salt stressed plants [54]. Ruan *et al.* [55] stated that SNP (a NO donor) significantly increased the activities of catalase (CAT) and superoxide dismutase (SOD) by reducing accumulation of H_2O_2 and O_2^- in wheat leaves under salt stress. NO was also found to enhance the accumulation of proline and in stimulation of PM H^+-ATPase and H^+-PPase activity to modulate ion homeostasis for salt tolerance. It further induced the adaptive capabilities of seedlings of wheat against salt stress by significantly increasing the K^+ content and K^+ to Na^+ ratio in the roots [56].

SNP application maintained ionic homeostasis during germination of wheat seeds under salt stress by decreasing Na^+ concentration and increasing K^+ concentration. SNP (NO donor) also increased the activities of catalase (CAT) and superoxide dismutase (SOD) whereas decreased the contents of hydrogen peroxide (H_2O_2), malondialdehyde (MDA) and release rate of superoxide anions (O_2^{--}) in the mitochondria [57]. Hasanuzzaman *et al.* [58] documented that pretreatment with SNP (NO donor) increased the glutathione (GSH) content, ascorbate (AsA) content, GSH/GSSG (glutathione/glutathione disulfide) ratio and also the activities of dehydroascorbate reductase (DHAR), monodehydroascorbate reductase (MDHAR), glutathione reductase (GR), glutathione peroxidase (GPX), glutathione S-transferase (GST), glyoxalase I (Gly I) and glyoxalase II (Gly II) in most of the wheat seedlings under salt stress. The pretreatment with NO significantly decreased hydrogen peroxide (H_2O_2) and malondialdehyde (MDA) content in salt-stressed seedlings.

SNP significantly increased activities of peroxidase (POD), superoxide dismutase (SOD) and catalase (CAT) by 15.4%, 89.1% and 5.5% respectively in the leaves of *Triticum aestivum*. Addition of SNP enhanced leaf chlorophyll content and reduced both root and H^+-ATPase activities of salt stressed plants Application of NO and $Ca(NO_3)_2$ improved ion homeostasis of plants under. salt stress by enhancing ion uptake and balancing the Ca/Na ratio [59]. A detailed summary of literature on nitric oxide mediated modulation of antioxidative responses under salinity stress in various plant species have been compiled in Table **1**.

Table 1. Role of nitric oxide (NO) in alleviation of salinity induced oxidative stress in different plants/plant parts.

S. No.	Name of the Plant and Family	Plant/ Plant Part	Treatment	Effects	Reference
1	*Zea Mays* L. (Maize) Poaceae	Seedlings	Seedlings of maize were treated with SNP at different concentrations and methylene blue (MB).	Increased dry matter, chlorophyll content, and relative water content and decreased membrane leakage from leaf cells.	[4]
2	*Cicer arietinum* L. (Chickpea) Fabaceae	Plants	Chickpea plants (50 days) were given different treatments of sodium nitroprusside (SNP), sodium chloride (NaCl) and putrescine.	Reduced lipid peroxidation.	[11]
3	*Aegiceras corniculatum* L. (Black mangrove) Primulaceae	Seedlings (leaves)	Sodium chloride (NaCl) and sodium nitroprusside (SNP)	Decreased hydrogen peroxide (H_2O_2) content, polyphenol oxidase (PPO) activity and lipid peroxidatio. Increased guaiacol peroxidase (GPX) activity and contents of reduced glutathione and polyphenols.	[14]
4	*Cicer arietinum* L. (Chickpea) Fabaceae	Plants	Sodium chloride (NaCl) and sodium nitroprusside (SNP)	Enhanced growth parameters, leaf relative water content (LRWC), chlorophyll content and accumulation osmolytes (glycine betaine, soluble proteins, proline and soluble sugars).	[15]
5	*Arabidopsis thaliana* L.	Plants	Sodium chloride (NaCl) and sodium nitroprusside (SNP)	Higher ratio of Na+ to K+ in shoots as compared to wild-type	[16]

	(Mouse-ear cress) Brassicaceae (Wild-type and mutant (Atnoa1))			plants thus combating oxidative stress.	
6	*Glycine max* L. (Soybean) Fabaceae	Plants	Different treatments of diethylenetriamine (DETA) and sodium chloride (NaCl) were given.	Improved growth of soybean plants under salt stress	[17]
7	*Cucumis sativus* L. (Cucumber) Cucurbitaceae	Seedlings (Roots, Shoots)	Sodium nitroprusside (SNP), sodium chloride (NaCl), sodium ferrocyanide ($Na_4Fe(CN)_6$) and haemoglobin at various concentrations.	Increase in dry weight of both stem and root. Activation of ROS-scavenging enzymes (SOD, CAT, GPX, APX, DHAR), reduced hydrogen peroxide (H_2O_2) content.	[18]
8	*Brassica juncea* L. (Mustard) Brassicaceae	Seedlings (Leaves)	SNP (sodium nitroprusside), $CaCl_2$ (calcium chloride) and cPTIO (2-(4-carboxyphenyl)-4,4,5,5-tetramethylimidazoline-1-oxyl-3-oxide).	Enhanced activities of ascorbate peroxidase (APX), superoxide dismutase (SOD), catalase (CAT), glutathione reductase (GR), peroxidase (POX) carbonic anhydrase (CA) and nitrate reductase (NR) enzymes. Leaf chlorophyll content, leaf relative water content, accumulation of osmolytes, potassium (K^+) and calcium (Ca^+) concentration of leaf were also enhanced.	[21]

(Table 1) cont.....

9	*Cucumis sativus* L. (Cucumber) Cucurbitaceae	Seedlings (Leaves and roots)	Sodium nitroprusside (SNP) and sodium chloride (NaCl)	Significant increase in plant height, stem thickness, fresh weight and dry weight and decrease in the content of free polyamine oxidase (PAO), putrescine (Put) and spermidine (Spd) activity in leaves.	[22]
10	*Hordeum vulgare* L. (Barley) Poaceae	Seedlings (Leaves)	Sodium nitroprusside (SNP) and sodium chloride (NaCl)	Decreased ion leakage and malondialdehyde (MDA), carbonyl, and hydrogen peroxide (H_2O_2) content.	[32]
11	*Lycopersicon esculentum* L. (Tomato) Solanaceae	Seedlings (Leaves)	Sodium nitroprusside (SNP) and sodium chloride (NaCl)	Increase in the activities of nitrate reductase (NR), proline content and antioxidant enzymes *i.e.* catalase (CAT), peroxidase (POX) and superoxide dismutase (SOD).	[40]
12	*Gossipium hirsutum* L. (Cotton) Malvaceae	Leaves and roots	Sodium nitroprusside (SNP)	Signiifcantly delayed leaf□ senescence, increase in photosynthetic rate, chlorophyll content and expression of LHCB gene. Signiifcantly enhanced the □⁺ content in leaves and roots and decreased Na^+ content in leaves and roots.	[45]
13	*Arabidopsis thaliana* L. (Mouse-ear cress) Brassicaceae	Seedling (Roots)	Sodium chloride (NaCl)	NaCl salt stress significantly down-regulated the expression of PIN genes and promoted AUXIN RESISTANT3 (AXR3)/IAA17 stabilization.	[46]

(Table 1) cont.....

14	*Jatropha curca* L. (Jatropha) Euphorbiacea e	Seedlings	Sodium chloride (NaCl) and nitric oxide (NO).	Improved salt tolerance by ameliorating oxidative damage and toxic ion accumulation.	[47]
15	*Kosteletzkya virginica* L. (Mallows) Malvaceae	Seedlings (Roots and leaves)	Sodium nitroprusside (SNP) and sodium chloride (NaCl)	Increased dry weight of both roots and leaves, activities of peroxidase (POD) in roots and leaves, superoxide dismutase (SOD) in leaves and catalase (CA) in roots and leaves.	[48]
16	*Lycopersicon esculentum* L. (Tomato) Solanaceae	Plants	Sodium nitroprusside (SNP) and sodium chloride (NaCl)	Growth parameters, relative water content, chlorophyll content, photosynthetic rate and activity of carbonic anhydrase enzyme were increased by SNP application.	[49]
17	*Oryza sativa* L. (Rice) Poaceae	Seedlings	Sodium Nitroprusside (SNP) and Hydrogen peroxide (H_2O_2)	SNP showed more inducible effects on the activities of CAT, POX and SOD than that of H_2O_2.	[50]
18	*Phragmities communis* L. Dune reed (DR) and Swamp reed (SR) Poaceae	Callus	NaCl, SNP, L-NNA (N^{ω}-nitro-l-arginine) or PTIO (2-phenyl-4,4,5,5-tetramethyl-imidazoline-1-oxyl-3-oxyde)	SNP restored relative water content (RWC) in callus (both DR and SR) to the normal status combating the salt stress.	[51]
19	*Populus euphratica* L. (Populus) Salicaceae	Callus	-NaCl and glucose/glucose oxidase (H_2O_2 donor), SNP (sodium nitroprusside, a NO donor), DPI	Application of glucose/glucose oxidase and SNP showed that both NO and H_2O_2 led to increase the K/Na ratio.	[52]

(Table 1) cont.....

			(diphenylene iodonium, an NADPH oxidase inhibitor), NMMA (N^G-monomethyl-l-Arg monoacetate, an NO synthase inhibitor), PTIO (2-phenyl4,4,5,5-tetramethyl-imidazoline -1-oxyl-3-oxyde, NO scavenger) or GSNO (S-nitrosoglutathione)		
20	*Solanum lycopersicum* L. (Tomato) Solanaceae	Plants (Leaves and roots)	Sodium nitroprusside (SNP) and sodium chloride (NaCl)	Increased antioxidant enzyme activities *i.e.* APX (ascorbate peroxidase), SOD (superoxide dismutase), POD (peroxidase) and GR (glutathione reductase) as well as enzymes involved in nitrogen metabolism *i.e.* NiR (nitrite reductase) and NR (nitrate reductase).	[53]
21	*Spinacia oleracea* L. (Spinach) Amaranthaceae	Plants	NaCl and NO	NO enhanced fresh and dry weights, growth photosynthetic rate and antioxidant enzyme activities. Decreased contents of hydrogen peroxide and malondialdehyde in leaves.	[54]
22	*Triticum aestivum* L. (Wheat) Poaceae	Leaves	Sodium nitroprusside (SNP) and sodium chloride (NaCl)	SNP treatment increased chlorophyll content and decreased MDA accumulation and decreased plasma membrane permeability	[55]

(Table 1) cont.....

23	*Triticum aestivum* L. (Wheat) Poaceae	Seedlings (Roots)	Sodium nitroprusside (SNP) and sodium chloride (NaCl)	SNP (NO donor) alleviated the water loss, growth inhibition and the decay of chlorophyll in seedlings of wheat caused by NaCl.	[56]
24	*Triticum aestivum* L. (Wheat) Poaceae	Seedlings	Sodium nitroprusside (SNP) and sodium chloride (NaCl)	SNP increased seed respiration rate and ATP synthesis, soluble sugar content and decreased seed starch content.	[57]
25	*Triticum aestivum* L. (Wheat) Poaceae	Seedlings	Sodium nitroprusside (SNP) and MG chloride (NaCl)	SNP application alleviated antioxidant defense and MG (methylglyoxal) detoxiifcation-systems.	[58]
26	*Triticum aestivum* L. (Wheat) Poaceae	Seedlings (Leaves)	Sodium chloride (NaCl), calcium nitrate (CaNO$_3$) and SNP.	Increased plant growth and relative water content (RWC). Decreased soluble sugar content, leaf electrolyte leakage, lipid peroxidation, H$_2$O$_2$ content and O$_2$$^-$ generation.	[59]

CONCLUSION

Salinity in soil or water is one of the major stressors for the food crops which can limit crop production by posing oxidative stress. Although, each plant has its own antioxidant enzyme system to counteract oxidative stress, but the working of this system exclusively depends on the balance between free radical generation and free radical sequestration. Thus, increasing the concentration of free radical scavengers can improve the antioxidant defense mechanism which can be attained by their exogenous application. In this line of study, the exogenous supply of nitric oxide was observed to be very effective in mitigating the salinity stress in various plant species. Thus, manipulation of endogenous NO content in various plants growing under salinity stress conditions, by application of various NO donors such as putrecine, sodium nitroprusside (SNP), 2-phenyl-4,4,5-5-tetremethy-limidazolinone1-oxyl 3-oxide and diethylenetriamine (DETA) can be a reliable strategy for salinity management in the modern period.

CONSENT FOR PUBLICATION

Not applicable.

CONFLICT OF INTEREST

The author declares no conflict of interest, financial or otherwise.

ACKNOWLEDGEMENTS

The authors are thankful to the University Grants Commission (UGC), New Delhi and Department of Botanical and Environmental Sciences, Guru Nanak Dev University, Amritsar for providing the necessary facilities.

REFERENCES

[1]   Pitman MG, LaEuchli A. Global impact of salinity and agricultural ecosystems. In: LaEuchli A, LuEttge U, Eds. Salinity Environment, Plants, Molecules 2002; 3-20.

[2]   Mahmood-ur-Rahman IM, Qamar S, Bukhari SA, Malik K. Abiotic stress signaling in rice crop. Advances in Rice Research for Abiotic Stress Tolerance. Elsevier Inc 2019; pp. 551-69.
[http://dx.doi.org/10.1016/B978-0-12-814332-2.00027-7]

[3]   Hu H, Dai M, Yao J, *et al.* Overexpressing a NAM, ATAF, and CUC (NAC) transcription factor enhances drought resistance and salt tolerance in rice. Proc Natl Acad Sci USA 2006; 103(35): 12987-92.
[http://dx.doi.org/10.1073/pnas.0604882103] [PMID: 16924117]

[4]   Zhang Y, Wang L, Liu Y, Zhang Q, Wei Q, Zhang W. Nitric oxide enhances salt tolerance in maize seedlings through increasing activities of proton-pump and Na+/H+ antiport in the tonoplast. Planta 2006; 224(3): 545-55.
[http://dx.doi.org/10.1007/s00425-006-0242-z] [PMID: 16501990]

[5]   Zhu JK. Plant salt stress. Encyclopedia of Life Sciences. Chichester, UK: Wiley 2007; pp. 1-3.
[http://dx.doi.org/10.1002/9780470015902.a0001300.pub2]

[6]   Munns R, Tester M. Mechanisms of salinity tolerance. Annu Rev Plant Biol 2008; 59: 651-81.
[http://dx.doi.org/10.1146/annurev.arplant.59.032607.092911] [PMID: 18444910]

[7]   Wang Y, Li K, Li X. Auxin redistribution modulates plastic development of root system architecture under salt stress in *Arabidopsis thaliana*. J Plant Physiol 2009; 166(15): 1637-45.
[http://dx.doi.org/10.1016/j.jplph.2009.04.009] [PMID: 19457582]

[8]   Ksouri R, Megdiche W, Debez A, Falleh H, Grignon C, Abdelly C. Salinity effects on polyphenol content and antioxidant activities in leaves of the halophyte *Cakile maritima*. Plant Physiol Biochem 2007; 45(3-4): 244-9.
[http://dx.doi.org/10.1016/j.plaphy.2007.02.001] [PMID: 17408958]

[9]   Parida AK, Jha B. Salt tolerance mechanisms in mangroves: a review. Trees (Berl) 2010; 24: 199-217.
[http://dx.doi.org/10.1007/s00468-010-0417-x]

[10]   Meneguzzo S, Navari-Izzo F, Izzo R. Antioxidative responses of shoots and roots of wheat to increasing NaCl concentrations. J Plant Physiol 1999; 155: 274-80.
[http://dx.doi.org/10.1016/S0176-1617(99)80019-4]

[11]   Sheokand S, Kumari A, Sawhney V. Effect of nitric oxide and putrescine on antioxidative responses under NaCl stress in chickpea plants. Physiol Mol Biol Plants 2008; 14(4): 355-62.
[http://dx.doi.org/10.1007/s12298-008-0034-y] [PMID: 23572902]

[12]   Verma S, Mishra SN. Putrescine alleviation of growth in salt stressed *Brassica juncea* by inducing antioxidative defense system. J Plant Physiol 2005; 162(6): 669-77.
[http://dx.doi.org/10.1016/j.jplph.2004.08.008] [PMID: 16008089]

[13] Wang YS, Yang ZM. Nitric oxide reduces aluminum toxicity by preventing oxidative stress in the roots of *Cassia tora* L. Plant Cell Physiol 2005; 46(12): 1915-23.
[http://dx.doi.org/10.1093/pcp/pci202] [PMID: 16179356]

[14] Chen J, Xiao Q, Wang C, *et al*. Nitric oxide alleviates oxidative stress caused by salt in leaves of a mangrove species, *Aegiceras corniculatum*. Aquat Bot 2014; 117: 41-7.
[http://dx.doi.org/10.1016/j.aquabot.2014.04.004]

[15] Ahmad P, AbdelLatef AA, Hashem A. AbdAllah EF, Gucel S, Tran LSP. Nitric oxide mitigates salt stress by regulating levels of osmolytes and antioxidant enzymes in chickpea. Front Plant Sci 2016; 7: 1-11.
[http://dx.doi.org/10.3389/fpls.2016.00347]

[16] Zhao MG, Tian QY, Zhang WH. Nitric oxide synthase-dependent nitric oxide production is associated with salt tolerance in *Arabidopsis*. Plant Physiol 2007; 144(1): 206-17.
[http://dx.doi.org/10.1104/pp.107.096842] [PMID: 17351048]

[17] Egbichi I, Keyster M, Ludidi N. Effect of exogenous application of nitric oxide on salt stress responses of soybean. S Afr J Bot 2014; 90: 131-6.
[http://dx.doi.org/10.1016/j.sajb.2013.11.002]

[18] Shi Q, Ding F, Wang X, Wei M. Exogenous nitric oxide protect cucumber roots against oxidative stress induced by salt stress. Plant Physiol Biochem 2007; 45(8): 542-50.
[http://dx.doi.org/10.1016/j.plaphy.2007.05.005] [PMID: 17606379]

[19] Zhu JK. Regulation of ion homeostasis under salt stress. Curr Opin Plant Biol 2003; 6(5): 441-5.
[http://dx.doi.org/10.1016/S1369-5266(03)00085-2] [PMID: 12972044]

[20] Hasegawa PM, Bressan RA, Zhu J-K, Bohnert HJ. Plant cellular and molecular responses to high salinity. Annu Rev Plant Physiol Plant Mol Biol 2000; 51: 463-99.
[http://dx.doi.org/10.1146/annurev.arplant.51.1.463] [PMID: 15012199]

[21] Khan MN, Siddiqui MH, Mohammad F, Naeem M. Interactive role of nitric oxide and calcium chloride in enhancing tolerance to salt stress. Nitric Oxide 2012; 27(4): 210-8.
[http://dx.doi.org/10.1016/j.niox.2012.07.005] [PMID: 22884961]

[22] Fan HF, Du CX, Guo SR. Nitric oxide enhances salt tolerance in cucumber seedlings by regulating free polyamine content. Environ Exp Bot 2013; 86: 52-9.
[http://dx.doi.org/10.1016/j.envexpbot.2010.09.007]

[23] Foyer CH, Noctor G. Oxygen processing in photosynthesis: regulation and signalling. New Phytol 2000; 146: 359-88.
[http://dx.doi.org/10.1046/j.1469-8137.2000.00667.x]

[24] Glenn EP, Brown JJ, Khan MJ. Mechanisms of salt tolerance in higher plants. In: Basra AS, Basra RK, Eds. Mechanisms of Environmental Stress Resistance in Plants. The Netherlands: Harwood Academic Publishers 1997; pp. 83-110.

[25] Munns R, James RA, Läuchli A. Approaches to increasing the salt tolerance of wheat and other cereals. J Exp Bot 2006; 57(5): 1025-43.
[http://dx.doi.org/10.1093/jxb/erj100] [PMID: 16510517]

[26] Krishnamurthy R. Amelioration of salinity effect in salt tolerant rice (*Oryza sativa* L.) by foliar application of putrescine. Plant Cell Physiol 1991; 32: 699-703.
[http://dx.doi.org/10.1093/oxfordjournals.pcp.a078133]

[27] Egbichi I, Keyster M, Jacobs A, Klein A, Ludidi N. Modulation of antioxidant enzyme activities and metabolites ratios bynitric oxide in short-term salt stressed soybean root nodules. S Afr J Bot 2013; 88: 326-33.
[http://dx.doi.org/10.1016/j.sajb.2013.08.008]

[28] Chen Q, Zhang W-H, Liu Y-L. Effect of NaCl, glutathione and ascorbic acid on function of tonoplast

vesicles isolated from barley leaves. J Plant Physiol 1999; 155: 685-90.
[http://dx.doi.org/10.1016/S0176-1617(99)80083-2]

[29]   Zhang JL, Shi H. Physiological and molecular mechanisms of plant salt tolerance. Photosynth Res 2013; 115(1): 1-22.
[http://dx.doi.org/10.1007/s11120-013-9813-6] [PMID: 23539361]

[30]   Bray EA, Bailey-Serres J, Weretilnyk E. Responses to abiotic stresses. In: Gruissem W, Buchannan B, Jones R, Eds. Biochemistry and Molecular Biology of Plants. Rockville, MD: American Society of Plant Physiologists 2000; pp. 1158-249.

[31]   Noctor G, Foyer CH. Ascorbate and glutathione: keeping active oxygen under control. Annu Rev Plant Physiol Plant Mol Biol 1998; 49: 249-79.
[http://dx.doi.org/10.1146/annurev.arplant.49.1.249] [PMID: 15012235]

[32]   Li QY, Niu HB, Yin J, *et al.* Protective role of exogenous nitric oxide against oxidative-stress induced by salt stress in barley (*Hordeum vulgare*). Colloids Surf B Biointerfaces 2008; 65(2): 220-5.
[http://dx.doi.org/10.1016/j.colsurfb.2008.04.007] [PMID: 18502620]

[33]   Rajaravindran M, Natarajan S. Effects of Salinity Stress on Growth and Biochemical Constituents of the Halophyte *Sesuvium portulacastrum.* Int J Res Biosci 2012; 2: 18-25.

[34]   Zhu Z, Chen J, Zheng HL. Physiological and proteomic characterization of salt tolerance in a mangrove plant, *Bruguiera gymnorrhiza* (L.) Lam. Tree Physiol 2012; 32(11): 1378-88.
[http://dx.doi.org/10.1093/treephys/tps097] [PMID: 23100256]

[35]   Chen J, Xiong DY, Wang WH, *et al.* Nitric oxide mediates root K+/Na+ balance in a mangrove plant, Kandelia obovata, by enhancing the expression of AKT1-type K+ channel and Na+/H+ antiporter under high salinity. PLoS One 2013; 8(8): e71543.
[http://dx.doi.org/10.1371/journal.pone.0071543] [PMID: 23977070]

[36]   Tester M, Davenport R. Na+ tolerance and Na+ transport in higher plants. Ann Bot 2003; 91(5): 503-27.
[http://dx.doi.org/10.1093/aob/mcg058] [PMID: 12646496]

[37]   Kopyra M, Gwozdz EA. The role of nitric oxide in plant growth regulation and responses to abiotic stress. Acta Physiol Plant 2004; 26: 459-72.
[http://dx.doi.org/10.1007/s11738-004-0037-4]

[38]   Garcia-Mata C, Lamattina L. Abscisic acid (ABA) inhibits light-induced stomatal opening through calcium- and nitric oxide-mediated signaling pathways. Nitric Oxide 2007; 17(3-4): 143-51.
[http://dx.doi.org/10.1016/j.niox.2007.08.001] [PMID: 17889574]

[39]   Chen J, Xiao Q, Wu F, *et al.* Nitric oxide enhances salt secretion and Na(+) sequestration in a mangrove plant, Avicennia marina, through increasing the expression of H(+)-ATPase and Na(+)/H(+) antiporter under high salinity. Tree Physiol 2010; 30(12): 1570-85.
[http://dx.doi.org/10.1093/treephys/tpq086] [PMID: 21030403]

[40]   Hayat S, Yadav S, Wani AS, Irfan M, Alyemeni MN, Ahmad A. Impact of sodium nitroprusside on nitrate reductase, proline content and antioxidant system in tomato under salinity stress. Hortic Environ Biotechnol 2012; 53: 362-7.
[http://dx.doi.org/10.1007/s13580-012-0481-9]

[41]   Jasid S, Simontacchi M, Puntarulo S. Exposure to nitric oxide protects against oxidative damage but increases the labile iron pool in sorghum embryonic axes. J Exp Bot 2008; 59(14): 3953-62.
[http://dx.doi.org/10.1093/jxb/ern235] [PMID: 18832188]

[42]   Kausar F, Shahbaz M. Interactive effect of foliar application of nitric oxide (NO) and salinity on wheat (*Triticum aestivum* L.). Pak J Bot 2013; 45: 67-73.

[43]   Ke X, Cheng Z, Ma W, Gong M. Nitric oxide enhances osmoregulation of tobacco (*Nicotiana tobacum* L.) cultured cells under phenylethanoid glycosides (PEG) 6000 stress by regulating proline metabolism. Afr J Biotechnol 2013; 12: 1257-66.

[44] Lu Y, Li N, Sun J, *et al.* Exogenous hydrogen peroxide, nitric oxide and calcium mediate root ion fluxes in two non-secretor mangrove species subjected to NaCl stress. Tree Physiol 2013; 33(1): 81-95.
[http://dx.doi.org/10.1093/treephys/tps119] [PMID: 23264032]

[45] Kong X, Wang T, Li W, Tang W, Zhang D, Dong H. Exogenous nitric oxide delays salt-induced leaf senescence in cotton (*Gossypium hirsutum* L.). Acta Physiol Plant 2016; 38: 1-9.
[http://dx.doi.org/10.1007/s11738-016-2079-9]

[46] Liu W, Li R-J, Han T-T, Cai W, Fu Z-W, Lu YT. Salt stress reduces root meristem size by nitric oxide-mediated modulation of auxin accumulation and signaling in Arabidopsis. Plant Physiol 2015; 168(1): 343-56.
[http://dx.doi.org/10.1104/pp.15.00030] [PMID: 25818700]

[47] Gadelha CG, Miranda RS, Alencar NLM, Costa JH, Prisco JT, Gomes-Filho E. Exogenous nitric oxide improves salt tolerance during establishment of *Jatropha curcas* seedlings by ameliorating oxidative damage and toxic ion accumulation. J Plant Physiol 2017; 212: 69-79.
[http://dx.doi.org/10.1016/j.jplph.2017.02.005] [PMID: 28278442]

[48] Guo Y, Tian Z, Yan D, Zhang J, Qin P. Effects of nitric oxide on salt stress tolerance in *Kosteletzkya virginica.* Life Sci J 2009; 6: 67-75.

[49] Hayat S, Yadav S, Alyemeni MN, Irfan M, Wani AS, Ahmad A. Alleviation of salinity stress with sodium nitroprusside in tomato. Int J Veg Sci 2013; 19: 164-76.
[http://dx.doi.org/10.1080/19315260.2012.697107]

[50] Uchida A, Jagendorf AT, Hibino T, Takabe T, Takabe T. Effects of hydrogen peroxide and nitric oxide on both salt and heat stress tolerance in rice. Plant Sci 2002; 163: 515-23.
[http://dx.doi.org/10.1016/S0168-9452(02)00159-0]

[51] Zhao L, Zhang F, Guo J, Yang Y, Li B, Zhang L. Nitric oxide functions as a signal in salt resistance in the calluses from two ecotypes of reed. Plant Physiol 2004; 134(2): 849-57.
[http://dx.doi.org/10.1104/pp.103.030023] [PMID: 14739346]

[52] Zhang F, Wang Y, Yang Y, Wu H, Wang D, Liu J. Involvement of hydrogen peroxide and nitric oxide in salt resistance in the calluses from *Populus euphratica.* Plant Cell Environ 2007; 30(7): 775-85.
[http://dx.doi.org/10.1111/j.1365-3040.2007.01667.x] [PMID: 17547650]

[53] Manai J, Kalai T, Gouia H, Corpas FJ. Exogenous nitric oxide (NO) ameliorates salinity-induced oxidative stress in tomato (*Solanum lycopersicum*) plants. J Soil Sci Plant Nutr 2014; 14: 433-46.
[http://dx.doi.org/10.4067/S0718-95162014005000034]

[54] Du ST, Liu Y, Zhang P, Liu HJ, Zhang XQ, Zhang RR. Atmospheric application of trace amounts of nitric oxide enhances tolerance to salt stress and improves nutritional quality in spinach (*Spinacia oleracea* L.). Food Chem 2015; 173: 905-11.
[http://dx.doi.org/10.1016/j.foodchem.2014.10.115] [PMID: 25466105]

[55] Ruan H, Shen W, Ye M, Xu L. Protective effects of nitric oxide on salt stress-induced oxidative damage to wheat (*Triticum aestivum* L.) leaves. Chin Sci Bull 2002; 47: 677-81.
[http://dx.doi.org/10.1360/02tb9154]

[56] Ruan HH, Shen WB, Xu LL. Nitric oxide modulates the activities of plasma membrane H+-ATPase and PPase in wheat seedling roots and promotes the salt tolerance against salt stress. Acta Bot Sin 2004; 46: 415-22.

[57] Zheng C, Jiang D, Liu F, *et al.* Exogenous nitric oxide improves seed germination in wheat against mitochondrial oxidative damage induced by high salinity. Environ Exp Bot 2009; 67: 222-7.
[http://dx.doi.org/10.1016/j.envexpbot.2009.05.002]

[58] Hasanuzzaman M, Hossain MA, Fujita M. Nitric oxide modulates antioxidant defense and the methyl glyoxal detoxification system and reduces salinity-induced damage of wheat seedlings. Plant Biotechnol Rep 2011; 5: 353-65.

[http://dx.doi.org/10.1007/s11816-011-0189-9]

[59]   Tian X, He M, Wang Z, Zhang J, Song Y. Application of nitric oxide and calcium nitrate enhances tolerance of wheat seedlings to salt stress. Plant Growth Regul 2015; 77: 343-56.
[http://dx.doi.org/10.1007/s10725-015-0069-3]

CHAPTER 7

Reactive Oxygen Species Metabolism and Antioxidant Defense in Plants under Stress

Sandeep Kaur[1,2], Samiksha[3], Jaskaran Kaur[1], Sharad Thakur[4,5], Neha Sharma[1], Kritika Pandit[1], Ajay Kumar[1], Shagun Verma[1], Satwinder Kaur Sohal[3] and Satwinderjeet Kaur[1,*]

[1] *Department of Botanical and Environmental Sciences, Guru Nanak Dev University, Amritsar, Punjab 143005, India*

[2] *PG Department of Botany, Khalsa College, Amritsar, Punjab 143005, India*

[3] *Department of Zoology, Guru Nanak Dev University, Amritsar, Punjab 143005, India*

[4] *Department of Molecular Biology and Biochemistry, Guru Nanak Dev University, Amritsar, Punjab 143005, India*

[5] *PG Department of Agriculture, Khalsa College, Amritsar, Punjab 143005, India*

Abstract: Plants are exposed to different types of environmental stressors throughout the different developmental stages. Reactive oxygen species (ROS) are found to play key roles in the maintenance of normal plant growth and improving their ability of stress tolerance. ROS as a secondary messenger performs crucial cellular functions, including the proliferation of cells, apoptosis, and necrosis. Both the external environmental factors and intrinsic genetic programs regulate the morphogenesis of plants. ROS are also considered as by-products of the aerobic metabolism of the plant and are formed in certain cellular compartments like mitochondria, chloroplasts, and peroxisomes. Plants form a huge number of ROS species under unfavorable circumstances that are involved in the regulation of different processes, including programmed cell death, pathogen defense, and stomatal behavior. These reactions often exert irreversible or profound effects on the development of organs and tissues, leading to abnormal death or plant growth. Several molecular approaches to understand the signaling and metabolism of ROS have opened novel avenues in comprehending its key role in abiotic stress. Plants possess their own enzymatic and non-enzymatic antioxidant defense system to encounter ROS. The interconnecting activities of these defensive antioxidants reduce oxidative load and regulate the detoxification of ROS in plants. This book chapter will highlight the importance of ROS metabolism and the role of the antioxidant defense mechanism of plants in combating the deleterious effect of oxidative stress under stressful conditions.

* **Corresponding Author Satwinderjeet Kaur:** Department of Botanical and Environmental Sciences, Guru Nanak Dev University, Amritsar, Punjab 143005, India; Tel: +91 828 3808 508; E-mail: satwinderjeet.botenv@gndu.ac.in

Keywords: Antioxidant defense system, Detoxification, Pathogen, Plant reactive oxygen species.

INTRODUCTION

In aerobic life, the production of reactive oxygen species (ROS) is regarded as an unavoidable chemical entity [1, 2]. The aerobic metabolic processes, including photosynthesis and respiration, unavoidably result in the formation of ROS in peroxisomes, chloroplasts, and mitochondria. The main characteristic feature of the various types of ROS is their capability to induce oxidative damage to DNA, proteins, and lipids [3]. ROS are the major aerobic metabolism byproducts, and their generation involves strong electron flow, which is confined to cellular compartments. About 2.7 billion years ago, in the earth's early reducing atmosphere, the molecular oxygen was first introduced by O_2^- evolving photosynthetic organisms that led to the advent of the unwanted byproducts as ROS [2]. Under favorable conditions, at basal levels, the ROS is being constantly generated. In plants, ROS exhibit a dual role, both acting as key regulators of development, growth, and defense pathways and as aerobic metabolism by-products which are toxic in nature. The disturbance between production and elimination of ROS leads to severe consequences within the cells [4]. A cell undergoes oxidative stress when the ROS levels surpass the range of the scavenging systems, resulting in oxidative change that causes cell damage and ultimately death. When the ROS production rate is small, the cells are in a reduced state, and ROS acts as second messengers in cell division, maintenance, organogenesis, differentiation, and abiotic and biotic maintenance [5]. Therefore, ROS levels must be kept within the correct range for plant safety. However, the different antioxidant mechanisms have the capability to scavenge the free radicals that protect from their damaging effects. Plants also produce ROS by the activation of several peroxidases and oxidases in response to various environmental changes [6, 7]. ROS includes superoxide radical ($O_2^{\cdot-}$), singlet oxygen ($^1O_2$), hydroxyl radical ($OH^{\cdot}$) and hydrogen peroxide (H_2O_2), *etc.* They exert numerous changes in the cellular structure of plants along with physiological responses and degrade proteins, enzymes, and nucleic acid [1]. The metabolism of redox and its related signaling act as crucial machinery at the time of abiotic stress [8].

In the evolutionary process, plants developed a high degree of control over the production of ROS and used them successfully as a signaling molecule [9]. The ROS family play a role of a double-edged sword by acting as secondary messengers in some physiological phenomena and under certain environmental stress conditions such as salinity, drought, UV irradiation, cold, heavy metals, *etc.*, they are able to induce oxidative damages when there is a disturbance in the delicate balance between the elimination and production of ROS, crucial for normal cellular homeostasis [10]. Plants have efficiently developed antioxidant machinery to ensure survival by involving two arms (i) non-enzymatic antioxidants like carotenoids, ascorbic acid (AA), reduced glutathione (GSH), α-tocopherol, flavonoids, and the osmolyte proline (ii) enzymatic components like monodehydroascorbate reductase (MDHAR), catalase (CAT), glutathione reductase (GR), guaiacol peroxidase (GPX), superoxide dismutase (SOD), ascorbate peroxidase (APX) and dehydroascorbate reductase (DHAR) [11].

To enhance the plants' tolerance against the harsh environmental condition, it is important to reinforce the comprehensive antioxidant systems and oxidative stress. The cells of the plant will remain in a state of "oxidative stress" if the level of ROS is raised than the inside antioxidant defense mechanisms. This shows retardation of growth under oxidative stress, such as root gravitropism, leaf, and flower abscission, seed germination, lignin biosynthesis in the cell wall, polar cell growth, and cell senescence [1, 8, 10]. In this chapter, the vital role of plants in controlling and regulating the excessive generation of ROS is elaborated.

REACTIVE OXYGEN SPECIES METABOLISM

The different types of ROS are produced within the plant cells in response to various abiotic, biotic, and other environmental cues. The levels of ROS are usually determined by a balance between the breakdown and production which is tightly controlled and achieved *via* highly complex and sophisticated antioxidant systems. It was found that both biotic and abiotic stresses cause the over-accumulation of ROS. Between the two types of stress, it is observed that the source of enhanced formation of ROS seems to be different. The accumulation of ROS is normally triggered primarily by the electron transport pathway damage in the mitochondria and the chloroplasts in case of abiotic stresses [12]. In the case of biotic stress, it was found that the infections of pathogens induce specific ROS-producing enzymes, including the cell wall peroxidases or plasma membrane NADPH oxidase complex [13]. In plants, ROS are present in molecular and/or ionic states. Ionic states comprise superoxide anions (O^{-2}) and hydroxyl radicals ($\cdot OH$), while molecular states chiefly include singlet oxygen ($^1O_2$) and hydrogen peroxide (H_2O_2) [3, 9, 14]. ROS generate various oxidative stressors that affect diverse biochemical and physiological reactions controlled by different genes in plants. Singlet oxygen ($^1O_2$) is generally produced by photosystem II (PSII) in chloroplast with high oxidizability potential. Though $^1O_2$ persists for a short time and is unstable in cells, it has an effect on photosynthesis. Superoxide anion (O^{-2}) is one of the antecedents of various ROS generation as randomness with high reducibility. O^{-2} might uphold the constancy of plant stem cells [15]. Excessive O^{-2} is also produced due to increased ROS levels that ultimately lead to cell death

[16]. In some plants adapted to the aquatic environments, their roots and stems which are emerged into the water, appear to be the key part of $O^{\cdot-2}$ generation. Among all molecules, H_2O_2 is known as an important redox molecule because of its extraordinary constancy within cells that leads to instantaneous oxidation of target proteins [17].

Production of ROS in Plants During Abiotic Stress

A large number of environmental factors such as high temperature, salinity, cold, drought, organic pollutants (OPs), heavy metal toxicity, pesticides, ultraviolet light, drought, salinity, UV, temperature, freezing stressors [18]. To respond to these abiotic and biotic stressors, plants show their defense system by generating reactive nitrogen species (RNS), reactive oxygen species (ROS), and reactive carbon species (RCS) [19]. ROS are aerobic metabolism byproducts in higher plants, and their site of production are chloroplast, mitochondria, peroxisomes, apoplast, and cellular compartments [20, 21]. ROS includes both free radicals such as perhydroxy radical ($HO_2^{\cdot}$), superoxide radical ($O_2^{\cdot}$), alkoxy radical ($RO^{\cdot}$), hydroxyl radical ($OH^{\cdot}$), and molecular forms like singlet oxygen ($^1O_2$) and hydrogen peroxide (H_2O_2) [22]. So, there is strong intercalation between abiotic stress and the generation of ROS in plants (Fig. **1**).

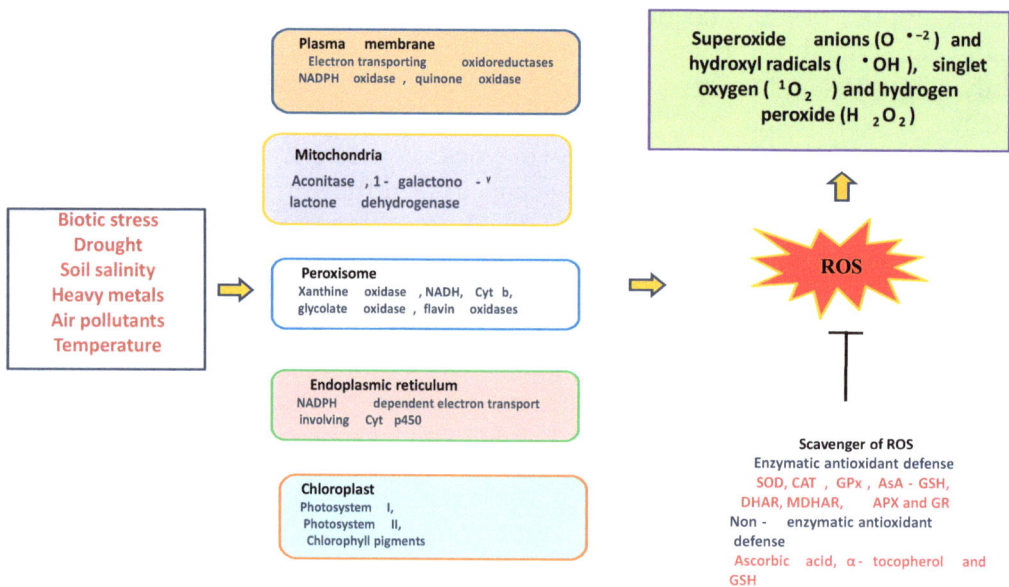

Fig. (1). Proposed mechanism for the production of ROS in plants.

Overproduction of ROS leads to degradation of cellular integrity like DNA damage, chlorophyll, proteins, membrane dysfunction, and inactivation of enzymes [10]. To maintain cellular functioning, plants activate their antioxidant system and ROS scavenging phenomenon by upregulating various enzyme activities [23]. An increase in temperature (heat) is one of the abiotic factors which increases stress for plants. It leads to the degradation of the photosynthetic machinery in plants, especially PSII (photosystem II) but the production of ROS (reactive oxygen species) absorbs excess light through LHC (light-harvesting complex). The excess generation of ROS during stress also leads to an imbalance in photosynthetic redox potential, resulting in photoinhibition [24]. In Arabidopsis, five heat sensors are associated with Ca^{2+} channel, cytosol (CPR) and endoplasmic reticulum (ER-UPR), red/far-red sensor in chloroplast and a histone sensor in the nucleus. Cyclic nucleotide-gated channel 2 (CNGC2) deficiencies in Arabidopsis lead to more production of H_2O_2 in the APX-dependent heat response pathway preventing the accumulation of its toxic level, which enhanced basal thermo tolerance in root and leaves during heat stress [25].

In cotton (*Gossypium* sp), the production of ROS regulates the development of cotton fibers and responds to abiotic stress by activating enzymes or integrating with other complex pathways. A GhCaM7 (calcium sensor) regulates the production of ROS and acts as a linker between ROS and Ca^{2+} signal pathways in the early stages of fiber development. More expression of ROS genes during abiotic stress (cold, heat, dehydration, and salt stress) in Gossypium sp. also revealed their significant effect in the regulation of stress conditions [26]. In mutant chickpea (CaGolS1 and CaGolS2), during abiotic stress, mainly heat and oxidative stress result in the production of ROS. Excessive accumulation of ROS damages their cellular integrity. Therefore RFOs (raffinose family oligosac-charides), mainly galactinol and raffinose, act as antioxidant molecules and show their radical scavenging activity and hence improve stress tolerance conditions in CaGolS1- and CaGolS2- by inhibiting accumulation of ROS and peroxidation of membrane lipid [27]. According to literature, biotic and abiotic factors stimulate non-enzymatic and enzymatic oxidant systems in plants in the form of production of H_2O_2, NO (nitric oxide) [28] and numerous molecular markers. For example, pre-treatment of heat or cold to maize (*Zea mays*) seedlings and cold or drought to tomato plants enhance the formation of H_2O_2 and thereby activate the expression of various genes related to defense system which provides cross-tolerance to plants towards heat, cold, and drought stress [29].

In spite of this, ROS also modulates the growth, development, reproduction [30], defense system, cell elongation, opening and closing of stomata and seed germination by cross-talk with various phytohormones like auxin [31, 32], brassinosteroids [33], abscisic acid, gibberellins [34], ethylene, salicylic acid, strigolactones and jasmonic acid by modulating enzymes which leads to the activation of sources and regulators of ROS [35]. In arabidopsis, LSU (low sulphur upregulation) is a protein network that gets upregulated/downregulated during biotic and abiotic stress circumstances. During high salinity conditions, LSU proteins trigger the production of ROS especially H_2O_2 which results in stomatal closure in response to these abiotic conditions. The generation of ROS through LSU is achieved by activation of superoxide dismutase (SOD) or iron (Fe)-dependent superoxide dismutase [36]. In many plant species such as tobacco and rice, tomato and *Arabidopsis*, overexpression of arginine decarboxylase initiates the production of ROS species which provide immunity to plants against drought, salinity and freezing conditions. According to literature, At*ARGAHs* (arginine aminohydrolase) plays a specific role in abiotic stress by maintaining expression of arginine synthase, endogenous polyamines, production of NO and ROS concentration [37]. Nitric oxide (NO) provides efficiency to plants to cope up with abiotic stress by activating various antioxidant enzymes including APX (ascorbate peroxidase), SOD (superoxide dismutase), GR (glutathione reductase), CAT (catalase), HO (heme oxygenase) and POD (peroxidase). Nitric oxide (NO) also regulates the ROS production and ROS scavenging enzymes which provides an additional immunity to plants to cope with stress conditions [38, 39].

In this way, ROS plays a significant role in plants to respond abiotic stress. ROS acts as a signaling molecule in plants which regulates the defense mechanism of plants by providing equilibrium between ROS production and their scavenging by activating enzymatic and non-enzymatic anti-oxidant pathway [26]. In spite of defense mechanism, ROS modulates expression of various genes that are involved in growth, development, PCD (programmed cell death), biotic and abiotic stress, stomata functioning, defense mechanism and signaling pathways in botanicals [3, 40].

Production of ROS in Plants During Biotic Stress

In the natural habitat, plants are exposed to lots of biotic stressors such as fungi, bacteria and many pathogens as well as the several bacterial infections might persuade oxidative stress in plants [28]. The oxidative stress triggers indirect generation and accumulation of ROS, which cause cell damage before elimination. In case of biotic stressors, a precise distinction was made between biotrophic and necrotrophic pathogens.

In those interactions of host-pathogen, where the necrotroph is a pathogen and observed to induced burst of oxidative stresses that normally promotes the development of disease. A study by Govrin and Levine [41], found that the increased ROS levels (using glucose/glucose oxidase and xanthine/xanthine oxidase treatments) in Arabidopsis caused elevation in fungal growth and necrotrophic Botrytis cinerea necroses. Suppression of ROS with diphenyleneiodonium (an inhibitor of NADPH oxidase) or with antioxidants inhibited the Botrytis induced fungal growth and necroses. Similarly, potato leaves were found to exhibit "extreme resistance" to Potato virus X. It was demonstrated that in potato exhibiting extreme resistance, very early the virus is killed after the onset of infection, and the hypersensitive response was not develop by the resistant host [42].

The hydrogen peroxide early application, after barley infection (before establishment) by powdery mildew, killed fungus and, as a result, the pathogen cannot induce hypersensitive response in the host. However, if 2 days after inoculation, the barley leaves were treated with ROS (hydrogen peroxide) (*i.e.* after infection establishment in the host), in the resistant plant, the hypersensitive response was increased, because sufficient time was given to the pathogen to cause the resistant barley leaves cell death (necroses) [43, 44].

ANTIOXIDANT DEFENSE MECHANISM IN PLANTS

ROS scavenging mechanisms can be divided into two types: non-enzymatic and enzymatic antioxidant protection systems that function to neutralize free radicals synergistically and interactively. The enzymatic processes primarily include peroxidase glutathione (GPX), catalase (CAT), peroxidase ascorbate (APX) and SOD. The enzymatic systems mainly include ascorbate peroxidase (APX), catalase (CAT), SOD and glutathione peroxidase (GPX) [3]. Among all the enzymatic systems, SOD can efficiently convert OH to H_2O_2, and the H_2O_2 produced is then converted by peroxidase and CAT to water and dioxygen [45]. The non-enzymatic systems are predominantly controlled by low molecular antioxidants, including flavonoids, glutathione and ascorbic acid (AsA) that are believed to eliminate singlet oxygen and hydroxyl radicals [46].

Enzymatic Antioxidant Defense

Antioxidants reduce or avoid substantial oxidation of oxidizable substrates such as free radicals when present in lower concentrations than the substrate [47]. The synchronized activity of antioxidants helps to reduce oxidative stress and detoxify ROS in plants [10]. Plants have evolved an effective antioxidant defense mechanism to escape the toxic effects of free radicals. The antioxidant system in plants consists of hybrid antioxidant enzymatic and non-enzymatic protection mechanisms. The enzymatic processes include superoxide dismutase (SOD), catalase (CAT), peroxidase glutathione (GPx), enzymes of ascorbate-glutathione (AsA-GSH) cycle such as dehydroascorbate reductase (DHAR), monodehydroascorbate reductase (MDHAR), ascorbate peroxidase (APX) and glutathione reductase (GR). The non-enzymatic metabolites include low molecular weight antioxidants (proline, carotenoids, ascorbic acid, glutathione, phenolic acids, flavonoids, *etc.*) and secondary metabolites with a high molecular weight such as tannins [48].

Among enzymatic antioxidant components, several antioxidant enzymes catalyze the ROS degradation processes. Superoxide dismutase (SOD, EC 1.15.1.1) are enzymes that catalyze the dismutation of O_2^- to H_2O_2 and are therefore a frontline defense against ROS [49].

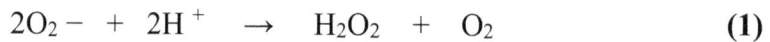

$$2O_2^- + 2H^+ \rightarrow H_2O_2 + O_2 \tag{1}$$

Catalases (CAT, EC 1.11.1.6) are known for eliminating H_2O_2 through a reduction of H_2O_2 to $2H_2O$ [50].

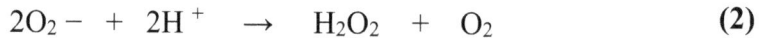

$$2O_2^- + 2H^+ \rightarrow H_2O_2 + O_2 \tag{2}$$

The nonheme thiol peroxidases, Glutathione peroxidases (GPX, EC 1.11.1.9) also catalyze the reduction of H_2O_2. In this reaction, glutathione (GSH) is used as a reducing agent [51].

$$2GSH + ROOH (H_2O_2) \rightarrow GSSG + ROH + H_2O (2H_2O) \tag{3}$$

Among the enzymes of the ascorbate-glutathione (AsA-GSH) cycle, ascorbate peroxidases (APX, EC 1.1.11.1) are enzymes that catalyze the conversion of H_2O_2 to $2H_2O$ and use ascorbate as a common donor of electrons [52].

$$H_2O_2 + 2AsA \rightarrow 2H_2O + 2MDHA \tag{4}$$

Dehydroascorbate reductase (DHAR) and monodehydroascorbate reductase (MDHAR) enzymes of AsA-GSH cycle are involved in the regeneration of ascorbic acid (AsA). MDHAR uses nicotinamide adenine dinucleotide phosphate (NAD(P)H) whereas DHAR uses GSH as an electron donor. Furthermore, the active enzyme of this cycle is glutathione reductase (GR) that checks the cell from damage caused by oxidative stress. It maintains ample amounts of reduced GSH by reduction of oxidized GSH at the expense of NADPH [53, 54].

Non-enzymatic Antioxidant Defense

The non-enzymatic molecules of the antioxidant defense system are vital to free radical scavengers by transfer of electron or hydrogen to plants. Ascorbic acid (vitamin C), α-tocopherol (vitamin E) and reduced glutathione (GSH) are major components of this system. α-tocopherol (vitamin E) is an important lipophilic, membrane-bound antioxidant that is known to have efficient scavenging potency for ROS and lipid radicals. According to literature, one α-tocopherol molecule can scavenge up to $120 \, _1O^2$ molecules *via* resonance energy transfer and it acts as a chain-breaking antioxidant [55, 56]. Moreover, an α-tocopherol radical may migrate to the membrane surface and react with ascorbate or GSH to regenerate α–tocopherol. The resulting radical ascorbate can recover ascorbate *via* GSH reduction in the AsA-GSH cycle [57]. Ascorbic acid (vitamin C) is a water-soluble component of this system that is distributed almost in all plants. It is an essential antioxidant that not only interacts with H_2O_2, but also with O_2^-, OH, and lipid peroxidases [58, 59]. Similarly, tripeptide glutathione (α-glutamyl cysteinyl glycine; GSH) maintains the redox homeostasis and is involved in various biosynthetic and detoxification, pathways [60]. Other than these, phenolics are numerous secondary metabolites (tannins, esters of hydroxycinnamate, lignin, and flavonoids) that contribute to plant defense mechanisms because of their antioxidant efficacy in response to various stress conditions [61].

ROS IN GROWTH AND DEVELOPMENT OF PLANTS

The plants are sessile organisms, have complex approaches to combat stressors. Before improving the crop yield the upgraded knowledge is needed for the synchronized development of plant tissues and organs. Plant development is controlled equally by intrinsic parameters as well as external environmental factors. ROS are generated as plant aerobic metabolism by-products and are produced in numerous cellular sections such as chloroplasts [62], mitochondria [63], and peroxisomes [64]. In stress conditions, plant responds by generating ROS that causes DNA damage and cell death and also produce significant signaling molecules that control normal plant growth.

Reactive oxygen species (ROS) are moderately basic molecules that reside in cells developing in aerobic conditions. ROS such as hydrogen peroxide (H_2O_2), superoxide (O_2^-), singlet oxygen ($^1O_2$), and hydroxyl radical (OH) were previously thought to be toxic to many cellular processes such as photosynthesis and aerobic respiration. Nevertheless, the investigation has reported that ROS has key physiological functions in the development and growth of plants [65]. The production of ROS in plants is found primarily in the mitochondria during mitochondrial respiration, in the chloroplast during photosynthesis and in peroxisomes during photorespiratory reactions. On the other hand, the cell membrane, endoplasmic reticulum, apoplast, and the cell wall act as secondary sites for ROS production [11]. ROS is important for seed viability for successive germination, at the time of desiccation, and for the early seedling growth. Various enzymes contribute to the production of apoplastic ROS, depending on the species. By germinating seeds, apoplastic ROS production is a direct microbicidal tool against the formation of invading pathogens [7]. Also, cellular ROS signaling triggers various reactions that are helpful for the survival of the seed and growth of seedlings. In a study, sweet chestnut seeds loose viability and are recalcitrant during desiccation. The excised embryos, when desiccated generate apoplastic O_2^- with a peak before the loss of viability, indicating that ROS serves as a signal triggering stress-related pathways. In a study, it was demonstrated that flowering time1 (PFT1)/MED25 and phytochrome restricts cell growth, while MED8 works independently of PFT1 to regulate the growth of organs and controls the balance of redox in roots [66]. Additionally, ROS are involved in controlling tip growth, such as the formation of root hairs. Loss of Root Hair Defective 2 (RHD2) mutant, encodes an NADPH oxidase or RbohC, does not produce ROS at the incipient root hair tip. As a consequence, the incipient root hair does not develop [67]. During seed germination of radish, both the embryos and the seed coats (living aleurone layer) produce many forms of ROS. In the seed coats, ROS production starts before any germination signs, while after the onset of the growth of radical, embryos begin to develop ROS. If the germination of radish seeds is inhibited by far-red light or abscisic acid, the production of ROS is also restricted in these organs [7]. Reactive oxygen species are also involved in the extension of leaf in maize, and in the leaf blades, expansion zone has more productivity [68]. ROS is also believed to play crucial roles in the growth of the branches and leaves, senescence and organ dormancy. *CAT2* expression (*Catalase 2*), for example, is decreased in Arabidopsis leaves after bolting. This results in the accumulation of H_2O_2 and promotes the WRKY53 gene expression, which is required for leaf senescence [69]. In rice, the homeobox gene *MADS3* was found to be important for the formation of stamens during early floral growth. At the anther development later stages, however, *MADS3* controls the abnormal MADS3 expression and ROS homeostasis that causes pollen sterility and the accumulation of O_2^- [70]. In rice (*Oryza sativa*) mutant *defective in tapetum cell death 1 (dtc1)*, ROS does not accumulate and displays delayed tapetum programmed cell death resulting in male-sterile plants [71]. *Arabidopsis* seeds germination requires the NADPH oxidase-dependent ROS. The freshly harvested seeds germination of the NADPH oxidase mutant (rbohD) is seriously decreased which is close to the treatment of the DPI test. NADPH oxidase-dependent ROS may induce germination by signaling activation of gibberellin [72]. Fruit ripening and post-ripening of seed are also based on ROS control. A low ROS storage condition delayed the fruit maturation while H_2O_2 treatment facilitated fruit ripening [73]. During the weakening of endosperm and radicle elongation, the production, as well as the action of •OH, increased which were inhibited by ABA. GA reversed these effects, indicating a positive role of •OH in cell wall loosening during seed germination [74]. In a study, it has been proposed that the accumulation of ROS at the tip of pollen tubes is important for their effective growth towards the female gametophyte [75]. In *Arabidopsis thaliana*, O_2^- accumulates primarily in the root tip's meristematic zone and is necessary for cell division, while H_2O_2 accumulates primarily in the elongation region, resulting in cell differentiation [76]. In a study, it was demonstrated that ROS also engages in breaking seed dormancy and germination by stimulating the oxidative pentose phosphate pathway [77]. ROS controls the primary development of the plant root, along with hormones and other signal molecules. ROS and auxin signaling are antagonistically regulated in the root apical meristem (RAM), to control root meristem development [78]. It was reported that ROS (H_2O_2) affects the proliferation of cortex, which is an inner cell type of the root. ROS also plays an important role in the development of the reproductive organs and tissues of plants. In a study, it was shown that glutathione/GRX systems played a vital role in flower development as determined by plant-specific class III CC-type GRXs, known as ROXYs [79].

CONCLUSION

The metabolism of reactive oxygen species was found to be beneficial for the growth and development of the plant as it plays an important role as a signaling molecule that controls the biological processes in all the cells. However, the overproduction of ROS under stressful circumstances leads to oxidative damage of various cellular biomolecules and may cause death. Although, plants have effective antioxidant machinery to combat the deleterious effect of ROS by scavenging and detoxifying the harmful pro-oxidants formed during ROS metabolism in order to maintain the redox homeostasis. Therefore, the antioxidative defense system of plants is effective in mitigating the harmful effects of excessive production of ROS.

CONSENT FOR PUBLICATION

Not applicable.

CONFLICT OF INTEREST

The author declares no conflict of interest, financial or otherwise.

ACKNOWLEDGEMENTS

We are thankful to the Indian Council of Medical Research (ICMR) [59/36/2011/BMS/TRM], New Delhi (India) for supporting this work.

REFERENCES

[1]   Halliwell B, Gutteridge JMC. Free radicals in biology and medicine. Gloucestershire UK: Clerendon Press 1989.

[2]   Halliwell B. Reactive species and antioxidants. Redox biology is a fundamental theme of aerobic life. Plant Physiol 2006; 141(2): 312-22.
        [http://dx.doi.org/10.1104/pp.106.077073] [PMID: 16760481]

[3]   Apel K, Hirt H. Reactive oxygen species: metabolism, oxidative stress, and signal transduction. Annu Rev Plant Biol 2004; 55: 373-99.
        [http://dx.doi.org/10.1146/annurev.arplant.55.031903.141701] [PMID: 15377225]

[4]   Foyer CH, Noctor G. Redox homeostasis and antioxidant signaling: a metabolic interface between stress perception and physiological responses. Plant Cell 2005; 17(7): 1866-75.
        [http://dx.doi.org/10.1105/tpc.105.033589] [PMID: 15987996]

[5]   Bolwell GP, Bindschedler LV, Blee KA, *et al.* The apoplastic oxidative burst in response to biotic stress in plants: a three-component system. J Exp Bot 2002; 53(372): 1367-76.
        [PMID: 11997382]

[6]   Schopfer P, Plachy C, Frahry G. Release of reactive oxygen intermediates (superoxide radicals, hydrogen peroxide, and hydroxyl radicals) and peroxidase in germinating radish seeds controlled by light, gibberellin, and abscisic acid. Plant Physiol 2001; 125(4): 1591-602.
        [http://dx.doi.org/10.1104/pp.125.4.1591] [PMID: 11299341]

[7]   Munné-Bosch S, Queval G, Foyer CH. The impact of global change factors on redox signaling underpinning stress tolerance. Plant Physiol 2013; 161(1): 5-19.
        [http://dx.doi.org/10.1104/pp.112.205690] [PMID: 23151347]

[8]   Mittler R, Vanderauwera S, Gollery M, Van Breusegem F. Reactive oxygen gene network of plants. Trends Plant Sci 2004; 9(10): 490-8.
        [http://dx.doi.org/10.1016/j.tplants.2004.08.009] [PMID: 15465684]

[9]   Choudhury S, Panda P, Sahoo L, Panda SK. Reactive oxygen species signaling in plants under abiotic stress. Plant Signal Behav 2013; 8(4): e23681.
        [http://dx.doi.org/10.4161/psb.23681] [PMID: 23425848]

[10]   Das K, Roychoudhury A. Reactive oxygen species (ROS) and response of antioxidants as ROS-scavengers during environmental stress in plants. Front Environ Sci 2014; 2: 1-14.
        [http://dx.doi.org/10.3389/fenvs.2014.00053]

[11]   Verma V, Ravindran P, Kumar PP. Plant hormone-mediated regulation of stress responses. BMC Plant Biol 2016; 16(1): 86.
        [http://dx.doi.org/10.1186/s12870-016-0771-y] [PMID: 27079791]

[12]   Prasad TK, Anderson MD, Stewart CR. Acclimation, hydrogen peroxide, and abscisic acid protect mitochondria against irreversible chilling injury in maize seedlings. Plant Physiol 1994; 105(2): 619-27.
        [http://dx.doi.org/10.1104/pp.105.2.619] [PMID: 12232229]

[13]   Mittler R. Oxidative stress, antioxidants and stress tolerance. Trends Plant Sci 2002; 7(9): 405-10.
        [http://dx.doi.org/10.1016/S1360-1385(02)02312-9] [PMID: 12234732]

[14]   Huang S, Van Aken O, Schwarzländer M, Belt K, Millar AH. The roles of mitochondrial reactive oxygen species in cellular signaling and stress response in plants. Plant Physiol 2016; 171(3): 1551-9.

[http://dx.doi.org/10.1104/pp.16.00166] [PMID: 27021189]

[15]   Sandalio LM, Romero-Puertas MC. Peroxisomes sense and respond to environmental cues by regulating ROS and RNS signalling networks. Ann Bot 2015; 116(4): 475-85.
[http://dx.doi.org/10.1093/aob/mcv074] [PMID: 26070643]

[16]   Blokhina O, Virolainen E, Fagerstedt KV. Antioxidants, oxidative damage and oxygen deprivation stress: a review. Ann Bot 2003; 91(Spec No): 179-94.
[http://dx.doi.org/10.1093/aob/mcf118] [PMID: 12509339]

[17]   Zeng J, Dong Z, Wu H, Tian Z, Zhao Z. Redox regulation of plant stem cell fate. EMBO J 2017; 36(19): 2844-55.
[http://dx.doi.org/10.15252/embj.201695955] [PMID: 28838936]

[18]   García-Cristobal J, García-Villaraco A, Ramos B, Gutierrez-Mañero J, Lucas JA. Priming of pathogenesis related-proteins and enzymes related to oxidative stress by plant growth promoting rhizobacteria on rice plants upon abiotic and biotic stress challenge. J Plant Physiol 2015; 188: 72-9.
[http://dx.doi.org/10.1016/j.jplph.2015.09.011] [PMID: 26439659]

[19]   Farooq MA, Niazi AK, Akhtar J, *et al.* Acquiring control: The evolution of ROS-Induced oxidative stress and redox signaling pathways in plant stress responses. Plant Physiol Biochem 2019; 141: 353-69.
[http://dx.doi.org/10.1016/j.plaphy.2019.04.039] [PMID: 31207496]

[20]   Waszczak C, Carmody M, Kangasjärvi J. Reactive oxygen species in plant signaling. Annu Rev Plant Biol 2018; 69: 209-36.
[http://dx.doi.org/10.1146/annurev-arplant-042817-040322] [PMID: 29489394]

[21]   Gill SS, Tuteja N. Reactive oxygen species and antioxidant machinery in abiotic stress tolerance in crop plants. Plant Physiol Biochem 2010; 48(12): 909-30.
[http://dx.doi.org/10.1016/j.plaphy.2010.08.016] [PMID: 20870416]

[22]   Czarnocka W, Karpiński S. Friend or foe? Reactive oxygen species production, scavenging and signaling in plant response to environmental stresses. Free Radic Biol Med 2018; 122: 4-20.
[http://dx.doi.org/10.1016/j.freeradbiomed.2018.01.011] [PMID: 29331649]

[23]   Gururani MA, Venkatesh J, Tran LSP. Regulation of photosynthesis during abiotic stress-induced photoinhibition. Mol Plant 2015; 8(9): 1304-20.
[http://dx.doi.org/10.1016/j.molp.2015.05.005] [PMID: 25997389]

[24]   Katano K, Honda K, Suzuki N. Integration between ROS regulatory systems and other signals in the regulation of various types of heat responses in plants. Int J Mol Sci 2018; 19(11): 3370.
[http://dx.doi.org/10.3390/ijms19113370] [PMID: 30373292]

[25]   Xu Y, Magwanga RO, Cai X, *et al.* Deep transcriptome analysis reveals reactive oxygen species (ROS) network evolution, response to abiotic stress, and regulation of fiber development in cotton. Int J Mol Sci 2019; 20(8): 1863.
[http://dx.doi.org/10.3390/ijms20081863] [PMID: 30991750]

[26]   Salvi P, Kamble NU, Majee M. Stress-inducible galactinol synthase of chickpea (CaGolS) is implicated in heat and oxidative stress tolerance through reducing stress-induced excessive reactive oxygen species accumulation. Plant Cell Physiol 2018; 59(1): 155-66.
[http://dx.doi.org/10.1093/pcp/pcx170] [PMID: 29121266]

[27]   Siddiqui MH, Al-Whaibi MH, Basalah MO. Role of nitric oxide in tolerance of plants to abiotic stress. Protoplasma 2011; 248(3): 447-55.
[http://dx.doi.org/10.1007/s00709-010-0206-9] [PMID: 20827494]

[28]   Hossain MA, Li ZG, Hoque TS, Burritt DJ, Fujita M, Munné-Bosch S. Heat or cold priming-induced cross-tolerance to abiotic stresses in plants: key regulators and possible mechanisms. Protoplasma 2018; 255(1): 399-412.
[http://dx.doi.org/10.1007/s00709-017-1150-8] [PMID: 28776104]

[29]   Kurusu T, Kuchitsu K. Autophagy, programmed cell death and reactive oxygen species in sexual reproduction in plants. J Plant Res 2017; 130(3): 491-9.
[http://dx.doi.org/10.1007/s10265-017-0934-4] [PMID: 28364377]

[30]   Khaksar G, Treesubsuntorn C, Thiravetyan P. Impact of endophytic colonization patterns on Zamioculcas zamiifolia stress response and in regulating ROS, tryptophan and IAA levels under airborne formaldehyde and formaldehyde-contaminated soil conditions. Plant Physiol Biochem 2017; 114: 1-9.
[http://dx.doi.org/10.1016/j.plaphy.2017.02.016] [PMID: 28246037]

[31]   Yuan HM, Liu WC, Jin Y, Lu YT. Role of ROS and auxin in plant response to metal-mediated stress. Plant Signal Behav 2013; 8(7): e24671.
[http://dx.doi.org/10.4161/psb.24671] [PMID: 23603941]

[32]   Xia XJ, Chen Z, Yu JQ. ROS mediate brassinosteroids-induced plant stress responses. Plant Signal Behav 2010; 5(5): 532-4.
[http://dx.doi.org/10.4161/psb.10989] [PMID: 20436298]

[33]   Xu N, Chu Y, Chen H, *et al.* Rice transcription factor OsMADS25 modulates root growth and confers salinity tolerance *via* the ABA-mediated regulatory pathway and ROS scavenging. PLoS Genet 2018; 14(10): e1007662.
[http://dx.doi.org/10.1371/journal.pgen.1007662] [PMID: 30303953]

[34]   Xia XJ, Zhou YH, Shi K, Zhou J, Foyer CH, Yu JQ. Interplay between reactive oxygen species and hormones in the control of plant development and stress tolerance. J Exp Bot 2015; 66(10): 2839-56.
[http://dx.doi.org/10.1093/jxb/erv089] [PMID: 25788732]

[35]   Sahay S, Gupta M. An update on nitric oxide and its benign role in plant responses under metal stress. Nitric Oxide 2017; 67: 39-52.
[http://dx.doi.org/10.1016/j.niox.2017.04.011] [PMID: 28456602]

[36]   Lim SD, Kim SH, Gilroy S, Cushman JC, Choi WG. Quantitative ROS bioreporters: A robust toolkit for studying biological roles of ROS in response to abiotic and biotic stresses. Physiol Plant 2019; 165(2): 356-68.
[http://dx.doi.org/10.1111/ppl.12866] [PMID: 30411793]

[37]   Pacheco JHL, Carballo MA, Gonsebatt ME. Antioxidants against environmental factor-induced oxidative stress. Nutritional Antioxidant Therapies. Treatments and Perspectives 2017; pp. 189-215.
[http://dx.doi.org/10.1007/978-3-319-67625-8_8]

[38]   Dietz KJ, Turkan I, Krieger-Liszkay A. Redox-and reactive oxygen species-dependent signaling into and out of the photosynthesizing chloroplast. Plant Physiol 2016; 171(3): 1541-50.
[http://dx.doi.org/10.1104/pp.16.00375] [PMID: 27255485]

[39]   Govrin EM, Levine A. The hypersensitive response facilitates plant infection by the necrotrophic pathogen Botrytis cinerea. Curr Biol 2000; 10(13): 751-7.
[http://dx.doi.org/10.1016/S0960-9822(00)00560-1] [PMID: 10898976]

[40]   Bendahmane A, Kanyuka K, Baulcombe DC. The Rx gene from potato controls separate virus resistance and cell death responses. Plant Cell 1999; 11(5): 781-92.
[http://dx.doi.org/10.1105/tpc.11.5.781] [PMID: 10330465]

[41]   El-Zahaby HM, Hafez YM, Király Z. Effect of reactive oxygen species on plant pathogens in planta and on disease symptoms. Acta Phytopathol Entomol Hung 2004; 39: 325-45.
[http://dx.doi.org/10.1556/APhyt.39.2004.4.2]

[42]   Hafez YM, Király Z. Role of hydrogen peroxide in symptom expression of barley susceptible and resistant to powdery mildew. Acta Phytopathol Entomol Hung 2003; 38: 227-36.
[http://dx.doi.org/10.1556/APhyt.38.2003.3-4.2]

[43]   Schippers JH, Foyer CH, van Dongen JT. Redox regulation in shoot growth, SAM maintenance and flowering. Curr Opin Plant Biol 2016; 29: 121-8.

[http://dx.doi.org/10.1016/j.pbi.2015.11.009] [PMID: 26799134]

[44]   Mhamdi A, Van Breusegem F. Reactive oxygen species in plant development. Development 2018; 145(15): dev164376.
[http://dx.doi.org/10.1242/dev.164376] [PMID: 30093413]

[45]   Glazener JA, Orlandi EW, Baker CJ. The active oxygen response of cell suspensions to incompatible bacteria is not sufficient to cause hypersensitive cell death. Plant Physiol 1996; 110(3): 759-63.
[http://dx.doi.org/10.1104/pp.110.3.759] [PMID: 12226215]

[46]   Gechev TS, Van Breusegem F, Stone JM, Denev I, Laloi C. Reactive oxygen species as signals that modulate plant stress responses and programmed cell death. BioEssays 2006; 28(11): 1091-101.
[http://dx.doi.org/10.1002/bies.20493] [PMID: 17041898]

[47]   Kasote DM, Katyare SS, Hegde MV, Bae H. Significance of antioxidant potential of plants and its relevance to therapeutic applications. Int J Biol Sci 2015; 11(8): 982-91.
[http://dx.doi.org/10.7150/ijbs.12096] [PMID: 26157352]

[48]   Ighodaro OM, Akinloye OA. First line defence antioxidants-superoxide dismutase (SOD), catalase (CAT) and glutathione peroxidase (GPX): Their fundamental role in the entire antioxidant defence grid. Alexandria Med J 2018; 54(4): 287-93.
[http://dx.doi.org/10.1016/j.ajme.2017.09.001]

[49]   Petrov VD, Van Breusegem F. Hydrogen peroxide-a central hub for information flow in plant cells. AoB Plants 2012; 2012: pls014.
[http://dx.doi.org/10.1093/aobpla/pls014] [PMID: 22708052]

[50]   Bela K, Horváth E, Gallé Á, Szabados L, Tari I, Csiszár J. Plant glutathione peroxidases: emerging role of the antioxidant enzymes in plant development and stress responses. J Plant Physiol 2015; 176: 192-201.
[http://dx.doi.org/10.1016/j.jplph.2014.12.014] [PMID: 25638402]

[51]   Gietler M, Nykiel M, Zagdańska BM. Changes in the reduction state of ascorbate and glutathione, protein oxidation and hydrolysis leading to the development of dehydration intolerance in *Triticum aestivum* L. seedlings. Plant Growth Regul 2016; 79(3): 287-97.
[http://dx.doi.org/10.1007/s10725-015-0133-z]

[52]   Venkatesh J, Park SW. Role of L-ascorbate in alleviating abiotic stresses in crop plants. Bot Stud (Taipei, Taiwan) 2014; 55(1): 38.
[http://dx.doi.org/10.1186/1999-3110-55-38] [PMID: 28510969]

[53]   Kapoor D, Singh S, Kumar V, Romero R, Prasad R, Singh J. Antioxidant enzymes regulation in plants in reference to reactive oxygen species (ROS) and reactive nitrogen species (RNS). Plant Gene 2019; 19: 100182.
[http://dx.doi.org/10.1016/j.plgene.2019.100182]

[54]   Ahmad P, Jaleel CA, Azooz MM, Nabi G. Generation of ROS and non-enzymatic antioxidants during abiotic stress in plants. BRI 2009; 2(1): 11-20.

[55]   El-Bahr SM. Biochemistry of free radicals and oxidative stress. Science international 2013; 1(5): 111-7.
[http://dx.doi.org/10.5567/sciintl.2013.111.117]

[56]   Jaleel CA, Riadh K, Gopi R, Manivannan P, Inès J, Al-Juburi HJ, *et al.* Antioxidant defense responses: physiological plasticity in higher plants under abiotic constraints. Acta Physiol Plant 2009; 31(3): 427-36.
[http://dx.doi.org/10.1007/s11738-009-0275-6]

[57]   Shao HB, Chu LY, Shao MA, Jaleel CA, Mi HM. Higher plant antioxidants and redox signaling under environmental stresses. C R Biol 2008; 331(6): 433-41.
[http://dx.doi.org/10.1016/j.crvi.2008.03.011] [PMID: 18510996]

[58]   Akram NA, Shafiq F, Ashraf M. Ascorbic acid-a potential oxidant scavenger and its role in plant

development and abiotic stress tolerance. Front Plant Sci 2017; 8: 613.
[http://dx.doi.org/10.3389/fpls.2017.00613] [PMID: 28491070]

[59] Passaia G, Margis-Pinheiro M. Glutathione peroxidases as redox sensor proteins in plant cells. Plant Sci 2015; 234: 22-6.
[http://dx.doi.org/10.1016/j.plantsci.2015.01.017] [PMID: 25804806]

[60] Caverzan A, Casassola A, Brammer SP. Reactive oxygen species and antioxidant enzymes involved in plant tolerance to stress. In: Shanker A, Shanker C, Eds. Abiotic and biotic stress in plants-Recent advances and future perspectives. London UK: IntechOpen 2016; pp. 463-80.

[61] Kärkönen A, Kuchitsu K. Reactive oxygen species in cell wall metabolism and development in plants. Phytochemistry 2015; 112: 22-32.
[http://dx.doi.org/10.1016/j.phytochem.2014.09.016] [PMID: 25446232]

[62] Garcia-Molina A, Altmann M, Alkofer A, Epple PM, Dangl JL, Falter-Braun P. LSU network hubs integrate abiotic and biotic stress responses *via* interaction with the superoxide dismutase FSD2. J Exp Bot 2017; 68(5): 1185-97.
[http://dx.doi.org/10.1093/jxb/erw498] [PMID: 28207043]

[63] Shi H, Ye T, Chen F, *et al.* Manipulation of arginase expression modulates abiotic stress tolerance in Arabidopsis: effect on arginine metabolism and ROS accumulation. J Exp Bot 2013; 64(5): 1367-79.
[http://dx.doi.org/10.1093/jxb/ers400] [PMID: 23378380]

[64] Arora D, Jain P, Singh N, Kaur H, Bhatla SC. Mechanisms of nitric oxide crosstalk with reactive oxygen species scavenging enzymes during abiotic stress tolerance in plants. Free Radic Res 2016; 50(3): 291-303.
[http://dx.doi.org/10.3109/10715762.2015.1118473] [PMID: 26554526]

[65] Xu R, Li Y. Control of final organ size by Mediator complex subunit 25 in *Arabidopsis thaliana.* Development 2011; 138(20): 4545-54.
[http://dx.doi.org/10.1242/dev.071423] [PMID: 21903673]

[66] Foreman J, Demidchik V, Bothwell JH, *et al.* Reactive oxygen species produced by NADPH oxidase regulate plant cell growth. Nature 2003; 422(6930): 442-6.
[http://dx.doi.org/10.1038/nature01485] [PMID: 12660786]

[67] Rodríguez AA, Grunberg KA, Taleisnik EL. Reactive oxygen species in the elongation zone of maize leaves are necessary for leaf extension. Plant Physiol 2002; 129(4): 1627-32.
[http://dx.doi.org/10.1104/pp.001222] [PMID: 12177475]

[68] Zimmermann P, Heinlein C, Orendi G, Zentgraf U. Senescence-specific regulation of catalases in *Arabidopsis thaliana* (L.) Heynh. Plant Cell Environ 2006; 29(6): 1049-60.
[http://dx.doi.org/10.1111/j.1365-3040.2005.01459.x] [PMID: 17080932]

[69] Hu L, Liang W, Yin C, *et al.* Rice MADS3 regulates ROS homeostasis during late anther development. Plant Cell 2011; 23(2): 515-33.
[http://dx.doi.org/10.1105/tpc.110.074369] [PMID: 21297036]

[70] Yi J, Moon S, Lee YS, *et al.* Defective tapetum cell death 1 (DTC1) regulates ROS levels by binding to metallothionein during tapetum degeneration. Plant Physiol 2016; 170(3): 1611-23.
[http://dx.doi.org/10.1104/pp.15.01561] [PMID: 26697896]

[71] Leymarie J, Vitkauskaité G, Hoang HH, *et al.* Role of reactive oxygen species in the regulation of Arabidopsis seed dormancy. Plant Cell Physiol 2012; 53(1): 96-106.
[http://dx.doi.org/10.1093/pcp/pcr129] [PMID: 21937678]

[72] Tian S, Qin G, Li B. Reactive oxygen species involved in regulating fruit senescence and fungal pathogenicity. Plant Mol Biol 2013; 82(6): 593-602.
[http://dx.doi.org/10.1007/s11103-013-0035-2] [PMID: 23515879]

[73] Müller K, Linkies A, Vreeburg RA, Fry SC, Krieger-Liszkay A, Leubner-Metzger G. *In vivo* cell wall loosening by hydroxyl radicals during cress seed germination and elongation growth. Plant Physiol

2009; 150(4): 1855-65.
[http://dx.doi.org/10.1104/pp.109.139204] [PMID: 19493972]

[74] Potocký M, Pejchar P, Gutkowska M, *et al.* NADPH oxidase activity in pollen tubes is affected by calcium ions, signaling phospholipids and Rac/Rop GTPases. J Plant Physiol 2012; 169(16): 1654-63.
[http://dx.doi.org/10.1016/j.jplph.2012.05.014] [PMID: 22762791]

[75] Tsukagoshi H, Busch W, Benfey PN. Transcriptional regulation of ROS controls transition from proliferation to differentiation in the root. Cell 2010; 143(4): 606-16.
[http://dx.doi.org/10.1016/j.cell.2010.10.020] [PMID: 21074051]

[76] Barba-Espín G, Diaz-Vivancos P, Job D, Belghazi M, Job C, Hernández JA. Understanding the role of H(2)O(2) during pea seed germination: a combined proteomic and hormone profiling approach. Plant Cell Environ 2011; 34(11): 1907-19.
[http://dx.doi.org/10.1111/j.1365-3040.2011.02386.x] [PMID: 21711356]

[77] Tognetti VB, Bielach A, Hrtyan M. Redox regulation at the site of primary growth: auxin, cytokinin and ROS crosstalk. Plant Cell Environ 2017; 40(11): 2586-605.
[http://dx.doi.org/10.1111/pce.13021] [PMID: 28708264]

[78] Cui H, Kong D, Wei P, *et al.* SPINDLY, ERECTA, and its ligand STOMAGEN have a role in redox-mediated cortex proliferation in the Arabidopsis root. Mol Plant 2014; 7(12): 1727-39.
[http://dx.doi.org/10.1093/mp/ssu106] [PMID: 25267734]

[79] Gutsche N, Thurow C, Zachgo S, Gatz C. Plant-specific CC-type glutaredoxins: functions in developmental processes and stress responses. Biol Chem 2015; 396(5): 495-509.
[http://dx.doi.org/10.1515/hsz-2014-0300] [PMID: 25781542]

CHAPTER 8

Role of Melatonin - A Signaling Molecule in Modulation of Antioxidant Defense System in Plants: Amelioration of Drought and Salinity Stress

Prabhjot Kaur[1], Davinder Singh[1], Farhana Rashid[1], Avinash Kumar[1], Harneetpal Kaur[1], Kirandeep Kaur[2], Atamjit Singh[2], Neena Bedi[2], Preet Mohinder Singh Bedi[2], Balbir Singh[2], Saroj Arora[1] and Shivani Attri[1,*]

[1] *Department of Botanical and Environmental Sciences, Guru Nanak Dev University, Amritsar, Punjab 143005, India*

[2] *Department of Pharmaceutical Sciences, Guru Nanak Dev University, Amritsar, Punjab 143005, India*

Abstract: Melatonin (N-acetyl-5-methoxytryptamine) is a small (232 daltons), non-toxic, indole molecule first isolated from the pineal gland of cows and later on found in different tissues of plants and bacteria. Melatonin acts as a signaling molecule/messenger, which plays an important role in coping with various biotic and abiotic stress conditions. Biosynthesis of melatonin involves two key cellular organelles *viz.* mitochondria and chloroplast. The endogenously produced melatonin serves as an antioxidant signaling molecule during the generation of reactive oxygen species (ROS) and reactive nitrogen species (RNS). The exogenous melatonin also functions in the same way during critical conditions by repairing mitochondria and deal with various stresses. The plants regulate the production of melatonin depending upon the conditions and regulate salt, drought, cold, heat, oxidative, and heavy metals-related stresses. Besides that, melatonin acts as a plant hormone and regulates numerous functions in plants, including growth, development, photoperiod, clearing oxygen species, rhizogenesis, photosynthesis, and enhances antioxidase activity. It acts as a multi regulatory molecule by regulating gene expressions as well as by cross-talks with phytohormones (auxin, cytokinin, salicylic acid, and abscisic acid) involved in plant growth and development. Therefore, understanding the mechanism of action of melatonin as a signaling molecule may serve as a novel strategy to combat various stresses in plants and animals. An attempt has been made in this chapter to discuss the important role of melatonin in modulating oxidative stress in plants during stressful conditions.

** **Corresponding Author Shivani Attri:** Department of Botanical and Environmental Sciences, Guru Nanak Dev University, Amritsar, Punjab 143005, India; Tel: +91-6284271258; E-mail: attrishivani1994@gmail.com*

Keywords: Chloroplast, Cytoplasm, Drought, Hormones, Melatonin, Mitochondria, N-acetyl serotonin methyltransferase, N-acetylserotonin, Oxidative stress, Plants, RNS, ROS, Salinity, Serotonin N-acetyl transferase, Serotonin, Stress, Tryptamine 5- hydroxylase, Tryptamine, Tryptophan decarboxylase, Tryptophan.

INTRODUCTION

Melatonin is a small, indole biological signaling molecule present in bacteria, unicellular eukaryotes, plants, and animals. First, it was isolated from the pineal glands of cows in 1958, where it plays an important role in reversing the darkening effect of melatonin by stimulating hormone (MSH) [1, 2]. In animals, it modulates numerous functions by regulating the activity of antioxidant enzymes, circadian rhythms, reproduction, coronary heart disease, immune system, mental status, and response to anticancer agents [3]. It also enhances the nighttime rise of serum levels [4]. After that, in 1995, melatonin was discovered in plants where it acts as a secondary messenger and regulates various functions by involving two key cellular organelles, mitochondria and chloroplast. It is distributed in different parts of the plants depending upon the type of plant species *i.e.*, *Lycopersicon esculentum* Mill. (Fruits), *Beta vulgaris* L. (leaf), *Fragaria magna* (root), *Brassica rapa* (corm), *Oryzasativum* (seeds) [4]. It is a multi regulatory molecule that modulates senescence, antioxidant activity, growth and development, circadian rhythms, root growth, and other biotic and abiotic stresses (salt, drought, cold, heat, oxidative, and heavy metals) [5, 6] in plants. Plants face more environmental stress due to their sessile nature. The concentration of melatonin in plants varies from species to species and also depends upon the growth rate, location, organ, and harvest time [7]. Its content in plant specimen can be analyzed by various methods such as RIA (radioimmunoassay), ELISA (enzyme-linked immuno-sorbent assay), GC-MS (gas chromatography-mass spectrometry), and HPLC (high-performance liquid chromatography) with electrochemical detection (HPLC-ECD), or HPLC-MS (high-performance liquid chromatography-mass spectrometry) [8]. It also acts as an antioxidant molecule by scavenging ROS (reactive oxygen species) and RNS (reactive nitrogen species) without depending on any kind of receptors [9]. In animals, MT1 and MT2 are the two major receptors present in the mitochondria, which communicate signaling pathways, whereas, in plants, only one melatonin receptor (CAND2/PMTR1) is identified to date. It modulates stomatal closure *via* H_2O_2 and Ca^{2+} signaling transduction pathway. Besides this, phytomelatonin improves stress tolerance by elevating the expression of antioxidant enzymes, regulating transcription factors involved in various stresses, and modulates photosynthetic rate by interacting with unknown receptors and activating the H_2O_2/NO signaling cascade [10]. The synthesis of phytomelatoninis is regulated in plants in a fashion to cope with any situation or stress.

MELATONIN: SYNTHESIS IN PLANTS

Melatonin is present in almost every living organism like bacteria, yeast, algae, plants, and animals. The production and secretion of melatonin occur in the pineal gland of animals, but there is no such organ in the case of plants. The synthesis pathways in plants are slightly different from animals; they are regulated by various environmental factors (light intensity, heat, cold, drought, and UV light) and developmental factors (fruit maturation, senescence, and leaf development) [4]. In plants, the chloroplast is the primary site of melatonin synthesis. In most plant species, tryptophan is the initiator of melatonin synthesis. The major enzymes are TDC, T5H, SNAT, CAMT, ASMT, and HIOMT, are involved in the biogenesis of melatonin. The first step enzyme involved in melatonin synthesis is TDC (tryptophan decarboxylase) that converts tryptophan into tryptamine. Then, tryptamine converts into serotonin involving T5H (tryptophan 5-hydroxylase). Serotonin is the intermediate molecule of melatonin synthesis [11, 12]. The final synthesis of melatonin from the serotonin involves a two-step mechanism, participation of enzyme SNAT, which converts serotonin into N-acetyl serotonin, and another enzyme ASMT (N-acetyl-serotonin methyltransferase), which converts N-acetyl serotonin into melatonin. Serotonin converts into melatonin by two pathways. In some plants, such as *Hypericum perforatum,* tryptophan 5-hydroxylate (TPH) catalyzes tryptophan into 5-hydroxytryptophan, and then modifies into serotonin by TDC/AADC (aromatic-L-amino-acid decarboxylase). Then the next two steps of conversion of serotonin into melatonin by involving SNAT/AANAT (arylalkylamine N-acetyltransferase) and ASMT/HIOMT (hydroxy indole-O-methyltransferase) is the same in both plants and animals. In some cases, the other pathway involves the conversion of serotonin into 5-methoxy-tryptamine by HIOMT, and finally, SNAT catalyzes 5-methox--tryptamine into melatonin [6, 13 - 18]. The enzymes involved in the synthesis are located in chloroplast and cytoplasm. SNAT, whose main function is to modulate the production of melatonin, is located in chloroplast and enzyme ASMT and T5H in the cytoplasm and endoplasmic reticulum, respectively [19]. In these pathways, the production of serotonin and melatonin proves to be useful for plants to cope with various biotic and abiotic stresses. The production of melatonin is enhanced or decreased according to the stress condition. Tryptophan begins to form various secondary metabolites during abiotic stresses in some plants. These secondary metabolites such as phytoalexins, indole glucosinolates, alkaloids, and serotonin provide a defense to plants by their spatial and temporal distribution (Fig. **1**) [20].

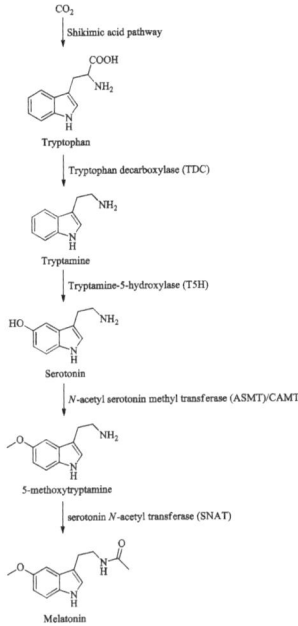

Fig. (1). Biosynthesis of melatonin from tryptophan in plants. The enzymes involve in this pathway are as follows: TDC (tryptophan decarboxylase); T5H (tryptophan 5-hydroxylase); ASMT (N-acetyl serotonin methyltransferase); SNAT (serotonin N-acetyltransferase) CAMT (caffeic acid o-methyltransferase) and HIOMT (hydroxy indole-*O*-methyltransferase).

CROSSTALK BETWEEN MELATONIN AND OTHER PHYTOHORMONES

Plants face lots of biotic (bacterias, viruses, fungi, nematodes, and other pathogens) and abiotic stresses (heat, cold, light intensity, UV radiations, and salinity) in their natural habitat [21]. To overcome all these stress conditions, the phytohormones like auxin (IAA), cytokinin (CKs), salicylic acid (SA), jasmonic acid (JA), ethylene (ET), brassinosteroids (BRs), abscisic acid (ABA), and melatonin crosstalk with each other through various signal transduction cascades [22, 23]. This signaling cascade of phytohormones provides immunity to plants to cope with any type of disaster. Melatonin does not only play various roles in animals but perform an important function in the immune / protective mechanism of plants. In Arabidopsis and tobacco leaves, the application of melatonin induces PR (Pathogenesis-related) genes and the number of other defense genes associated with SA (salicylic acid) and ET (ethylene) pathways, resulting in the downregulation of the multiplication mechanism of bacteria [24]. Auxin, salicylic acid (SA), and melatonin interactions play a very crucial role in the ripening of climacteric and non-climacteric fruits. Interestingly, these hormones have a common precursor, "chorismate" [28]. Auxin initiates the fruit ripening by promoting cell division in combination with cytokinins and regulates cell expansion in the combination of gibberellins. In this way, it plays a significant role in the growth and development of plants [25]. SA provides a defense to plants against abiotic stresses, attack of pathogens, flowering, and cell cycle regulation. Currently, the role of melatonin in fruit ripening, reproductive development [26], growth, and development exerted by auxin (indole-3-acetic acid). It also promotes grape berry ripening by showing interaction with ABA, ethylene, and hydrogen peroxide. In addition to this, melatonin also regulates the ripening of tomatoes [27], pears, and bananas by regulating the expression of enzymes. In this way interactions between hormones provide a fine shape and weight to climacteric and non-climacteric fruits [28].

Melatonin is also known as bio stimulator, because of its anti-senescence and anti-stress activity. This effect of melatonin is due to the opposite actions of kinetin and ABA on leaf senescence. An increase in photosynthetic rate, reduction in chlorophyll degradation and enhancing activities of ROS-scavenging enzymes on melatonin treatment reduced the adverse effects of water stress [29]. There is one such example to show the relationship between melatonin and auxin hormone in *Arabidopsis* roots. A higher concentration of melatonin downregulates the expression of auxin by reducing auxin transport and decreasing PIN1,3,7 protein expression and resulting decrease in root meristem size [30]. Similarly, crosstalk between serotonin and melatonin modulates various functions in plants by involving other components (IAA, GA, CK, ABA, SA or NO, ethylene and jasmonates) in their signaling cascade [31]. During abiotic and biotic stress, the generation of ROS/NO changes the expressions of phytohormones [32]. More production of melatonin in stress conditions increases the production of NO by upregulating expression of nitrate reductase. Melatonin and NO production increased/decreased the functioning of other hormones by effecting their synthesis, catabolic enzymes, transcription factors and hormone signaling which determine their anti-stress efficiency [32]. Due to the significant roles of melatonin in growth, development, senescence, antioxidant activity, regulation of opening and closing of stomata, circadian rhythms maintain the permeability of membranes by regulating ion-exchange mechanism, seed germination, osmoprotectant and phototropism, and numerous other functions, it is not wrong to say that melatonin plays a very significant role in botanicals by associating with other phytohormones [33 - 35].

ANTIOXIDANT POTENTIAL OF MELATONIN

The maintenance of cellular redox equilibrium requires a stable balance between reactive oxygen species and antioxidants [36]. Reactive oxygen species (ROS) are by-products of normal plant cellular metabolism. Among these chemically reactive species, examples of oxygen radicals include superoxide (O_2^-), hydroxyl

(OH˙), and peroxyl (ROO˙) [37]. Non-radical forms of ROS include hydrogen peroxide (H_2O_2), singlet oxygen ($^1O_2$), and ozone (O_3). Reactive oxygen species are of immense significance for plants. In adaptation processes, a limited number of radicals serve as a signal to induce response towards abiotic stresses, whereas overproduction causes oxidative damage to essential macromolecules [38].

External stresses such as drought, salinity, freezing, metal toxicity, UV-B radiation and pathogens attack lead to increased ROS production in plants due to cellular homeostasis disturbance [36, 39]. To counteract, the adverse effects of elevated levels of ROS, plants have evolved a wide spectrum of antioxidant defense systems. The antioxidant protection mechanism includes enzymatic and nonenzymatic components responsible for either detoxifying or scavenging reactive oxygen species upto a certain level [40]. ROS development, however, exceeds the efficiency of the antioxidant protection system under adverse or unfavorable conditions that result in oxidative stress [41]. Among antioxidant systems, the potent enzymatic metabolites are superoxide dismutase (SOD), catalase (CAT), guaiacol peroxidase (GPX), enzymes of ascorbate-glutathione (AsA-GSH) cycle such as ascorbate peroxidase (APX), monodehydroascorbate reductase (MDHAR), dehydroascorbate reductase (DHAR), and glutathione reductase (GR). However, ascorbate (AsA), glutathione (GSH), carotenoids, tocopherols, and phenolics are nonenzymatic antioxidants within the cells [42].

Melatonin is omnipresent in various parts of a plant and responsible for inducing many physiochemical responses to adverse environmental conditions [43]. Melatonin is also known as an abiotic and biotic anti-stressor [44]. The literature shows that within the antioxidant network, ROS and melatonin are closely related [45]. The melatonin family is known to follow a free radical scavenging cascade pathway. As a result of this cascade, one melatonin molecule can scavenge up to four or more reactive species. Therefore, Melatonin is very effective as an antioxidant [46]. The antioxidant activity tends to work through the following pathways: (i) by direct scavenging of free radicals, (ii) by stimulating antioxidant enzymes such as peroxidase (POD), superoxide dismutase (SOD), and catalase (CAT) (iii) by boosting the function of other antioxidants, such as ascorbic acid (AsA), soluble sugars, and flavonoids (iv) by protecting antioxidant enzymes against oxidative damage and (v) by improving the performance of the mitochondrial electron transport chain, thus simplifying electron leakage and reducing free radical content [46, 47].

The key enzyme of an antioxidant system is sodium dismutase (SOD) which provides the first line of defense against reactive oxygen species by dismutation of O_2^- into oxygen and hydrogen peroxide (H_2O_2). In plants, three nuclear-encoded isozymes of SOD are present that is, copper/zinc SOD, manganese SOD, and iron SOD [48]. Also, the normal aerobic metabolism in plants produces hydrogen peroxide (H_2O_2)⁻ an essential relatively stable non-radical reactive oxygen species (ROS). Furthermore, stress factors cause increased development of H_2O_2 in plants as the elevated and non-metabolized cellular H_2O_2 causes severe damage to biomolecules. To scavenge H_2O_2, plants are equipped with H_2O_2-metabolizing enzymes such as catalases (CAT), ascorbate peroxidases (APX), some peroxyredoxins, peroxidase glutathione/thioredoxin and glutathione sulfotransferases (GST). As result of the action of SOD, CAT and APX subsequently H_2O_2 is converted to H_2O [49]. The Catalase (CAT) catalyzes the dismutation of H_2O_2 molecules into O_2 and $2H_2O$. It is a tetrameric enzyme that comprises heme molecules and is located mainly in peroxisomes. However, it is also present in cytosol, mitochondria and chloroplasts. CAT prefers H_2O_2 molecules strongly over organic peroxides. Whereas, ascorbate peroxidase (APX) has a higher affinity for H_2O_2 than CAT. In chloroplast, cytosol, mitochondria and peroxisomes as well as in an apoplastic vacuum, APX reduces H_2O_2 to H_2O *via* ascorbate-glutathione (AsA-GSH) cycle [50]. It uses ascorbate as a common donor of electrons. Other than this, H_2O_2 is detoxified by Glutathione peroxidase (GPX) and glutathione sulfotransferase (GST) using glutathione (GSH) as a reducing agent [51]. Along with APX, monodehydroascorbate reductase (MDHAR), dehydroascorbate reductase (DHAR), and glutathione reductase (GR) are the other enzymes involved in the ascorbate-glutathione (AsA-GSH) cycle. MDHAR regenerates ascorbic acid (AsA) using NADPH as a reducing agent and DHAR transforms dehydroascorbate (DHA) into AsA using reduced glutathione (GSH) as a donor of electrons [52]. It has been reported that exogenous melatonin treatment positively controls MDHAR and DHAR at both the protein and mRNA levels [41]. Glutathione is involved in both degradations of the enzymatic and non-enzymatic H_2O_2. The ratio of GSH / GSSG is known as a determinant of oxidative stress. Glutathione disulfide (GSSG) accumulates into the cells during oxidative stress, and the ratio of GSH to GSSG decreases. The essential enzyme for the recovery of GSH is glutathione reductase (GR), a possible enzyme for AsA-GSH [53] (Fig. **2**).

The potential antioxidant effects of melatonin have been confirmed in various food crops under different abiotic stresses. In literature, SOD activity is known to be elevated in plants under harsh abiotic stress conditions such as drought, salinity, heat stress, cold stress, metal toxicity, oxidative stress, *etc.* [54]. It has also been documented that the exogenous application of melatonin further increases the activity of SOD enzymes under severe conditions. For instance, the behavior of SOD was observed by exogenous melatonin concentrations (4 µM, 20 µM, and 100 µM) under salt, drought, and cold stress conditions. Their findings showed that melatonin had major effects on SOD activity, alleviating the accumulation of abiotic stress-induced ROS and related oxidative damage in

Bermuda grass [55]. Thus, pretreatment with melatonin improves SOD development, which in turn increases crop tolerance to different abiotic stresses.

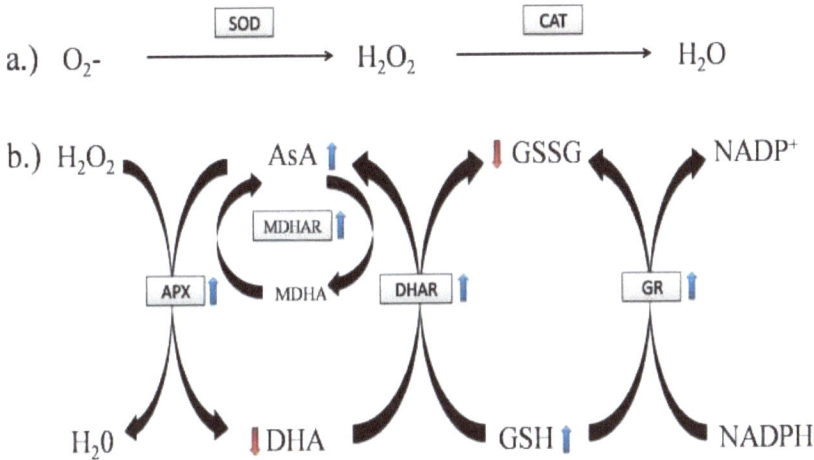

Fig. (2). a) Superoxide (O_2-) is reduced to hydrogen peroxide (H_2O_2) by superoxide dismutase (SOD) enzyme and H_2O_2 is detoxified to water (H_2O) with the help of catalase (CAT). b) Effect of exogenous melatonin on ascorbate-glutathione (AsA-GSH) cycle: H_2O_2 is metabolized *via* ascorbate peroxide (APX) to H_2O using ascorbate as an electron donor and the reaction is assisted with monodehydroascorbate reductase (MDHAR), dehydroascorbate reductase (DHAR), and glutathione reductase (GR). AsA is oxidized to monodehydroascorbate (MDHA) which is, further either reduced to AsA by MDHAR or reacts to form dehydroascorbate (DHA). DHA is reduced by DHAR to AsA. The reduced form of glutathione (GSH) is oxidized to glutathione disulfide (GSSG) in this reaction. Then, glutathione reductase (GR) reduces GSSG to GSH. During the reduction of MDHA and GSSG, the electron acceptor nicotinamide adenine dinucleotide phosphate (NADP) is regenerated by the respective enzymes. The electron acceptor nicotinamide adenine dinucleotide phosphate (NADP) is regenerated with the respective enzymes during the reduction of MDHA and GSSG.

But, SOD activity in naked oat seedlings treated with melatonin has been reported to be lower than that in non-melatonin treated seedlings [56]. Similarly, melatonin pretreated tomato seedlings having a combination of heat and salinity stresses showed decreased levels of SOD activity [57]. Increased antioxidant activity of the enzyme catalase is known as an indication of plant abiotic stress [43]. The impact of exogenous melatonin on CAT activity under abiotic stress conditions was also studied and the results indicated that it differs (*i.e.*, increases, decreases, or does not affect CAT activity) under various abiotic stress conditions [41].

The upregulation or downregulation of different components of the antioxidant system has been observed to vary depending upon the type of crop as well as the stress faced. It was found that 1μM melatonin-treated stored broccoli florets retained higher SOD content and decreased POD activity compared to 3-day storage control. This result suggested that melatonin could delay the senescence of broccoli flowering by controlling the enzyme antioxidant system [58]. Melatonin pretreatment soothes the growth inhibition and oxidative damage to

Malus hupehensis by directly scavenging H_2O_2 and boosting antioxidant enzyme activity of APX, CAT and POD to maintain homeostasis [59]. Also, exogenous melatonin pretreatment decreased oxidative damage under salinity stress in cucumber by direct scavenging of H_2O_2 or enhancing antioxidant enzyme activity (including SOD, POD, CAT, APX) and antioxidant concentrations (ascorbic acid and glutathione) [60]. Similarly, alleviation of drought-induced oxidative damage has been reported in maize (*Zea mays* L.) by increased enzymatic activity, including SOD, CAT, APX and POD, as well as DPPH-radical scavenging capacity, and decreased accumulation of H_2O_2 and Malondialdehyde (MDA) [61]. Exogenous use of melatonin increased the production of CAT and POD in apple leaves under stress from drought [62]. However, in the case of wheat seedlings and cucumber, studies have shown that melatonin under chilling stress increased the activities of SOD, Glutathione Peroxidase (GPX), APX, and Glutathione Reductase (GR), but not CAT, suggesting that CAT might not be responsive to melatonin treatment [63, 64]. Although many studies have been conducted still the specific processes involved in melatonin regulation of antioxidant enzymes have not been published yet, thus further research in this field is needed.

MELATONIN AND ITS PROTECTIVE EFFECT IN DROUGHT STRESS

Water scarcity is among the most severe adversities. Water availability is a restricting factor for the climate and severely affecting the plants 'homeostatic equilibrium. In general, responses to water deficit continue with an increased endogenous synthesis of abscisic acid (ABA), that triggers increased expression of many water stress response genes and triggers a signaling cascade resulting in reduced stomatal conductivity (gs) and, subsequently, the internal concentration of carbon dioxide (CO_2) [65]. Along with this, the carboxylation activity of ribulose bisphosphate carboxylase (Rubisco) also decreased due to the depletion of its substrate CO_2 [66]. Hence NADPH accumulates and $NADP^+$ remains unavailable for electron acceptance at the end of the photochemical stage [67]. These result in rapid accumulation of molecular oxygen that generate superoxide radicals and damage photochemical apparatus, cause peroxidation of lipid membranes and chlorophylls [68]. This reduced photosynthetic efficiency causes suppressed synthesis of sucrose along with deprivation of sugars for the maintenance of metabolism [69]. This stress leads to a change in membrane fluidity and its composition, loss of turgor pressure and modulation of protein-lipid and protein-protein interactions.

Melatonin has been known for its beneficial effect in various abiotic stress, *i.e.* heat stress, heavy metal stress, UV radiations, salt stress and drought stress [70 - 73]. Melatonin limits the production of reactive oxygen species (ROS) by enhancing the antioxidant defense system and by lowering lipid peroxidation and hydrogen peroxide (H_2O_2) levels [73]. It enhances the gene expression of several stress response genes, *i.e.* drought response element-binding factors (DREB 1), antioxidant enzymes and reduces the levels of malondialdehyde and H_2O_2 levels [74, 75]. Moreover, it enhances the levels of ascorbate, a secondary antioxidant acting as a retrograde signaling molecule during low photosynthetic rate [59]. Interestingly the growth rate of melatonin treated plant is observed high as compared to non-treated that might be due to enhanced sugar metabolism and photosynthetic rate [76].

Melatonin protects the chlorophyll from degradation during drought stress [77, 78]. There are specific enzymes that catalyze the degradation process of chlorophyll *i.e.* chlorophyllase (Chlase), pheophorbide-a-oxygenase (PAO), chlorophyll degrading peroxidase (Chl-PRX) and pheophytinase (PPH) [79 - 81]. Melatonin treatment successfully down-regulates the expression of these genes [79]. The protection of chloroplast structure from oxidative damage is another reason behind the restoration of photosynthesis following the melatonin treatment [82, 83]. It prevents shortening of chloroplast length and disruption of the membrane, grana, stroma, lamellae and thylakoid under water deficit conditions. Recently, it also observed that the melatonin treatment regulates the carbon fixation pathway at the molecular level by up-regulating the expression of transcripts, *i.e. RUBISCO*, *PGK* (phosphoglycerate kinase), *FBP* (fructose-1,--bisphosphatase), *TIM* (triosephosphate isomerase), *RPK* (phosphoribulokinase), *TKT* (transketolase), *FBA* (fructose bis-phosphate aldolase), *RPI* (ribose 5-phosphate isomerase), *SEBP* (sedoheptulose-1,7-bisphosphatase), *GAPA* (glyceraldehydes-3-phosphate dehydrogenase) [84].

Melatonin protects plants from drought stress-induced oxidative stress by enhancing the scavenging or by controlling the production of free radicals [55, 83, 84]. It enhances the detoxification of H_2O_2, harmful hydroxyl radicals and other aldehydes involved in oxidative damage during drought stress [85, 86]. Melatonins down-regulates the biosynthesis of ABA and enhance the gene expression of enzymes used in ABA catabolism [85]. All of these activities are verified using plant model *Arabidopsis thaliana* by over-expressing the *TaCOMT* gene responsible for the biosynthesis of melatonin in drought conditions [87]. It also regulates the scavenging of ROS through an enzymatic cycle known as the Asada-Halliwell pathway [88]. In addition to this, it regulates the AsA-GSH pathway for the detoxification of ROS. This enhances the AsA/DHA and GSH/GSSG ratio an indication of an improved antioxidant defense system [88]. Drought stress induces a negative impact on nitrogen metabolic enzymes. In-reverse, melatonin enhances the expression of *NR*, *NiR*, *GS*, *GOGAT* enzymes under water-deficient conditions for proper metabolism of nitrogen and *AMT* (ammonium transporter), *NRT* (a nitrate transporter) for nitrogen uptake [77].

Cuticle, is a waxy layer present on plant leaves that are important for their growth under drought to reduce water retention. Melatonin treatment enhances the biosynthesis of wax by up-regulating the gene expression of *KCS1*, *CER3*, *TTS1* and *LTP1* [89]. Overall, melatonin protects the plants from drought stress by regulating multiple survivor pathways including photosynthesis, ROS scavenging, antioxidant defense system, and carbon and nitrogen metabolism (Table **1**).

Table 1. Summary table explaining the effect of melatonin on various parameters under the drought stress.

S. No.	Plant	Concentration	Parameters Modulated	References
1	*Triticum aestivum*	100 μM	↑(APX, *MDHAR, DHAR*, GPX, GST, GSH, AsA), ↓H_2O_2	[83]
2	*Malus prunifolia, Malus hupehensis*	100 μM	↓(*MdNCED3*, H_2O_2), ↑(*MdCYP707A1, MdCYP707A2, MdTDC1, MdAANAT2, Mdt5H4, MdASMT1*)	[85]
3	*Arabidopsis thaliana (transgenic)*	Endogenous	↓ROS	[20]
4	*Solanum lycopersicum*	0.1 mM	↑(Chlorophyll, transpiration rate, stomatal conductance, ΦPSII, Fv/Fm, SOD, POD, CAT, APX, GR, AsA), ↓(MDA, O_2^-)	[88]
5	*Zea mays* L.	1 mM	↓Lipid hydroperoxide, ↑Fv/Fm	[90]
6	*Zea mays* L.	100 μM	↑photosynthetic rate, stomatal conductance, transpiration rate, SOD, POD, CAT, APX), ↓(Melanodialdehyde, hydrogen peroxide)	[60]
7	*Malus domestica* Borkh.	100 μM	↓(*SAG12, PAO*), ↑(ΦPSII, H_2O_2)	[61]
8	*Malus domestica* Borkh.	100 μM	↓(NR, NiR, GS, GOGAT), ↑(*Fd-GOGAT, NADH-GOGAT, AMT1;2, AMT1;5, AMT1;6, AMT2;1, NRT1;1, NRT2;4, NRT2;5, NRT2;7*)	[77]
9	*A. chinensis*	100 μM	↓(lipid peroxidation, pigment degradation), ↑(*Rubisco, PGK, GAPA, FBA, FBP, TIM, TKT, RPK, SEBP, RPI, RPE*)	[84]
10	*Avenanuda* L.	100 μM	↓(H_2O_2, O_2^-, *Aamap1, Aspk11*), ↑(SOD, POD, CAT, APX, *WRKY1, DREB2, MYB*)	[91]
11	*Brassica napus* L.	50 μM	↓(H_2O_2), ↑(SOD, POD, CAT, APX, WSP, WSG, Pro)	[92]
12	*Agrostis stolonifera*	20 μM	↓(*Chlase, PPH, Chl-PRX, TDC1, SNAT1, COMT*), ↑(*JUB1, DREB2A*)	[79]
13	*Cucumis sativus* L.	100 μM	↑(SOD, CAT, APX, photosynthetic rate, chlorophyll content, stomatal conductance)	[93]

(Table 1) cont.....

S. No.	Plant	Concentration	Parameters Modulated	References
14	*Coffea arabica*	300 μM	↓(MDA, proline), ↑(Photosynthetic rate, stomatal conductance, transpiration rate, carboxylation efficiency, SOD, CAT)	[55]
15	*Dracocephalum moldavica*	100 μM	↓(Lipid peroxidation, H_2O_2), ↑(SOD, GPX, APX)	[86]

MELATONIN: SALINITY STRESS TOLERANCE

Salinity stress is a major issue in plants associated with their reduced growth. It is generally of two types; osmotic stress and ionic stress. During osmotic stress, soil has an abundant concentration of salts which results in loss of water due to high osmotic pressure. On the other hand, during ionic stress, a plant has a high concentration of ions, *i.e.* Na^+ and Cl^- which leads to functional and structural abnormalities in plants such as reduced root and shoot growth, inadequate intercellular solute level, *etc.* Some small genes encode peptides that behave similarly to phytohormones are overexpressed/knockdown during the salinity stress conditions. Moreover, the resultant mutations due to this stress condition in these small peptides lead to tolerance for the respective condition [94, 95]. Melatonin is a key molecule that increased stress conditions to cop-up the salinity stress level. It acts in two ways either directly reducing the reactive oxygen species or by indirect way *i.e.* by magnifying the activity of antioxidant enzymes, increasing the metabolic content, photosynthetic efficiency by modulating the stress-associated transcription factors [96 - 99]. Various mechanisms underlying the salinity stress modulation by melatonin is given in Fig. (**1**). Besides, some phytochemicals have been reported to modulate the level of melatonin to enhance salinity stress tolerance. Nitrogen oxide (NO) regulates the downstream of melatonin to increase the tolerance ability [100]. In the recent past, various plants such as Rapeseed, Arabidopsis, Rice, Maize, Tomato plant and Watermelon [56, 101 - 104] have been used by various researchers in the field to investigate the effect of melatonin on stress tolerance.

CONCLUSION AND FUTURE PERSPECTIVES

From the above data, melatonin proved to be a pleiotropic molecule that shows its activity by binding to membrane, cytosolic and nuclear receptors or act independently. The synthesis of melatonin occurs in chloroplast and mitochondria involving various enzymes. The melatonin present in different plant organs but their concentration is not stable. Its amount changes according to the stress (biotic and abiotic) conditions. The crosstalk of melatonin to other phytohormones shows its role in growth, development, stomata regulation, and defense mechanism. The networking intercalation of ROS, NO, RNS and melatonin proved its antioxidant activity by modulating enzymes which involve in ROS production and their scavenging activity. So, melatonin maintains equilibrium between ROS production and free radicals scavenging by their production and metabolism. Therefore, mass production of melatonin during abiotic (drought and salinity) and biotic stress provides immunity to plants to cope/respond towards adverse conditions. To conclude, melatonin a non-toxic marker/molecule proved useful in botanical sciences.

Besides the above data, how the concentration of melatonin changes in the same plant during different environmental conditions? Which mechanism or proteins to be involved to change in melatonin content? The role of melatonin against pathogens, bacteria and viruses are further needed to be explored in the future.

CONSENT FOR PUBLICATION

Not applicable.

CONFLICT OF INTEREST

The author declares no conflict of interest, financial or otherwise.

ACKNOWLEDGEMENTS

We are grateful to the DST-INSPIRE (Department of Science & Technology); dated (04/10/2018) and CSIR-JRF (Council of Scientific & Industrial Research); dated (01/01/2019) for providing fellowship and special thanks to the Department of Botanical and Environmental Sciences for providing necessary facilities and support.

REFERENCES

[1]   Tan DX, Manchester LC, Esteban-Zubero E, Zhou Z, Reiter RJ. Melatonin as a potent and inducible endogenous antioxidant: synthesis and metabolism. Molecules 2015; 20(10): 18886-906.
[http://dx.doi.org/10.3390/molecules201018886] [PMID: 26501252]

[2]   Kopin IJ, Pare CM, Axelrod J, Weissbach H. The fate of melatonin in animals. J Biol Chem 1961; 236(3072): 3072-5.
[http://dx.doi.org/10.1016/S0021-9258(19)76431-X] [PMID: 14458327]

[3]   Arendt J. Melatonin. Clinical Endocrinology 1988; 29(2): 205-29.

[4]   Mayo JC, Sainz RM, González-Menéndez P, He D, Cernuda-Cernuda R. Melatonin transport into mitochondria. Cell Mol Life Sci 2017; 74(21): 3927-40.
[http://dx.doi.org/10.1007/s00018-017-2616-8] [PMID: 28828619]

[5]   Fan J, Xie Y, Zhang Z, Chen L. Melatonin: A multifunctional factor in plants. Int J Mol Sci 2018; 19(5): 1-14.

[http://dx.doi.org/10.3390/ijms19051528] [PMID: 29883400]

[6]   Arnao MB, Hernández-Ruiz J. Functions of melatonin in plants: a review. J Pineal Res 2015; 59(2): 133-50.
        [http://dx.doi.org/10.1111/jpi.12253] [PMID: 26094813]

[7]   Van Tassel DL, Roberts N, Lewy A, O'Neill SD. Melatonin in plant organs. J Pineal Res 2001; 31(1): 8-15.
        [http://dx.doi.org/10.1034/j.1600-079X.2001.310102.x] [PMID: 11485009]

[8]   Sharif R, Xie C, Zhang H, *et al.* Melatonin and its effects on plant systems. Molecules 2018; 23(9): 1-20.
        [http://dx.doi.org/10.3390/molecules23092352] [PMID: 30223442]

[9]   Lee K, Back K. Melatonin-deficient rice plants show a common semidwarf phenotype either dependent or independent of brassinosteroid biosynthesis. J Pineal Res 2019; 66(2): e12537.
        [http://dx.doi.org/10.1111/jpi.12537] [PMID: 30403303]

[10]   Tan DX, Reiter RJ. Mitochondria: the birth place, battle ground and the site of melatonin metabolism in cells. Melatonin Research 2019; 2(1): 44-66.
        [http://dx.doi.org/10.32794/mr11250011]

[11]   Zhao D, Yu Y, Shen Y, *et al.* Melatonin synthesis and function: evolutionary history in animals and plants. Front Endocrinol (Lausanne) 2019; 10: 249.
        [http://dx.doi.org/10.3389/fendo.2019.00249] [PMID: 31057485]

[12]   Back K, Tan DX, Reiter RJ. Melatonin biosynthesis in plants: multiple pathways catalyze tryptophan to melatonin in the cytoplasm or chloroplasts. J Pineal Res 2016; 61(4): 426-37.
        [http://dx.doi.org/10.1111/jpi.12364] [PMID: 27600803]

[13]   Posmyk MM, Janas KM. Melatonin in plants. Acta Physiol Plant 2009; 31(1): 1-11.
        [http://dx.doi.org/10.1007/s11738-008-0213-z]

[14]   Arnao MB, Hernández-Ruiz J. Melatonin: a new plant hormone and/or a plant master regulator? Trends Plant Sci 2019; 24(1): 38-48.
        [http://dx.doi.org/10.1016/j.tplants.2018.10.010] [PMID: 30446305]

[15]   Reiter RJ, Tan DX, Fuentes-Broto L. Melatonin: a multitasking molecule. In Progress in brain research 2010; 181: 127-51.
        [http://dx.doi.org/10.1016/S0079-6123(08)81008-4]

[16]   Hardeland R. Melatonin in plants and other phototrophs: advances and gaps concerning the diversity of functions. J Exp Bot 2015; 66(3): 627-46.
        [http://dx.doi.org/10.1093/jxb/eru386] [PMID: 25240067]

[17]   Debnath B, Islam W, Li M, *et al.* Melatonin mediates enhancement of stress tolerance in plants. Int J Mol Sci 2019; 20(5): 1-17.
        [http://dx.doi.org/10.3390/ijms20051040] [PMID: 30818835]

[18]   Tan DX, Manchester LC, Liu X, Rosales-Corral SA, Acuna-Castroviejo D, Reiter RJ. Mitochondria and chloroplasts as the original sites of melatonin synthesis: a hypothesis related to melatonin's primary function and evolution in eukaryotes. J Pineal Res 2013; 54(2): 127-38.
        [http://dx.doi.org/10.1111/jpi.12026] [PMID: 23137057]

[19]   Wang L, Feng C, Zheng X, *et al.* Plant mitochondria synthesize melatonin and enhance the tolerance of plants to drought stress. J Pineal Res 2017; 63(3): e1-e27.
        [http://dx.doi.org/10.1111/jpi.12429] [PMID: 28599069]

[20]   Byeon Y, Lee HY, Lee K, Park S, Back K. Cellular localization and kinetics of the rice melatonin biosynthetic enzymes SNAT and ASMT. J Pineal Res 2014; 56(1): 107-14.
        [http://dx.doi.org/10.1111/jpi.12103] [PMID: 24134674]

[21]   Kaur H, Mukherjee S, Baluska F, Bhatla SC. Regulatory roles of serotonin and melatonin in abiotic

stress tolerance in plants. Plant Signal Behav 2015; 10(11): e1049788.
[http://dx.doi.org/10.1080/15592324.2015.1049788] [PMID: 26633566]

[22]   Verma V, Ravindran P, Kumar PP. Plant hormone-mediated regulation of stress responses. BMC Plant Biol 2016; 16(1): 86.
[http://dx.doi.org/10.1186/s12870-016-0771-y] [PMID: 27079791]

[23]   Kohli A, Sreenivasulu N, Lakshmanan P, Kumar PP. The phytohormone crosstalk paradigm takes center stage in understanding how plants respond to abiotic stresses. Plant Cell Rep 2013; 32(7): 945-57.
[http://dx.doi.org/10.1007/s00299-013-1461-y] [PMID: 23749097]

[24]   Yang YXJ, Ahammed GJ, Wu C, Fan SY, Zhou YH. Crosstalk among jasmonate, salicylate and ethylene signaling pathways in plant disease and immune responses. Curr Protein Pept Sci 2015; 16(5): 450-61.
[http://dx.doi.org/10.2174/1389203716666150330141638] [PMID: 25824390]

[25]   Schaller GE, Bishopp A, Kieber JJ. The yin-yang of hormones: cytokinin and auxin interactions in plant development. Plant Cell 2015; 27(1): 44-63.
[http://dx.doi.org/10.1105/tpc.114.133595] [PMID: 25604447]

[26]   Edmonds KE. Melatonin, But not auxin, affects postnatal reproductive development in the marsh rice rat (Oryzomys palustris). Zool Sci 2013; 30(6): 439-45.
[http://dx.doi.org/10.2108/zsj.30.439] [PMID: 23721467]

[27]   Sun Q, Zhang N, Wang J, *et al.* Melatonin promotes ripening and improves quality of tomato fruit during postharvest life. J Exp Bot 2015; 66(3): 657-68.
[http://dx.doi.org/10.1093/jxb/eru332] [PMID: 25147270]

[28]   Pérez-Llorca M, Muñoz P, Müller M, Munné-Bosch S. Biosynthesis, metabolism and function of auxin, salicylic acid and melatonin in climacteric and non-climacteric fruits. Front Plant Sci 2019; 10: 136.
[http://dx.doi.org/10.3389/fpls.2019.00136] [PMID: 30833953]

[29]   Arnao MB, Hernández-Ruiz J. Melatonin: plant growth regulator and/or biostimulator during stress? Trends Plant Sci 2014; 19(12): 789-97.
[http://dx.doi.org/10.1016/j.tplants.2014.07.006] [PMID: 25156541]

[30]   Wang Q, An B, Wei Y, *et al.* Melatonin regulates root meristem by repressing auxin synthesis and polar auxin transport in Arabidopsis. Front Plant Sci 2016; 7: 1882.
[http://dx.doi.org/10.3389/fpls.2016.01882] [PMID: 28018411]

[31]   Mukherjee S. Novel perspectives on the molecular crosstalk mechanisms of serotonin and melatonin in plants. Plant Physiol Biochem 2018; 132: 33-45.
[http://dx.doi.org/10.1016/j.plaphy.2018.08.031] [PMID: 30172851]

[32]   Nawaz F, Shabbir RN, Shahbaz M, *et al.* Cross talk between nitric oxide and phytohormones regulate plant development during abiotic stresses. Phytohormones-signaling mechanisms and crosstalk in plant development and stress responses 2017; 117-41.

[33]   Arnao MB, Hernández-Ruiz J. Melatonin and its relationship to plant hormones. Ann Bot 2018; 121(2): 195-207.
[http://dx.doi.org/10.1093/aob/mcx114] [PMID: 29069281]

[34]   Agathokleous E, Kitao M, Calabrese EJ. New insights into the role of melatonin in plants and animals. Chem Biol Interact 2019; 299: 163-7.
[http://dx.doi.org/10.1016/j.cbi.2018.12.008] [PMID: 30553720]

[35]   Kim YS, Park SI, Kim JJ, *et al.* Expression of Heterologous OsDHAR Gene Improves Glutathione (GSH)-Dependent Antioxidant System and Maintenance of Cellular Redox Status in *Synechococcus* elongatus PCC 7942. Front Plant Sci 2020; 11: 1-14.
[http://dx.doi.org/10.3389/fpls.2020.00231]

[36]   Sharma P, Jha AB, Dubey RS, Pessarakli M. Reactive oxygen species, oxidative damage, and antioxidative defense mechanism in plants under stressful conditions. J Bot 2012; 2012: 1-27.
[http://dx.doi.org/10.1155/2012/217037]

[37]   Hasanuzzaman M, Bhuyan MHMB, Anee TI, *et al*. Regulation of ascorbate-glutathione pathway in mitigating oxidative damage in plants under abiotic stress. Antioxidants 2019; 8(9): 1-50.
[http://dx.doi.org/10.3390/antiox8090384] [PMID: 31505852]

[38]   He JD, Zou YN, Wu QS, Kuča K. Mycorrhizas enhance drought tolerance of trifoliate orange by enhancing activities and gene expression of antioxidant enzymes. Sci Hortic (Amsterdam) 2020; 262: 1-8.
[http://dx.doi.org/10.1016/j.scienta.2019.108745]

[39]   Halliwell B, Gutteridge JM. Antioxidant defences: endogenous and diet derived. Free Radic Biol Med 2007; 4: 79-186.

[40]   Kapoor D, Singh S, Kumar V, Romero R, Prasad R, Singh J. Antioxidant enzymes regulation in plants in reference to reactive oxygen species (ROS) and reactive nitrogen species (RNS). Plant Gene 2019; 19: 1-13.
[http://dx.doi.org/10.1016/j.plgene.2019.100182]

[41]   Fimognari L, Dölker R, Kaselyte G, *et al*. Simple semi-high throughput determination of activity signatures of key antioxidant enzymes for physiological phenotyping. Plant Methods 2020; 16(1): 42.
[http://dx.doi.org/10.1186/s13007-020-00583-8] [PMID: 32206082]

[42]   Khan A, Numan M, Khan AL, *et al*. Melatonin: awakening the defense mechanisms during plant oxidative stress. Plants (Basel) 2020; 9(4): 1-22.
[http://dx.doi.org/10.3390/plants9040407] [PMID: 32218185]

[43]   Miao H, Zeng W, Zhao M, Wang J, Wang Q. Effect of melatonin treatment on visual quality and health-promoting properties of broccoli florets under room temperature. Food Chem 2020; 319: 126498.
[http://dx.doi.org/10.1016/j.foodchem.2020.126498] [PMID: 32169761]

[44]   Tan DX, Reiter RJ, Manchester LC, *et al*. Chemical and physical properties and potential mechanisms: melatonin as a broad spectrum antioxidant and free radical scavenger. Curr Top Med Chem 2002; 2(2): 181-97.
[http://dx.doi.org/10.2174/1568026023394443] [PMID: 11899100]

[45]   Liang D, Shen Y, Ni Z, *et al*. Exogenous melatonin application delays senescence of kiwifruit leaves by regulating the antioxidant capacity and biosynthesis of flavonoids. Front Plant Sci 2018; 9: 426.
[http://dx.doi.org/10.3389/fpls.2018.00426] [PMID: 29675031]

[46]   Reiter RJ, Ma Q, Sharma R. Melatonin in mitochondria: mitigating clear and present dangers. Physiology (Bethesda) 2020; 35(2): 86-95.
[http://dx.doi.org/10.1152/physiol.00034.2019] [PMID: 32024428]

[47]   Kumar A, Khushboo , Pandey R, Sharma B. Modulation of Superoxide Dismutase Activity by Mercury, Lead, and Arsenic. Biol Trace Elem Res 2020; 196(2): 654-61.
[http://dx.doi.org/10.1007/s12011-019-01957-3] [PMID: 31925741]

[48]   Saxena I, Srikanth S, Chen Z. Cross talk between H_2O_2 and interacting signal molecules under plant stress response. Front Plant Sci 2016; 7: 570.
[http://dx.doi.org/10.3389/fpls.2016.00570] [PMID: 27200043]

[49]   Sofo A, Scopa A, Nuzzaci M, Vitti A. Ascorbate peroxidase and catalase activities and their genetic regulation in plants subjected to drought and salinity stresses. Int J Mol Sci 2015; 16(6): 13561-78.
[http://dx.doi.org/10.3390/ijms160613561] [PMID: 26075872]

[50]   Venkatesh J, Park SW. Role of L-ascorbate in alleviating abiotic stresses in crop plants. Bot Stud (Taipei, Taiwan) 2014; 55(1): 1-20.
[http://dx.doi.org/10.1186/1999-3110-55-38]

[51]   Gill SS, Tuteja N. Reactive oxygen species and antioxidant machinery in abiotic stress tolerance in crop plants. Plant Physiol Biochem 2010; 48(12): 909-30.
[http://dx.doi.org/10.1016/j.plaphy.2010.08.016] [PMID: 20870416]

[52]   Gietler M, Nykiel M, Zagdańska BM. Changes in the reduction state of ascorbate and glutathione, protein oxidation and hydrolysis leading to the development of dehydration intolerance in *Triticum aestivum* L. seedlings. Plant Growth Regul 2016; 79(3): 287-97.
[http://dx.doi.org/10.1007/s10725-015-0133-z]

[53]   Wei J, Li DX, Zhang JR, *et al.* Phytomelatonin receptor PMTR1-mediated signaling regulates stomatal closure in *Arabidopsis thaliana*. J Pineal Res 2018; 65(2): e12500.
[http://dx.doi.org/10.1111/jpi.12500] [PMID: 29702752]

[54]   Shi H, Jiang C, Ye T, *et al.* Comparative physiological, metabolomic, and transcriptomic analyses reveal mechanisms of improved abiotic stress resistance in bermudagrass [*Cynodon dactylon* (L). Pers.] by exogenous melatonin. J Exp Bot 2015; 66(3): 681-94.
[http://dx.doi.org/10.1093/jxb/eru373] [PMID: 25225478]

[55]   Campos CN, Ávila RG, de Souza KR, Azevedo LM, Alves JD. Melatonin reduces oxidative stress and promotes drought tolerance in young Coffeaarabica L. plants. Agric Water Manage 2019; 211: 37-47.
[http://dx.doi.org/10.1016/j.agwat.2018.09.025]

[56]   Martinez V, Nieves-Cordones M, Lopez-Delacalle M, *et al.* Tolerance to stress combination in tomato plants: new insights in the protective role of melatonin. Molecules 2018; 23(3): 1-20.
[http://dx.doi.org/10.3390/molecules23030535] [PMID: 29495548]

[57]   KolevaGudeva L, ButleskaGjoroska V, Trajkova F, Mihajlov L. Activity of enzyme catalase in alfalfa (*Medicago sativa* L.) as an indicator for abiotic stress. J Agricul Plant Sci 2019; 17(1): 45-52.

[58]   Li C, Wang P, Wei Z, *et al.* The mitigation effects of exogenous melatonin on salinity-induced stress in *Malus hupehensis*. J Pineal Res 2012; 53(3): 298-306.
[http://dx.doi.org/10.1111/j.1600-079X.2012.00999.x] [PMID: 22507106]

[59]   Wang LY, Liu JL, Wang WX, Sun Y. Exogenous melatonin improves growth and photosynthetic capacity of cucumber under salinity-induced stress. Photosynthetica 2016; 54(1): 19-27.
[http://dx.doi.org/10.1007/s11099-015-0140-3]

[60]   Ye J, Wang S, Deng X, Yin L, Xiong B, Wang X. Melatonin increased maize (*Zea mays* L.) seedling drought tolerance by alleviating drought-induced photosynthetic inhibition and oxidative damage. Acta Physiol Plant 2016; 38(2): 48.
[http://dx.doi.org/10.1007/s11738-015-2045-y]

[61]   Wang P, Sun X, Li C, Wei Z, Liang D, Ma F. Long-term exogenous application of melatonin delays drought-induced leaf senescence in apple. J Pineal Res 2013; 54(3): 292-302.
[http://dx.doi.org/10.1111/jpi.12017] [PMID: 23106234]

[62]   Marta B, Szafrańska K, Posmyk MM. Exogenous melatonin improves antioxidant defense in cucumber seeds (*Cucumis sativus* L.) germinated under chilling stress. Front Plant Sci 2016; 7: 575.
[http://dx.doi.org/10.3389/fpls.2016.00575] [PMID: 27200048]

[63]   Turk H, Erdal S, Genisel M, Atici O, Demir Y, Yanmis D. The regulatory effect of melatonin on physiological, biochemical and molecular parameters in cold-stressed wheat seedlings. Plant Growth Regul 2014; 74(2): 139-52.
[http://dx.doi.org/10.1007/s10725-014-9905-0]

[64]   Osakabe Y, Osakabe K, Shinozaki K, Tran LS. Response of plants to water stress. Front Plant Sci 2014; 5: 86.
[http://dx.doi.org/10.3389/fpls.2014.00086] [PMID: 24659993]

[65]   Xu L, Yu J, Han L, Huang B. Photosynthetic enzyme activities and gene expression associated with drought tolerance and post-drought recovery in Kentucky bluegrass. Environ Exp Bot 2013; 89: 28-35.
[http://dx.doi.org/10.1016/j.envexpbot.2012.12.001]

[66]   Porcar-Castell A, Tyystjärvi E, Atherton J, *et al.* Linking chlorophyll a fluorescence to photosynthesis for remote sensing applications: mechanisms and challenges. J Exp Bot 2014; 65(15): 4065-95.
       [http://dx.doi.org/10.1093/jxb/eru191] [PMID: 24868038]

[67]   Nishiyama Y, Allakhverdiev SI, Murata N. Protein synthesis is the primary target of reactive oxygen species in the photoinhibition of photosystem II. Physiol Plant 2011; 142(1): 35-46.
       [http://dx.doi.org/10.1111/j.1399-3054.2011.01457.x] [PMID: 21320129]

[68]   Hammond JP, White PJ. Sucrose transport in the phloem: integrating root responses to phosphorus starvation. J Exp Bot 2008; 59(1): 93-109.
       [http://dx.doi.org/10.1093/jxb/erm221] [PMID: 18212031]

[69]   Shi H, Chan Z. The cysteine2/histidine2-type transcription factor zinc finger of *Arabidopsis thaliana* 6-activated C-repeat-binding factor pathway is essential for melatonin-mediated freezing stress resistance in Arabidopsis. J Pineal Res 2014; 57(2): 185-91.
       [http://dx.doi.org/10.1111/jpi.12155] [PMID: 24962049]

[70]   Posmyk MM, Kuran H, Marciniak K, Janas KM. Presowing seed treatment with melatonin protects red cabbage seedlings against toxic copper ion concentrations. J Pineal Res 2008; 45(1): 24-31.
       [http://dx.doi.org/10.1111/j.1600-079X.2007.00552.x] [PMID: 18205729]

[71]   Afreen F, Zobayed SM, Kozai T. Melatonin in *Glycyrrhiza uralensis*: response of plant roots to spectral quality of light and UV-B radiation. J Pineal Res 2006; 41(2): 108-15.
       [http://dx.doi.org/10.1111/j.1600-079X.2006.00337.x] [PMID: 16879315]

[72]   Wang P, Sun X, Xie Y, *et al.* Melatonin regulates proteomic changes during leaf senescence in *Malus hupehensis*. J Pineal Res 2014; 57(3): 291-307.
       [http://dx.doi.org/10.1111/jpi.12169] [PMID: 25146528]

[73]   Reiter RJ, Tan DX, Zhou Z, Cruz MH, Fuentes-Broto L, Galano A. Phytomelatonin: assisting plants to survive and thrive. Molecules 2015; 20(4): 7396-437.
       [http://dx.doi.org/10.3390/molecules20047396] [PMID: 25911967]

[74]   Shi H, Qian Y, Tan DX, Reiter RJ, He C. Melatonin induces the transcripts of CBF/DREB1s and their involvement in both abiotic and biotic stresses in Arabidopsis. J Pineal Res 2015; 59(3): 334-42.
       [http://dx.doi.org/10.1111/jpi.12262] [PMID: 26182834]

[75]   Jiang C, Cui Q, Feng K, Xu D, Li C, Zheng Q. Melatonin improves antioxidant capacity and ion homeostasis and enhances salt tolerance in maize seedlings. Acta Physiol Plant 2016; 38(4): 1-9.
       [http://dx.doi.org/10.1007/s11738-016-2101-2]

[76]   Zhao H, Su T, Huo L, *et al.* Unveiling the mechanism of melatonin impacts on maize seedling growth: sugar metabolism as a case. J Pineal Res 2015; 59(2): 255-66.
       [http://dx.doi.org/10.1111/jpi.12258] [PMID: 26122919]

[77]   Liang B, Ma C, Zhang Z, *et al.* Long-term exogenous application of melatonin improves nutrient uptake fluxes in apple plants under moderate drought stress. Environ Exp Bot 2018; 155: 650-61.
       [http://dx.doi.org/10.1016/j.envexpbot.2018.08.016]

[78]   Karaca P, Cekic FÖ. Exogenous melatonin-stimulated defense responses in tomato plants treated with polyethylene glycol. Int J Veg Sci 2019; 25(6): 601-9.
       [http://dx.doi.org/10.1080/19315260.2019.1575317]

[79]   Ma X, Zhang J, Burgess P, Rossi S, Huang B. Interactive effects of melatonin and cytokinin on alleviating drought-induced leaf senescence in creeping bentgrass (*Agrostis stolonifera*). Environ Exp Bot 2018; 145: 1-11.
       [http://dx.doi.org/10.1016/j.envexpbot.2017.10.010]

[80]   Sharma A, Yuan H, Kumar V, *et al.* Castasterone attenuates insecticide induced phytotoxicity in mustard. Ecotoxicol Environ Saf 2019; 179: 50-61.
       [http://dx.doi.org/10.1016/j.ecoenv.2019.03.120] [PMID: 31026750]

[81]   Sharma A, Kumar V, Yuan H, *et al.* Jasmonic acid seed treatment stimulates insecticide detoxification in *Brassica juncea* L. Front Plant Sci 2018; 9: 1609.
       [http://dx.doi.org/10.3389/fpls.2018.01609] [PMID: 30450109]

[82]   Meng JF, Xu TF, Wang ZZ, Fang YL, Xi ZM, Zhang ZW. The ameliorative effects of exogenous melatonin on grape cuttings under water-deficient stress: antioxidant metabolites, leaf anatomy, and chloroplast morphology. J Pineal Res 2014; 57(2): 200-12.
       [http://dx.doi.org/10.1111/jpi.12159] [PMID: 25039750]

[83]   Cui G, Zhao X, Liu S, Sun F, Zhang C, Xi Y. Beneficial effects of melatonin in overcoming drought stress in wheat seedlings. Plant Physiol Biochem 2017; 118: 138-49.
       [http://dx.doi.org/10.1016/j.plaphy.2017.06.014] [PMID: 28633086]

[84]   Liang D, Ni Z, Xia H, *et al.* Exogenous melatonin promotes biomass accumulation and photosynthesis of kiwifruit seedlings under drought stress. Sci Hortic (Amsterdam) 2019; 246: 34-43.
       [http://dx.doi.org/10.1016/j.scienta.2018.10.058]

[85]   Li C, Tan DX, Liang D, Chang C, Jia D, Ma F. Melatonin mediates the regulation of ABA metabolism, free-radical scavenging, and stomatal behaviour in two *Malus* species under drought stress. J Exp Bot 2015; 66(3): 669-80.
       [http://dx.doi.org/10.1093/jxb/eru476] [PMID: 25481689]

[86]   Kabiri R, Hatami A, Oloumi H, Naghizadeh M, Nasibi F, Tahmasebi Z. Foliar application of melatonin induces tolerance to drought stress in Moldavian balm plants (*Dracocephalum moldavica*) through regulating the antioxidant system. Folia Hortic 2018; 30(1): 155-67.
       [http://dx.doi.org/10.2478/fhort-2018-0016]

[87]   Yang WJ, Du YT, Zhou YB, *et al.* Overexpression of TaCOMT improves melatonin production and enhances drought tolerance in transgenic Arabidopsis. Int J Mol Sci 2019; 20(3): 1-16.
       [http://dx.doi.org/10.3390/ijms20030652] [PMID: 30717398]

[88]   Liu J, Wang W, Wang L, Sun Y. Exogenous melatonin improves seedling health index and drought tolerance in tomato. Plant Growth Regul 2015; 77(3): 317-26.
       [http://dx.doi.org/10.1007/s10725-015-0066-6]

[89]   Ding F, Wang G, Wang M, Zhang S. Exogenous melatonin improves tolerance to water deficit by promoting cuticle formation in tomato plants. Molecules 2018; 23(7): 1-10.
       [http://dx.doi.org/10.3390/molecules23071605] [PMID: 30004432]

[90]   Fleta-Soriano E, Díaz L, Bonet E, Munné-Bosch S. Melatonin may exert a protective role against drought stress in maize. J Agron Crop Sci 2017; 203(4): 286-94.
       [http://dx.doi.org/10.1111/jac.12201]

[91]   Gao W, Zhang Y, Feng Z, Bai Q, He J, Wang Y. Effects of melatonin on antioxidant capacity in naked oat seedlings under drought stress. Molecules 2018; 23(7): 1-14.
       [http://dx.doi.org/10.3390/molecules23071580] [PMID: 29966243]

[92]   Li J, Zeng L, Cheng Y, *et al.* Exogenous melatonin alleviates damage from drought stress in *Brassica napus* L.(rapeseed) seedlings. Acta Physiol Plant 2018; 40(3): 1-11.
       [http://dx.doi.org/10.1007/s11738-017-2601-8]

[93]   Zhang N, Zhao B, Zhang HJ, *et al.* Melatonin promotes water-stress tolerance, lateral root formation, and seed germination in cucumber (*Cucumis sativus* L.). J Pineal Res 2013; 54(1): 15-23.
       [http://dx.doi.org/10.1111/j.1600-079X.2012.01015.x] [PMID: 22747917]

[94]   Nakaminami K, Okamoto M, Higuchi-Takeuchi M, *et al.* AtPep3 is a hormone-like peptide that plays a role in the salinity stress tolerance of plants. Proc Natl Acad Sci USA 2018; 115(22): 5810-5.
       [http://dx.doi.org/10.1073/pnas.1719491115] [PMID: 29760074]

[95]   Sako K, Sunaoshi Y, Tanaka M, Matsui A, Seki M. The duration of ethanol-induced high-salinity stress tolerance in *Arabidopsis thaliana*. Plant Signal Behav 2018; 13(8): e1500065.
       [http://dx.doi.org/10.1080/15592324.2018.1500065] [PMID: 30067446]

[96]   Guo J, Li Y, Han G, Song J, Wang B. NaCl markedly improved the reproductive capacity of the euhalophyte Suaeda salsa. Funct Plant Biol 2018; 45(3): 350-61.
[http://dx.doi.org/10.1071/FP17181] [PMID: 32290958]

[97]   Zhao KF, Song J, Fan H, Zhou S, Zhao M. Growth response to ionic and osmotic stress of NaCl in salt-tolerant and salt-sensitive maize. J Integr Plant Biol 2010; 52(5): 468-75.
[http://dx.doi.org/10.1111/j.1744-7909.2010.00947.x] [PMID: 20537042]

[98]   Liu S, Wang W, Li M, Wan S, Sui N. Antioxidants and unsaturated fatty acids are involved in salt tolerance in peanut. Acta Physiol Plant 2017; 39(9): 1-10.
[http://dx.doi.org/10.1007/s11738-017-2501-y]

[99]   Liu Q, Liu R, Ma Y, Song J. Physiological and molecular evidence for Na+ and Cl− exclusion in the roots of two Suaeda salsa populations. Aquat Bot 2018; 146: 1-7.
[http://dx.doi.org/10.1016/j.aquabot.2018.01.001]

[100] Li J, Liu J, Zhu T, Zhao C, Li L, Chen M. The role of melatonin in salt stress responses. Int J Mol Sci 2019; 20(7): 1-15.
[http://dx.doi.org/10.3390/ijms20071735] [PMID: 30965607]

[101] Zhao G, Zhao Y, Yu X, *et al.* Nitric oxide is required for melatonin-enhanced tolerance against salinity stress in rapeseed (*Brassica napus* L.) seedlings. Int J Mol Sci 2018; 19(7): 1-22.
[http://dx.doi.org/10.3390/ijms19071912] [PMID: 29966262]

[102] Chen Z, Xie Y, Gu Q, *et al.* The AtrbohF-dependent regulation of ROS signaling is required for melatonin-induced salinity tolerance in Arabidopsis. Free Radic Biol Med 2017; 108: 465-77.
[http://dx.doi.org/10.1016/j.freeradbiomed.2017.04.009] [PMID: 28412199]

[103] Hwang OJ, Back K. Melatonin deficiency confers tolerance to multiple abiotic stresses in rice *via* decreased brassinosteroid levels. Int J Mol Sci 2019; 20(20): 1-17.
[http://dx.doi.org/10.3390/ijms20205173] [PMID: 31635310]

[104] Li H, Chang J, Chen H, *et al.* Exogenous melatonin confers salt stress tolerance to watermelon by improving photosynthesis and redox homeostasis. Front Plant Sci 2017; 8: 295.
[http://dx.doi.org/10.3389/fpls.2017.00295] [PMID: 28298921]

CHAPTER 9

Phenylpropanoid Biosynthesis and its Protective Effects against Plants Stress

Sharad Thakur[1,4], **Kritika Pandit**[2], **Ajay Kumar**[2], **Jaskaran Kaur**[2] and **Sandeep Kaur**[2,*]

[1] *CSIR-Institute of Himalayan Bioresource Technology, Palampur, H.P. 176061, India*

[2] *Department of Botanical and Environmental Sciences, Guru Nanak Dev University, Amritsar, Punjab 143005, India*

[3] *PG Department of Botany, Khalsa College, Amritsar, Punjab 143005, India*

[4] *PG Department of Agriculture, Khalsa College, Amritsar, Punjab 143005, India*

Abstract: Phenylpropanoids are a class of secondary metabolites in plants that are derived from aromatic amino acids like tyrosine and phenylalanine. It mainly includes stilbenes, monolignols, coumarins, flavonoids, and phenolic acids. These are considered to play a crucial role in protecting the plants against both abiotic and biotic stress by quenching the generation of reactive oxygen species (ROS) through a wide range of mechanisms. Phenylpropanoids are found widely in the plant kingdom and serve an essential role in the development of plant by acting as an important cell wall component, floral pigments to mediate the interactions of plant–pollinator, antibiotics (phytoalexins) against pathogens and herbivores, and protectants against UV radiation and high light. Several phenylpropanoids are helpful for the plant to fight against microbial diseases and thereby show broad-spectrum antimicrobial activity. The biosynthetic pathway of phenylpropanoid is mostly activated under abiotic stress conditions including salinity, heavy metal, ultraviolet radiations, high/low temperature, and drought and results in the accumulation of different phenolic compounds that are helpful in scavenging the deleterious effect of ROS. A series of enzymes involved in the activation of the biosynthetic pathway are reductases, transferases, lyases, oxygenases, and ligases. Among these, many are encoded by superfamilies of genes, like NADPH-dependent reductase gene family, the 2-oxoglutarate dependent dioxygenase (2-ODD) gene family, the cytochrome P450 membrane-bound monooxygenase (P450) gene family, and the type III polyketide synthase (PKS III) gene family. Thus, keeping in view the importance of phenylpropanoids in plant defense, the present book chapter is focused on unraveling the role of these essential compounds in ameliorating the stressful conditions in plants.

Keywords: Antimicrobial, Biosynthetic pathway, Phenylpropanoids, Plant, ROS.

---

* **Corresponding Author Sandeep Kaur:** Department of Botanical and Environmental Sciences, Guru Nanak Dev University, Amritsar, Punjab 143005, India; Tel: +91-9877168954; E-mail: soniasandeep4@gmail.com

Tajinder Kaur & Saroj Arora (Eds.)

INTRODUCTION

In continuously evolving environmental conditions, plants are subjected to numerous abiotic stresses that are unfavorable for growth and development. Such abiotic stresses include water stress (drought and flooding), heavy metals, salinity, nutritional surplus or deficiency, high and low temperature (chilling and freezing), intense light periods (high and low), radiations (UV-A and UV-B ultraviolet), ozone layer depletion, sulfur dioxide and other mild plant stressors [4]. The adaptive response of plants to the abiotic stresses is the synthesis and accumulation of the phenolic compounds in plant tissues [1]. Plant synthesizes an enormous number of primary as well as secondary metabolites that is an essential adaptation strategy adopted by plants to their immediate environment. Primary metabolites are distributed everywhere in plants and are required by the plants for growth and development. These include sugars, nucleic acids, fatty acids, and amino acids. Secondary metabolites are not essential for the plants as such but are synthesized in the changing environment scenario. Moreover, secondary metabolites are structurally and chemically more diverse than primary metabolites [2]. Other than polyphenols, the most widely occurring group of secondary metabolites with substantial physiological and morphological importance in plants are the phenylpropanoids. The phenylpropanoid pathway is an essential and ubiquitous pathway in plants leading to the bio-synthesis of many important signaling and protection molecules.

In agricultural crops, the losses due to the abiotic stresses are manifold, and hence an adequate understanding of the plant adaptation to the environment is necessary to minimize the yield losses in crop plants [3]. Hence the understanding of the bio-synthesis of phenylpropanoids is of utmost importance for the selection of useful cultivars with the ability to survive under harmful conditions and for improving the nutraceutical value of food. Since the secondary metabolites of plants like phenylpropanoids also have antioxidant properties, hence their accumulation in the fruits is a health benefit for humans. To neutralize the environmental stresses and environmental irritants, phenylpropanoid pathways are one of the main defense mechanisms employed by the plants. Phenylpropanoid compounds are basically phenols with one or more hydroxyl groups. Cinnamic acid is the precursor molecule for the phenylpropanoids having a simple carbon skeleton with C_3-C_6 carbon chain and complex carbon skeleton with C_7-C_{22} carbon chain [5]. The precursor molecule of the phenylpropanoid pathway, cinnamic acid, is synthesized from phenylalanine by undergoing acylation, cyclization, condensation, glycosylation, methylation, hydroxylation, prenylation, and dehydration reactions. The various enzymes involved in the synthesis of phenylpropanoids are phenylalanine ammonia lyase, chalcone synthase, chalcone isomerase, dihydroflavonol reductase, and anthocyanidin synthase [6].

The main compounds belonging to the phenylpropanoids are benzoquinones, napthoquinones, acetophenones, coumaric acids, flavonoids, bioflavonoids, isoflavonoids, and tannins. The biosynthesis of all the above-mentioned compounds under abiotic and biotic stress conditions in plants is essential to ameliorate the reactive oxygen species (ROS) synthesized by the plants in response to the external stress. The oxidation and reduction potential of the bio-compounds synthesized in phenylpropanoid pathway is known to have excellent scavenging potential for the reactive oxygen species [7]. This chapter will explore the protective role of phenylpropanoids in the growth and development of plants.

SECONDARY METABOLITES PRODUCED BY PHENYLPROPANOID METABOLISM

Phenylpropanoid pathway is considered as the major frequently evaluated metabolic route for the synthesis of the secondary metabolites. Certain simple phenylpropanoids (having basic carbon skeleton C6-C3 of phenylalanine) are generated from cinnamate *via* a series of methylation, hydroxylation, and dehydration reactions, including caffeic, sinapic acids, p-coumaric, ferulic acid, and simple coumarins [8]. The phenylpropanoid pathway products are involved in several aspects of plant growth and response to the stimuli associated inextricably with the land life as well as structural support. They play a key role in stress response upon mineral shortage [9], light variations [10] as well as act as crucial mediators of the interactions of plants with other organisms [11, 12].

The diverse aromatic metabolites produced by the phenylpropanoid biosynthesis in plants play a number of important biological functions. The shikimate pathway of the plant is the entry to the phenylpropanoids biosynthesis [13]. The general metabolism of phenylpropanoid produces a wide array of secondary metabolites based on the few shikimate pathway intermediates as the core unit. The three enzymatic activities of the central phenylpropanoid pathway is defined by (i) the deamination of phenylalanine to the trans-cinnamic acid by phenylalanine ammonia-lyase (PAL), (ii) the hydroxylation of trans-cinnamic acid by the activity of cinnamic acid 4-hydroxylase (C4H) activity to the 4-coumarate, and finally (iii) the 4-coumarate conversion by 4-coumarate-CoA ligase (4CL) to the 4-coumaroyl-CoA [14]. Phenylpropanoid-based polymers such as condensed tannins, suberin, and lignin substantially contribute to the robustness and stability of angiosperms and gymnosperms towards environmental or mechanical damage like wounding or drought. When plants are stressed under sub-optimal photosynthesis, the low concentration of shikimate pathway intermediates may re-direct the whole phenylpropanoid pathway for the formation of anthocyanins, volatiles, phytoalexins, flavonoids, and de novo proteins synthesis [15, 16].

On planet earth, lignin is regarded as the highly prominent polymer besides cellulose. Lignin is synthesized exclusively from phenylpropanoid units derived from the hydroxycinnamoyl alcohol derivatives by oxidative polymerization. The hydrolyzable tannins, like ellagitannins and galloyl-glucose esters, similar to the non-hydrolyzable tannins like proanthocyanidins are usually characterized by the oxidative coupling of neighboring galloyl groups. They show a strong ability of radical scavenging. They precipitate and cross-link enzymes due to the presence of multiple aromatic hydroxyl groups, making them prominent defense compounds when present in high concentrations, specifically in sumac or in oat [17, 18].

In all plant species, all the classes of compounds of phenylpropanoid are not present. Although the flavonoid and hydroxycinnamic acid are ubiquitously present in higher plants, with specific substitution patterns, the members of these classes may be peculiar to some species or genera. Other phenylpropanoid classes, including stilbenes and isoflavonoids, are limited to only some plant families. Their structural variations are large involving the complexity and number of substituents on the different oxidation levels of the heterocycle, 3-phenylchroman framework, and the presence of additional heterocyclic rings. Mostly, isoflavonoids are limited to the leguminosae subfamily, papilionoideae. Stilbenes, isoflavonoids, and terpenophenolics are limited to only a few species, but they have gained major interest due to health-promoting potential in mammals [19, 20].

ROLE OF PHENYLPROPANOIDS IN THE GROWTH AND DEVELOPMENT OF PLANTS

Phenylpropanoids (PPs) are the largest group of secondary metabolites produced by plants in response to biotic and/or non-biotic pressures, such as infections, wounds, UV irradiations, ozone exposure, and pollutants, as well as other harsh environmental conditions. PPs belong to a large class of plant phenols synthesized through the shikimic acid pathway [21]. It includes various plant-derived phenolic compounds like flavonoids, isoflavonoids, coumarines, and lignans [8]. PPs plays many vital roles in the growth and development of plants. The distribution of auxin, a major growth-controlling hormone, is negatively modulated in plants by flavonols and isoflavones. Flavonoid-deficient mutants display a wide variety of phenotypic changes such as root hair production, root formation, length of root hair, lateral hair density, seed organ count, number of shoot organ, increased inflorescence, plant stature and fertility [22]. There are also different colors which are exhibited by flavonoids.

Anthocyanins for example, have different shades of colors, from orange to pink, red and purple. The shades of flavonols and flavones are usually light yellow or colorless. Pigmentation is supplied by them to seeds, fruits and flowers to attract different kinds of pollinators [23, 24]. Flavonoids also play an important function in leguminoseae family. They not only act as chemo-attractants for symbiotic bacteria, but also play significant direct roles in root nodule organogenesis [25, 26]. Flavonoids can also shield plants from a number of abiotic stresses, including UV-B radiation, sugar stress, temperature, water deficit, light abundance or deficiency, nutrient deficiency and also acts as antimicrobial agents [27]. In plant evolution, flavonoids perform a significant role in the UV-B irradiation response from marine algal ancestors to terrestrial plants. UV-B irradiation can cause DNA, fatty acid and protein damage. It also leads to reactive oxygen (ROS) production that damages plant growth [28]. The effects of UV-B radiation on oxidative stress and on the role of PPs in the plant as antioxidants have also been studied. It has been found that many plants respond to enhanced UV radiations by the growth of smaller and thicker leaves, by increasing the thickness of the epidermal wall and cutin layer and further by raising UV-absorbing compounds concentrations in the epidermis, waxes and leaf hairs and by activation of the antioxidant defense system [29].

Flavonoids including flavones, flavonol glycosides, isoflavonoids, anthocyanins, as well as other phenylpropanoids including coumarins, stilbenes, and sinapate esters are induced to act as UV attenuators to scavange reactive oxygen species (ROS) as plants are exposed to UV-B irradiation [30]. There are several pathogen protection mechanisms based on phenylpropanoids, for instance, the formation of structural barriers contains lignins which prevent pathogens from entering the plant tissues. The application of phytoalexin and scopoletin as a wide-ranging antibiotic is another method. Scopoletin can also be an effective peroxidase substrate and thus eliminate or minimize the oxidative damage to infected plant cells as a scavenger of reactive oxygen species [31]. Phenolic compounds as coumarin, trans-cinnamic acid, p-hydroxybenzoic acid and benzoic acid, if accumulated in large concentrations can prevent germination and growth of seedlings, and found potentially phytotoxic [32]. PPs accumulation is very important in plants to counteract the harmful impacts of drought stress. Accumulation of flavonoids in cytoplasm can effectively detoxify harmful H_2O_2 molecules produced by drought stress and ascorbic acid mediated re-conversion of flavonoids to primary metabolites at the end of flavonoid oxidation [33].

In a study, enhancement in phenolic compounds like ferulic acid, benzoic acid, caffeic acid, 4-hydroxybenzoic acid, coumaric acid, gallic acid, homovanillic acid, cinnamic acid, salicylic acid and vanillic acid has been noted in *Festuca trachyphylla* plants developing under heat stress [34]. In *Glycine max* under saline conditions flavone biosynthesis was enhanced and it was found that salinity up-regulates the expression of flavone synthase genes, GmFNSII-1 and GmFNSII-2 [35]. Plants that grow under stressful conditions show potential to biosynthesize more phenolic compounds compared to plants that grow under normal conditions. These compounds have antioxidant properties and are capable of scavenging free radicals, reducing cell membrane peroxidation and thus shielding plant cells from the ill effects of oxidative stress [36].

In addition, phenolic compounds improve nutrient uptake by chelating metallic ions, improving active absorption sites and soil porosity with enhanced mobilization of elements such as magnesium (Mg), potassium (K), calcium (Ca), iron (Fe), zinc (Zn), and manganese (Mn) [37]. After shading with UV-proof film the transcript level of the FLS4 gene (related to flavonol biosynthesis) is lowered. Eventually, a recent study based on the synergistic activity between temperature and light of the anthocyanin accumulation in grape berry skin was carried out [38]. It is proposed that in carrots, phenolics such as caffeic acid, coumaric acid and anthocyanins prevent heat-induced oxidative damage by increasing their aggregation [39]. Phenolic compounds such as lignin or suberin under chilling stress start accumulating in plant cell walls which helps in enhancing resistance against chilling stress [40]. Transcriptomic and metabolomic analysis carried out on Arabidopsis plants revealed that enhanced accumulation of flavonoids under drought stress is very helpful in providing resistance [41].

RESPONSE AND ROLE OF ENDOGENOUS PHENYLPROPANOIDS IN PLANTS AGAINST ABIOTIC STRESS

Biosynthesis of secondary metabolites, including Phenylpropanoids, is usually enhanced in plants in reaction to abiotic stress. Phenylpropanoids (PPs) are among the largest community of plant-produced secondary metabolites, mainly in response to biotic or abiotic stress such as pathogens, wounding, UV irradiation, ozone exposure, contaminants, and other aggressive environmental factors. Phenylpropanoids have a huge class of secondary metabolites, such as coumarins, aurones, catechin, isoflavonoids, lignin, lignocellulose, flavonoids, and stilbenes [42]. Phenylpropanoids are present in fruits, vegetables, medicinal herbs (seeds, root, and leaves), and nuts [43].

Phenylpropanoids has antioxidants activity and are intricate in fighting to cancer too. These compounds have anti-inflammatory, antiviral and antibacterial activities with proven properties in wound healing. Due to the basic differences, the role of phenylpropanoid derivatives are quite complex. Phenolic compounds originated from phenylalanine are also called as phenylpropanoids. The temporal dynamics of phenylpropanoid pathway events in growth and the spatial spread across individual plant organs and cell types differ greatly due to the broad structural and functional variety of these compounds. This study suggested visions into the role of abiotic stress on phenylpropanoid production in plants [44].

Salinity

Anthropogenic practices have caused the increase of soil salinisation which have an effect on the sustainability of medicinal plants and the production of the bioactive compounds they generate [45]. Different environmental factors (especially drought and salinity) affect the production, growth and yield of plants. The changes can be caused by environmental components that involve local geo-climatic and seasonal shifts, environmental conditions of temperature, wind, humidity and developmental cycles, affect on biomass growth and biosynthesis of plant secondary metabolites (PSMs). Physiological, biochemical, morphological and biosynthetic parameters get highly changed by salinity in terms of profile of plant-natural products through oxidative stress and defensive reaction pathways in medicinal plants. These compounds have antioxidative properties and are capable of scavenging free radicals, resulting in reduction of cell membrane peroxidation hence protecting plant cells from ill effects of oxidative stress [46].

Biosynthesis of phenolics under stressful environments is regulated by the altered activities of various key enzymes of phenolic biosynthetic pathways like PAL and CHS (chalcone synthase). Plant development of metabolites is considered an adaptive capacity to cope with stressful constraints during difficult and evolving growth environments that can require the creation of diverse chemical forms and interactions in structural and functional stability across signaling processes and pathways. Salinity is one of the main obstacles in growing agricultural land development worldwide, and can significantly restrict crop output, especially in arid and semi-arid regions [12]. Salinity impairs seed germination, inhibits nodule production, retards the growth of plants and decreases crop yields. This can contribute to water depletion due to osmotic stress induced by salt accumulation in higher plant. Salt stress contributes to the production of ROS such as superoxide anions, hydrogen peroxide, and hydroxyl ions [47] and includes the activation of well-orchestrated and finely balanced antioxidant system in plants to contrast ROS dissemination [48].

Phenolic compounds have strong antioxidant properties and aid in scavenging of toxic ROS under salt stress [49]. In addition, phenylpropanoid biosynthetic pathways are activated in response to salt stress and result in the development of various phenolic compounds with high antioxidant potential [50]. Misra and Gupta [51], reported that *Catharanthus roseous*, in salinity stress can induce conditions of oxidative stress. Cuong *et al* [52], reported the accumulation of phenylpropanoid in NaCl stressed plants of *Triticum aestivum* L.

Light

Plants vary in their biological responses on exposure to light in the form of photoperiods or short durations associated with the development of secondary metabolites during conditions of *in vivo* and *in vitro* growth [53]. The light is often considered among restricting factors influencing plant growth and development during both *in vivo* and *in vitro* environments, and may influence the output of SMs, depending on species or genotype, stage of development, form of light and period of exposure [54]. For example, production of SMs under diverse temperatures, light intensities and phonological cycle during greenhouse growth of *H. perforatum* exhibited inconsistency for each of the specific compounds evaluated [55].

Chen *et al* [56], reported that exposure of different light intensity to *Peucedanum japonicum callus* cultures exhibited various somatic differentiations and a number of secondary metabolites including phenylpropanoid. When plants are exposed to UV-B radiations (280–320 nm) it can cause damage to DNA, lipid and proteins. It also induces the production of reactive oxygen species (ROS), which damages the plant growth. UV-B irradiation, flavonoids, as well as other phenylpropanoids are encouraged to act as UV attenuators to scavenge ROS.

Heavy Metals

Heavy metals increase the oxidative stress in plants by activating the production of destructive ROSs that eventually cause damage and reduce growth of plants [57]. An increased biosynthesis of secondary metabolites in plants in response to heavy metals supports the plants by shielding from oxidative stress [58]. Secondary metabolites can increase the metal chelation progression which aids in minimizing the presence of harmful hydroxyl radical in plant cells [59] and this showed that the levels of secondary metabolites in plants have found to be higher by heavy metal stress condition [60].

Under heavy metal stress, the generation of phenylpropanoids play key role in aiding the plant's defense mechanism [61]. Aggregation of phenolic compounds is attributable to the up-regulation of the phenylpropanoid enzyme biosynthesis, which in fact relies on the control of transcript rates of genes encoding biochemical enzymes under heavy metal stress [62].

Drought

Secondary metabolites are very useful and are reported to increase in response to various environmental stresses in plants [63]. Metabolomic analysis in *Arabidopsis* plants showed that increase in secondary metabolites under drought stress is very cooperative to provide resistance [41]. Biosynthesis and accumulation of secondary metabolites occur under water deficient conditions and also help to develop resistance against drought stress [63, 64]. Flavonoids and anthocyanins are reported to have protective function in plants plants under drought conditions [64].

Drought stress also maintain the biosynthetic pathways of secondary metabolites for enhanced generation of these compounds which act as antioxidants and prevent plants from adverse effects of water deficit conditions. For example, the increase in concentration of flavonoids like kaempferol and quercetin in tomato plants help in enhancing drought tolerance [65]. The major cause for accumulation of phenolic compounds induced by drought is the modulation of phenylpropanoid biosynthetic pathway. Drought regulates many major genes encoding key enzymes of phenylpropanoid pathway, resulting in stimulated biosynthesis of phenolic compounds [66]. Phenylpropanoids generation in cytoplasm can competently clean the damaging H_2O_2 molecules produced as a result of drought stress [33].

CONCLUSION

The stress conditions at the molecular level induce synthesis of phenylpropanoids, which play an important role in the maintenance of physiological and metabolic homeostasis. Phenylpropanoids reduce the oxidative stress of plant tissues and increase the recovery rate in the post stress events by accumulating the phenylpropanoids in leaves, flowers and roots. The molecular changes at the genomic level occurring in plants subjected to the stress conditions, could be an interesting area for further research and could be useful in the manipulation of the plants at the genetic level for their better adaptation to the stressful conditions.

CONSENT FOR PUBLICATION

Not applicable.

CONFLICT OF INTEREST

The author declares no conflict of interest, financial or otherwise.

ACKNOWLEDGEMENTS

We are thankful to DST-PURSE, Programme, Department of Science and Technology (DST), New Delhi (India) for supporting this work.

REFERENCES

[1]   Weaver LM, Herrmann KM. Dynamics of the shikimate pathway in plants. Trends Plant Sci 1997; 2(9): 346351.
[http://dx.doi.org/10.1016/S1360-1385(97)84622-5]

[2]   Dixon RA, Achnine L, Kota P, Liu CJ, Reddy MS, Wang L. The phenylpropanoid pathway and plant defence-a genomics perspective. Mol Plant Pathol 2002; 3(5): 371-90.
[http://dx.doi.org/10.1046/j.1364-3703.2002.00131.x] [PMID: 20569344]

[3]   Boyer JS. Plant productivity and environment. Science 1982; 218(4571): 443-8.
[http://dx.doi.org/10.1126/science.218.4571.443] [PMID: 17808529]

[4]   Mariani L, Ferrante A. Agronomic management for enhancing plant tolerance to abiotic stresses—drought, salinity, hypoxia and lodging. Horticulturae 2017; 3: 52.
[http://dx.doi.org/10.3390/horticulturae3040052]

[5]   Balasundram N, Sundram K, Samman S. Phenolic compounds in plants and agri-industrial by-products: antioxidant activity, occurrence, and potential uses. Food Chem 2006; 99(1): 191203.
[http://dx.doi.org/10.1016/j.foodchem.2005.07.042]

[6]   Pereira DM. Valenta~ OP, Pereira JA, Andrade PB. Phenolics: from chemistry to biology. Molecules 2009; 14: 22022211.
[http://dx.doi.org/10.3390/molecules14062202]

[7]   Grace SC, Logan BA. Energy dissipation and radical scavenging by the plant phenylpropanoid pathway. Philos Trans R Soc Lond B Biol Sci 2000; 355(1402): 1499-510.
[http://dx.doi.org/10.1098/rstb.2000.0710] [PMID: 11128003]

[8]   Fellenberg C, Vogt T. Evolutionarily conserved phenylpropanoid pattern on angiosperm pollen. Trends Plant Sci 2015; 20(4): 212-8.
[http://dx.doi.org/10.1016/j.tplants.2015.01.011] [PMID: 25739656]

[9]   Lanfranco L, Bonfante P, Genre A. The mutualistic interaction between plants and arbuscular Mycorrhizal fungi. Microbiol Spectr 2016; 4(6): 1-20.
[PMID: 28087942]

[10]   Dixon RA, Paiva NL. Stress-induced phenylpropanoid metabolism. Plant Cell 1995; 7(7): 1085-97.
[http://dx.doi.org/10.2307/3870059] [PMID: 12242399]

[11]   Clemens S, Weber M. The essential role of coumarin secretion for Fe acquisition from alkaline soil. Plant Signal Behav 2016; 11(2): e1114197.
[http://dx.doi.org/10.1080/15592324.2015.1114197] [PMID: 26618918]

[12]   Yang L, Wen KS, Ruan X, Zhao YX, Wei F, Wang Q. Response of plant secondary metabolites to environmental factors. Molecules 2018; 23(4): E762.
[http://dx.doi.org/10.3390/molecules23040762] [PMID: 29584636]

[13]   Naoumkina MA, Zhao Q, Gallego-Giraldo L, Dai X, Zhao PX, Dixon RA. Genome-wide analysis of phenylpropanoid defence pathways. Mol Plant Pathol 2010; 11(6): 829-46.
[PMID: 21029326]

[14]   Liu CW, Murray JD. The role of flavonoids in nodulation host-range specificity: an update. Plants (Basel) 2016; 5(3): E33.
[http://dx.doi.org/10.3390/plants5030033] [PMID: 27529286]

[15]   Yu S-i, Kim H, Yun D-J. Suh M C, Lee B-ha. Post-translational and transcriptional regulation of phenylpropanoid biosynthesis pathway by Kelch repeat F-box protein SAGL1. Plant Mol Biol 2018; 99: 135-48.
[http://dx.doi.org/10.1007/s11103-018-0808-8] [PMID: 30542810]

[16]   Olsen KM, Lea US, Slimestad R, Verheul M, Lillo C. Differential expression of four Arabidopsis PAL genes; PAL1 and PAL2 have functional specialization in abiotic environmental-triggered flavonoid synthesis. J Plant Physiol 2008; 165(14): 1491-9.
[http://dx.doi.org/10.1016/j.jplph.2007.11.005] [PMID: 18242769]

[17]   Abdulrazzak N, Pollet B, Ehlting J, *et al.* A coumaroyl-ester-3-hydroxylase insertion mutant reveals the existence of nonredundant meta-hydroxylation pathways and essential roles for phenolic precursors in cell expansion and plant growth. Plant Physiol 2006; 140(1): 30-48.
[http://dx.doi.org/10.1104/pp.105.069690] [PMID: 16377748]

[18]   Schoch G, Morant M, Abdulrazzak N, *et al.* The meta-hydroxylation step in the phenylpropanoid pathway: a new level of complexity in the pathway and its regulation. Environ Chem Lett 2006; 4: 127-36.
[http://dx.doi.org/10.1007/s10311-006-0062-1]

[19]   Barbehenn RV, Jones C-P, Hagerman AE, Karonen M, Salminen J-P. Ellagitannins have greater oxidative activities than condensed tannins and galloyl glucoses at high pH: potential impact on caterpillars. J Chem Ecol 2006; 32(10): 2253-67.
[http://dx.doi.org/10.1007/s10886-006-9143-7] [PMID: 17019621]

[20]   Gross GG. From lignins to tannins: forty years of enzyme studies on the biosynthesis of phenolic compounds. Phytochemistry 2008; 69(18): 3018-31.
[http://dx.doi.org/10.1016/j.phytochem.2007.04.031] [PMID: 17559893]

[21]   Veitch NC. Isoflavonoids of the leguminosae. Nat Prod Rep 2009; 26(6): 776-802.
[http://dx.doi.org/10.1039/b616809b] [PMID: 19471685]

[22]   Stevens JF, Page JE. Xanthohumol and related prenylflavonoids from hops and beer: to your good health! Phytochemistry 2004; 65(10): 1317-30.
[http://dx.doi.org/10.1016/j.phytochem.2004.04.025] [PMID: 15231405]

[23]   Korkina LG. Phenylpropanoids as naturally occurring antioxidants: from plant defense to human health. Cell Mol Biol 2007; 53(1): 15-25.
[PMID: 17519109]

[24]   Buer CS, Djordjevic MA. Architectural phenotypes in the transparent testa mutants of *Arabidopsis thaliana.* J Exp Bot 2009; 60(3): 751-63.
[http://dx.doi.org/10.1093/jxb/ern323] [PMID: 19129166]

[25]   Buer CS, Imin N, Djordjevic MA. Flavonoids: new roles for old molecules. J Integr Plant Biol 2010; 52(1): 98-111.
[http://dx.doi.org/10.1111/j.1744-7909.2010.00905.x] [PMID: 20074144]

[26]   Shi MZ, Xie DY. Biosynthesis and metabolic engineering of anthocyanins in *Arabidopsis thaliana.* Recent Pat Biotechnol 2014; 8(1): 47-60.
[http://dx.doi.org/10.2174/1872208307666131218123538] [PMID: 24354533]

[27]   Wasson AP, Pellerone FI, Mathesius U. Silencing the flavonoid pathway in *Medicago truncatula* inhibits root nodule formation and prevents auxin transport regulation by rhizobia. Plant Cell 2006; 18(7): 1617-29.
[http://dx.doi.org/10.1105/tpc.105.038232] [PMID: 16751348]

[28]   Zhang J, Subramanian S, Stacey G, Yu O. Flavones and flavonols play distinct critical roles during nodulation of *Medicago truncatula* by Sinorhizobium meliloti. Plant J 2009; 57(1): 171-83.
[http://dx.doi.org/10.1111/j.1365-313X.2008.03676.x] [PMID: 18786000]

[29]   Petrussa E, Braidot E, Zancani M, *et al.* Plant flavonoids--biosynthesis, transport and involvement in

stress responses. Int J Mol Sci 2013; 14(7): 14950-73.
[http://dx.doi.org/10.3390/ijms140714950] [PMID: 23867610]

[30]   Frohnmeyer H, Staiger D. Ultraviolet-B radiation-mediated responses in plants. Balancing damage and protection. Plant Physiol 2003; 133(4): 1420-8.
[http://dx.doi.org/10.1104/pp.103.030049] [PMID: 14681524]

[31]   Turunen M, Latola K. UV-B radiation and acclimation in timberline plants. Environ Pollut 2005; 137(3): 390-403.
[http://dx.doi.org/10.1016/j.envpol.2005.01.030] [PMID: 16005753]

[32]   Heijde M, Ulm R. UV-B photoreceptor-mediated signalling in plants. Trends Plant Sci 2012; 17(4): 230-7.
[http://dx.doi.org/10.1016/j.tplants.2012.01.007] [PMID: 22326562]

[33]   Bednarek P, Schneider B, Svatoš A, Oldham NJ, Hahlbrock K. Structural complexity, differential response to infection, and tissue specificity of indolic and phenylpropanoid secondary metabolism in Arabidopsis roots. Plant Physiol 2005; 138(2): 1058-70.
[http://dx.doi.org/10.1104/pp.104.057794] [PMID: 15923335]

[34]   Baleroni CRS, Ferrarese MLL, Souza NE, Ferrarese-Filho O. Lipid accumulation during canola seed germination in response to cinnamic acid derivatives. Biol Plantarum 2000; 43(2): 313-6.
[http://dx.doi.org/10.1023/A:1002789218415]

[35]   Hernández I, Alegre L, Van Breusegem F, Munné-Bosch S. How relevant are flavonoids as antioxidants in plants? Trends Plant Sci 2009; 14(3): 125-32.
[http://dx.doi.org/10.1016/j.tplants.2008.12.003] [PMID: 19230744]

[36]   Wang J, Yuan B, Huang B. Differential heat-induced changes in phenolic acids associated with genotypic variations in heat tolerance for hard fescue. Crop Sci 2019; 59(2): 667-74.
[http://dx.doi.org/10.2135/cropsci2018.01.0063]

[37]   Yan J, Wang B, Jiang Y, Cheng L, Wu T. GmFNSII-controlled soybean flavone metabolism responds to abiotic stresses and regulates plant salt tolerance. Plant Cell Physiol 2014; 55(1): 74-86.
[http://dx.doi.org/10.1093/pcp/pct159] [PMID: 24192294]

[38]   Schroeter H, Boyd C, Spencer JP, Williams RJ, Cadenas E, Rice-Evans C. MAPK signaling in neurodegeneration: influences of flavonoids and of nitric oxide. Neurobiol Aging 2002; 23(5): 861-80.
[http://dx.doi.org/10.1016/S0197-4580(02)00075-1] [PMID: 12392791]

[39]   Seneviratne G, Jayasinghearachchi HS. Mycelial colonization by bradyrhizobia and azorhizobia. J Biosci 2003; 28(2): 243-7.
[http://dx.doi.org/10.1007/BF02706224] [PMID: 12711817]

[40]   Azuma A, Yakushiji H, Koshita Y, Kobayashi S. Flavonoid biosynthesis-related genes in grape skin are differentially regulated by temperature and light conditions. Planta 2012; 236(4): 1067-80.
[http://dx.doi.org/10.1007/s00425-012-1650-x] [PMID: 22569920]

[41]   Commisso M, Toffali K, Strazzer P, *et al.* Impact of phenylpropanoid compounds on heat stress tolerance in carrot cell cultures. Front Plant Sci 2016; 7: 1439.
[http://dx.doi.org/10.3389/fpls.2016.01439] [PMID: 27713760]

[42]   Griffith M, Yaish MW. Antifreeze proteins in overwintering plants: a tale of two activities. Trends Plant Sci 2004; 9(8): 399-405.
[http://dx.doi.org/10.1016/j.tplants.2004.06.007] [PMID: 15358271]

[43]   Nakabayashi R, Yonekura-Sakakibara K, Urano K, *et al.* Enhancement of oxidative and drought tolerance in Arabidopsis by overaccumulation of antioxidant flavonoids. Plant J 2014; 77(3): 367-79.
[http://dx.doi.org/10.1111/tpj.12388] [PMID: 24274116]

[44]   Hemm MR, Rider SD, Ogas J, Murry DJ, Chapple C. Light induces phenylpropanoid metabolism in Arabidopsis roots. Plant J 2004; 38(5): 765-78.
[http://dx.doi.org/10.1111/j.1365-313X.2004.02089.x] [PMID: 15144378]

[45] Shahidi F, Ambigaipalan P. Phenolics and polyphenolics in foods, beverages and spices: Antioxidant activity and health effects–A review. J Funct Foods 2015; 18: 820-97.
[http://dx.doi.org/10.1016/j.jff.2015.06.018]

[46] Cuong DM, Jeon J, Morgan AMA, *et al.* Accumulation of charantin and expression of triterpenoid biosynthesis genes in bitter melon (*Momordica charantia*). J Agric Food Chem 2017; 65(33): 7240-9.
[http://dx.doi.org/10.1021/acs.jafc.7b01948] [PMID: 28737900]

[47] Seo BN, Ryu JM, Yun SP, *et al.* Delphinidin prevents hypoxia-induced mouse embryonic stem cell apoptosis through reduction of intracellular reactive oxygen species-mediated activation of JNK and NF-κB, and Akt inhibition. Apoptosis 2013; 18(7): 811-24.
[http://dx.doi.org/10.1007/s10495-013-0838-2] [PMID: 23584725]

[48] Gupta B, Huang B. Mechanism of salinity tolerance in plants: physiological, biochemical, and molecular characterization. Int J Genomics 2014; 2014: 701596.
[http://dx.doi.org/10.1155/2014/701596] [PMID: 24804192]

[49] Taïbi K, Taïbi F, Abderrahim LA, Ennajah A, Belkhodja M, Mulet JM. Effect of salt stress on growth, chlorophyll content, lipid peroxidation and antioxidant defence systems in *Phaseolus vulgaris* L. S Afr J Bot 2016; 105: 306-12.
[http://dx.doi.org/10.1016/j.sajb.2016.03.011]

[50] Martinez V, Mestre TC, Rubio F, *et al.* Accumulation of flavonols over hydroxycinnamic acids favors oxidative damage protection under abiotic stress. Front Plant Sci 2016; 7: 838.
[http://dx.doi.org/10.3389/fpls.2016.00838] [PMID: 27379130]

[51] Chen S, Wu F, Li Y, *et al.* NtMYB4 and NtCHS1 are critical factors in the regulation of flavonoid biosynthesis and are involved in salinity responsiveness. Front Plant Sci 2019; 10: 178.
[http://dx.doi.org/10.3389/fpls.2019.00178] [PMID: 30846995]

[52] Al-Ghamdi AA, Elansary HO. Synergetic effects of 5-aminolevulinic acid and *Ascophyllum nodosum* seaweed extracts on Asparagus phenolics and stress related genes under saline irrigation. Plant Physiol Biochem 2018; 129: 273-84.
[http://dx.doi.org/10.1016/j.plaphy.2018.06.008] [PMID: 29906777]

[53] Mishra MR, Srivastava RK, Akhtar N. Abiotic stresses of salinity and water to enhance alkaloids production in cell suspension culture of *Catharanthus roseus*. GJBB 2019; 9(1): 7-14.

[54] Cuong DM, Kwon SJ, Jeon J, Park YJ, Park JS, Park SU. Identification and characterization of phenylpropanoid biosynthetic genes and their accumulation in bitter melon (*Momordica charantia*). Molecules 2018; 23(2): 469.
[http://dx.doi.org/10.3390/molecules23020469] [PMID: 29466305]

[55] Ghosh S, Watson A, Gonzalez-Navarro OE, *et al.* Speed breeding in growth chambers and glasshouses for crop breeding and model plant research. Nat Protoc 2018; 13(12): 2944-63.
[http://dx.doi.org/10.1038/s41596-018-0072-z] [PMID: 30446746]

[56] Isah T, Umar S. Influencing *in vitro* clonal propagation of *Chonemorpha fragrans* (moon) Alston by culture media strength, plant growth regulators, carbon source and photo periodic incubation. J For Res 2020; 31(1): 27-43.
[http://dx.doi.org/10.1007/s11676-018-0794-3]

[57] Sanchez DH, Lippold F, Redestig H, *et al.* Integrative functional genomics of salt acclimatization in the model legume *Lotus japonicus*. Plant J 2008; 53(6): 973-87.
[http://dx.doi.org/10.1111/j.1365-313X.2007.03381.x] [PMID: 18047558]

[58] Chen CC, Agrawal DC, Lee MR, *et al.* Influence of LED light spectra on *in vitro* somatic embryogenesis and LC-MS analysis of chlorogenic acid and rutin in *Peucedanum japonicum Thunb.*: a medicinal herb. Bot Stud (Taipei, Taiwan) 2016; 57(1): 9.
[http://dx.doi.org/10.1186/s40529-016-0124-z] [PMID: 28597418]

[59] Williams RJ, Spencer JP, Rice-Evans C. Flavonoids: antioxidants or signalling molecules? Free Radic

Biol Med 2004; 36(7): 838-49.
[http://dx.doi.org/10.1016/j.freeradbiomed.2004.01.001] [PMID: 15019969]

[60]   Kısa D, Elmastaş M, Öztürk L, Kayır Ö. Responses of the phenolic compounds of *Zea mays* under heavy metal stress. Appl Biol Chem 2016; 59(6): 813-20.
[http://dx.doi.org/10.1007/s13765-016-0229-9]

[61]   Leng X, Jia H, Sun X, *et al.* Comparative transcriptome analysis of grapevine in response to copper stress. Sci Rep 2015; 5: 17749.
[http://dx.doi.org/10.1038/srep17749] [PMID: 26673527]

[62]   Naikoo MI, Dar MI, Raghib F. Role and regulation of plants phenolics in abiotic stress tolerance: an overview. Plant Signaling Molecules 2019; 157-68.
[http://dx.doi.org/10.1016/B978-0-12-816451-8.00009-5]

[63]   Ballizany WL, Hofmann RW, Jahufer MZZ, Barrett BA. Multivariate associations of flavonoid and biomass accumulation in white clover (*Trifolium repens*) under drought. Funct Plant Biol 2012; 39(2): 167-77.
[http://dx.doi.org/10.1071/FP11193] [PMID: 32480771]

[64]   Nichols SN, Hofmann RW, Williams WM. Physiological drought resistance and accumulation of leaf phenolics in white clover interspecific hybrids. Environ Exp Bot 2015; 119: 40-7.
[http://dx.doi.org/10.1016/j.envexpbot.2015.05.014]

[65]   Kapoor D, Bhardwaj S, Landi M, Sharma A, Ramakrishnan M, Sharma A. The impact of drought in plant metabolism: how to exploit tolerance mechanisms to increase crop production. Appl Sci (Basel) 2020; 10(16): 5692.
[http://dx.doi.org/10.3390/app10165692]

[66]   Akula R, Ravishankar GA. Influence of abiotic stress signals on secondary metabolites in plants. Plant signaling & behavior 6(11): 1720-31.
[http://dx.doi.org/10.4161/psb.6.11.17613]

CHAPTER 10

Role of Plant Growth Regulators in Abiotic Stress Tolerance

Sakshi Sharma[1], Inderpreet Kaur[2,*] and **Avinash Kaur Nagpal[3,*]**

[1] *Department of Botany, DAV College, Amritsar, Punjab 143005, India*

[2] *Department of Chemistry, Centre for Advanced Studies, Guru Nanak Dev University, Amritsar, Punjab 143005, India*

[3] *Department of Botanical and Environmental Sciences, Guru Nanak Dev University, Amritsar, Punjab 143005, India*

Abstract: Plants are frequently exposed to different types of stressful environmental conditions, which have adverse effects on their growth, development, and productivity. These conditions, such as salinity, drought, floods, chilling, freezing, UV exposure, pollution, nutritional deficiencies, metal toxicity, *etc.*, are collectively known as abiotic stressors and hinder plants from fully expressing their genetic potential. With advancements in scientific fields such as genetics and molecular biology, it has become easier to understand that under abiotic stress, a myriad of responses are triggered in plants. These changes include alterations in gene expression to changes in cell metabolism to avoid or tolerate the stress. The intensity of these plant responses depends on affected tissue, age of the plant, type of stress posed, duration and severity of stress, *etc.* It has been observed that plant growth regulators such as auxins, abscisic acid, cytokinins, ethylene, gibberellins, jasmonic acid, brassinosteroids, salicylic acid, polyamines, strigolactones, *etc.*, which influence the growth and differentiation in plants, also have very important roles in regulating the stress tolerance in plants. This chapter is a comprehensive account of literature based on the role of different plant growth regulators in the regulation of tolerance of plants towards abiotic stressors. The contents of this chapter include a brief discussion about different types of abiotic stressors, their effects on plants, and responses developed in plants against them. There is also a detailed discussion about plant growth regulators, their role in the normal functioning of plants, followed by their contribution and underlying mechanisms in building abiotic stress tolerance in plants.

Keywords: Abscisic acid, Auxins, Brassinosteroids, Cytokinins, Ethylene, Gibberellins, Jasmonic acid, Polyamines, Salicylic acid, Strigolactones.

---

* **Corresponding Author Avinash Kaur Nagpal & Inderpreet Kaur:** Department of Botanical and Environmental Sciences, Guru Nanak Dev University, Amritsar, Punjab 143005, India; Tel: +91-94174-26060, 2258802-09 Ext. 3423 (O); Fax: 0183-2258819, 20; E-mail:avnagpal@yahoo.co.in and Department of Chemistry, Centre for Advanced Studies, Guru Nanak Dev University, Amritsar, Punjab 143005, India; Tel: +91-8427662766, +91-183-2258802 to 9, Extn.-3285(O); E-mail:inderpreet11@yahoo.co.in

Tajinder Kaur & Saroj Arora (Eds.)

INTRODUCTION

Abiotic stress in plants is a cumulative adverse effect of non-living environmental factors on the plants affecting their physiology, growth, development, and yield [1]. Plants are immobile in nature and highly susceptible to ever-changing environmental and climatic conditions, including abiotic stress. The latter results in the malfunctioning of different physiological, biochemical, and molecular mechanisms, ultimately decreasing the plant yield [2]. Demand for agricultural products has increased enormously with the increasing human population. However, the impact of different abiotic stressors has posed a grave danger to plant growth, fertility, and agricultural productivity [3].

Abiotic stress on plants is posed by environmental conditions (physical or chemical) such as drought, water logging, UV exposure, excessive heat, chilling, frosting, soil salinity, mineral toxicity, pollution, and nutritional deficiencies (Fig. 1) [1, 4, 5]. Plants have developed various methods to adapt to these abiotic stressors, in turn causing a reduction in biomass and yield [3]. It is extremely important to manage abiotic stress on plants as it limits the agricultural yield affecting the economy of countries, and the livelihood of the farmers is dependent on these crops [5]. To improve the yield and sustainability of crops, it is necessary to understand the factors and mechanisms behind plant responses to different abiotic stressors [3]. The severity of plant responses is dependant on different factors such as affected part or tissue of the plant, growth stage, and conditions of plant, species or ecotypes of plant, type and intensity of abiotic stress, duration of exposure to the stress, *etc* [6, 7].

Plants are able to perceive even minor unfavourable alterations in environmental conditions and start a myriad of stress responses for coping with the situation and establishing homeostasis [8, 9]. Plant growth regulators, which play a vital role in the regulation of the growth and development of plants, are instrumental in tolerating and combating abiotic stress in plants [10]. Different plant growth regulators, such as abscisic acid, cytokinins, auxins, gibberellins, ethylene, jasmonic acid, salicylic acid, brassinosteroids, polyamines, strigolactones, *etc.* have an extensive role in developing and maintaining abiotic stress tolerance in plants [11 - 13]. Considering these facts, this chapter has been written as an attempt to compile information on adverse effects of different types of abiotic stressors on plants, responses of plants to these stressors, the role of different plant growth regulators in stress tolerance and crop improvement, regulation of tolerance of plants towards abiotic stressors.

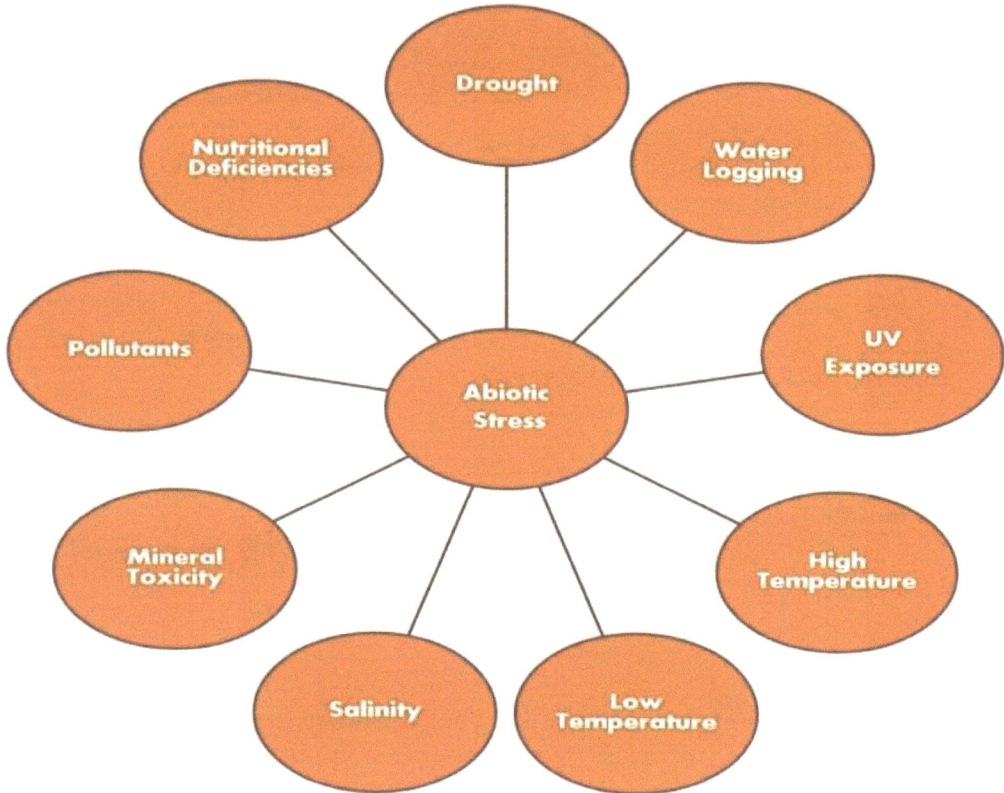

Fig. (1). Different types of abiotic stressors posed to plants [Based on: 1, 4, 5].

ABIOTIC STRESS AND PLANT RESPONSES

Abiotic stress is a major factor responsible for hindrance in the full expression of the genetic potential of plants. It triggers a variety of responses in plants, including changes in cellular metabolism, signal transduction, gene expression, plant growth rate, and crop yield [14]. According to Onaga and Wydra, under abiotic stress, membrane receptors in plants perceive different initial stress signals such as changes in membrane fluidity and osmotic effects. Then, signal transduction triggers transcription, which is controlled by different hormones, miRNAs, transcription factor binding proteins, and transcription factors. It activates defense response in plant cells to fix damaged membranes and proteins and to reinstate ion homeostasis in cells (Fig **2**) [15]. This defense response includes various changes such as:

- Increased production of lignin and callose for hardening of cell walls in order to increase the mechanical strength of plants.
- Reduced secondary metabolic activities to save energy.
- Induction of changes in proteins involved in growth signalling.
- Elevated transmembrane activity for maintenance of osmotic and ion homeostasis.
- Improved primary metabolism to meet energy requirements of cell.
- Prevention of oxidative damage by increased scavenging of reactive oxygen species (ROS).
- Increased activity of chaperon proteins such as heat shock proteins (HSPs) and late embryogenesis abundant (LEA) proteins.
- Increased cell cycle activity for maintenance of cell size and repair of cell damage.

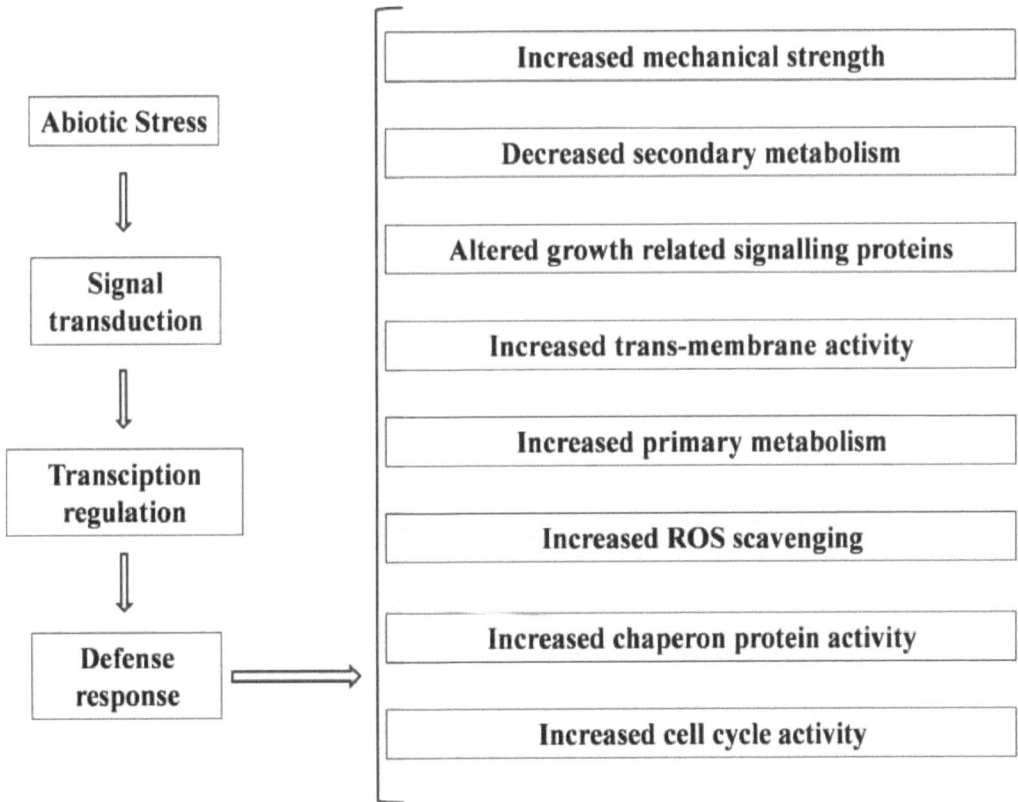

Abiotic Stress		Increased mechanical strength
⇓		Decreased secondary metabolism
Signal transduction		Altered growth related signalling proteins
⇑		Increased trans-membrane activity
Transciption regulation		Increased primary metabolism
⇓		Increased ROS scavenging
Defense response	⇒	Increased chaperon protein activity
		Increased cell cycle activity

Fig. (2). Plant responses to abiotic stress posed to plants [Based on: 15].

Abiotic stressors had caused major harm to crops and reduced the overall worldwide yield of cereal crops by 9 to 10% during 1964 to 2007 [2, 16].

Excessive abiotic stress reduces crop yield by affecting morphology, physiology, vegetative growth, formation and functioning of reproductive organs, photosynthesis and seed germination, along with increase in denaturation of proteins, alteration in membrane fluidity, generation of reactive oxygen species (ROS) and inactivation of enzymatic activities in chloroplasts and mitochondria and reduction in electron transport [17 - 20]. According to Nadeem *et al.*, various physiological mechanisms in plants help in avoiding the membrane damage and maintain transpiration, photosynthesis and respiration [20]. Nitric oxide (NO) and calcium (Ca^{2+}) help in controlling stress signalling in plants, modulating the expression of genes, regulation of biochemical reactions, *etc* [21]. According to He *et al.*, plants have a variety of defences against abiotic stressors including formation of cuticle, unsaturated fatty acids, reactive oxygen species, heat shock proteins, plant growth regulators, gasotransmitters, calcium, *etc*, making plants capable to strive in unfavourable conditions [22]. The role of plant growth regulators in combating abiotic stressors is discussed in detail in the following section.

PLANT GROWTH REGULATORS AND THEIR ROLE IN ABIOTIC STRESS TOLERANCE

Plant growth regulators (PGRs) influence the physiological processes in plants even when present in minute quantities [23]. Abiotic stressors such as, excessive water, drought, UV exposure, nutritional deficiencies, chilling, frosting, excessive heat, soil salinity and pollution, initialise different complex and explicit responses in plants to deal with stress and increase the possibilities of survival under unfavourable conditions. PGRs, play a major role in modulation of different plant responses to abiotic stressors [24, 25]. PGRs, alter different physiological, cellular and molecular mechanisms in plants in response to abiotic stressors [26]. Various metabolic adaptations such as accumulation of different organic solutes and activation of free-radical scavengers also take place under stress [26, 27]. The roles of plant growth regulators such as, auxins, abscisic acid, cytokinins, ethylene, gibberellins, jasmonic acid, brassinosteroids, salicylic acid, polyamines *etc*, in plant growth, development and abiotic stress tolerance are discussed below.

Auxins

Auxins are first discovered plant hormones [28]. There are different types of active auxins such as, indole-3-acetic acid (IAA), 4-chloroindole-3-acetic acid (4-Cl-IAA), phenylacetic acid (PAA); natural inactive precursors of auxins such as, indole-3-pyruvic acid (IPyA), indole-3-acetaldehyde (IAAld), indole--acetaldoxime (IAOx), indoleacetamine (IAM) and indole-3-acetonitrile (IAN); natural storage forms of auxins such as, indole-3-butyric acid (IBA) and methyl-

IAA (MeIAA); and auxins attached to sugars or amino acids [29]. Structure of IAAand IBA are presented in Fig. **(3)**. Auxins are essential for the inititation of DNA synthesis and also induce qualitative changes in proteins and RNA molecules as a response to wounding without affecting their rates of synthesis [30]. Auxins play very important role in plant development such as, formation of auxillary buds, differentiation of vascular tissues, root growth, apical dominance, and development of various parts of flowers [31]. Auxins also help in tolerance against abiotic stressors [32].

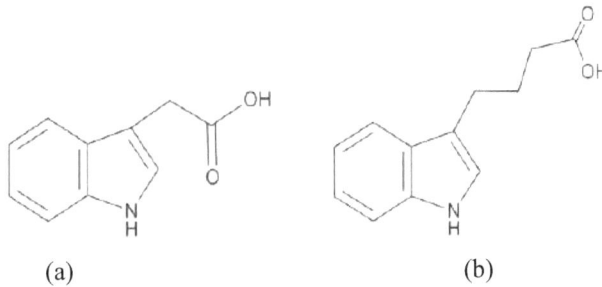

(a)   (b)

Fig. (3). Chemical structures of auxins **(a)** indole-3-acetic acid (IAA) and **(b)** indole-3-butyric acid (IBA).

Auxins are produced in the shoot apical meristems, flower buds and young leaves, and then transported throughout the plant *via* phloem tissue [33]. According to Zhao, auxin (IAA) synthesis in plants occur in two distinct steps, which are highly conserved throughout the plant kingdom. First step involves removal of an amino group from amino acid "Tryptophan" by the transaminases of the family of "Tryptophan aminotransferase of the *Arabidopsis*" (TAA) in order to produce indole-3-pyruvate (IPyA). In the second step, oxidative decarboxylation of IPyA takes place to convert it into Indole-3-acetic acid. This process is catalyzed by the YUCCA (YUC) family of flavin monooxygenases [34]. A variety of factors such as F-box auxin receptor (TIR1/AFB) (nuclear factors), ARFs (transcriptional activators) and Aux/IAAs (repressors) have been reported to recognize the auxins and regulate signalling pathways; and high affinity binding of receptors to auxin requires a co-receptor complex of F- box protein and an Aux/IAA [35].

Under stress conditions IAA levels get modified in the plant parts *via* different mechanisms such as, by modifications in expression of gene responsible for polar transport of auxins or by restricted auxin polar transport due to presence of some compounds accumulated in plants due to stress conditions [36]. Auxins are instrumental in regulating the plasticity of roots in plants during salt or water stressors, hence making plants able to combat abiotic stressors [37]. High IAA content in reproductive organs of capsicum plants helps in lowering of abscission and induces higher tolerance towards temperature stress [38].

Abscisic Acid

Abscisic acid (ABA) is a sesquiterpene phytohormone, which inhibits plant growth and metabolism along with initiation of destructive changes such as abscission, ripening and senescence [39, 40] (Fig. **4**). It is produced primarily in vascular tissues and transported to other parts (shoots and roots) of plant *via* both phloem and xylem [41]. Upreti and Sharma explained that ABA is produced in cytosol *via* carotenoid biosynthetic pathway [38]. Family of receptors such as PYR1 (Pyrabactin Resistance 1), PYL1-PYL13 (PYR1-like 1–13) and RCAR1-RCAR14 (Regulatory Component of ABA Receptor) is instrumental in perception and signal transduction of ABA [42]. Signal transduction pathway of ABA consists of a cascade of signals including different components such as ABA receptors and enzymes such as PP2Cs (type 2C protein phosphatases) and SnRK2s (Snf1-related protein kinases 2) for regulation of abiotic stress resistance, senescence, stomatal movements, development of roots and seed germination, stomatal closure [43]. According to Minocha and Dibona, ABA is capable of inhibiting nucleic acid synthesis in plants [30]. Abscisic acid helps in maintaining stomatal movements, hydraulic conductivity in roots, photosynthesis, osmolyte production, production of stress responsive genes and proteins, *etc* [24].

ABA has a very important role as a chemical signal during plant response to salinity [44]. ABA acts as a messenger, which is produced endogenously, for regulation of water content in plant, therefore plays an important role in plant growth and defense of plant during water stress [45]. According to Abobatta, under drought stress, in order to reduce water loss, plants synthesise higher amounts of abscisic acid (ABA) leading to stomatal closure and hence reduction in transpiration [46]. ABA irreversibly degrades into its hydroxylated products such as, dihydrophaseic acid (DPA) and phaseic acid (PA) and contents of DPA and PA increase with ABA [47]. Further, it was stated that even after ABA attains a plateau under stress, contents of DPA and PA still keep increasing. Moreover, when stress conditions are no more present, ABA content comes back to pre-stressed levels, however, DPA and PA contents increase or remain more or less unaltered [24, 47]. Abscisic acid along with other plant hormones has been found to control expression of various proteins such as, late embryogenesis abundant proteins and dehydrins, under stress conditions [48]. External application of ABA also increases the capacity of plants to deal with stress conditions [49, 50].

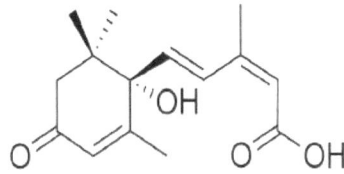

Fig. (4). Chemical structure of abscisic acid (ABA).

Cytokinins

Cytokinins are plant hormones, which are synthesized primarily in roots besides shoot apex and other tissues in plants. They have a very important role in regulating growth of plants, morphogenesis, seed germination, nutrient uptake, sink/source relationships, biogenesis of chloroplasts, abscission of fruits and leaves, apical dominance, stomatal movements and communication from roots to shoots in plants [51, 52]. Cytokinins are natural N^6-substituted adenine derivatives [53]. There are different forms of cytokinins such as, kinetin, *trans*-zeatin, *cis*-zeatin, *etc.* as represented in [51] (Fig. **5**).

According to Upreti and Sharma, cytokinins are biosynthesized in plants by two different pathways, *i.e.*, t-RNA pathway and *de novo* biosynthetic pathway, latter being the pathway responsible for most of the biologically active cytokinins [24]. In the t-RNA pathway, degradation of t-RNA occurs and its isomer *"cis-zeatin"* is formed in the presence of enzyme *"cis-trans* isomerase" [24]. One of the major steps in the cytokinin formation *via de novo* biosynthestic pathway is the synthesis of N^6-(Δ^2-isopentenyl) adenosine-5'-mono phosphate from Δ^2-isopentenyl pyrophosphate and adenosine-5'-monophosphate *via* catalytic reactions aided by isopentenyl transferase (ipt) [54]. Rao *et al.* explained that regulation of cytokinins is done by converting them to derivatives to ribosides and N- and O-linked glycosides, which further release free cytokinins in the presence of enzyme β-glucosidase during different developmental stages in plants [52]. When not required in plants, cytokinins are irreversibly rendered inactive by the action of enzyme cytokinin oxidase. According to Auyoma and Oka (2003), in *Arabidopsis*, histidine kinases (CRE1, AHK2 and AHK3) are hybrid-type sensors which act as receptors and sense different cytokinins and lead to phosphorylation of conserved residues of histidine [55]. Then these sensor histidine kinases transport the phosphoryl group *via* the C-terminal receiver domains of the sensors to the *Arabidopsis* histidine-containing phosphotransfer (HPt) factors, AHPs [55]. Further, phosphoryl group is shifted to the nucleus by the AHPs and transferred to transcription-factor-type response regulators (ARR1), which helps in transactivation of various cytokinin-responsive genes [55].

Cytokinin concentration decreases in plants under stress due to reduction in cytokinin biosynthesis and/ or increased cytokinin degradation, however, their content gets restored under normal conditions [24, 47, 56]. Cytokinin content in xylem sap decreases under drought stress [57]. Exogenous cytokinin application is capable to reverse the abscission in leaves and fruits induced by water stress or abscisic acid [44].

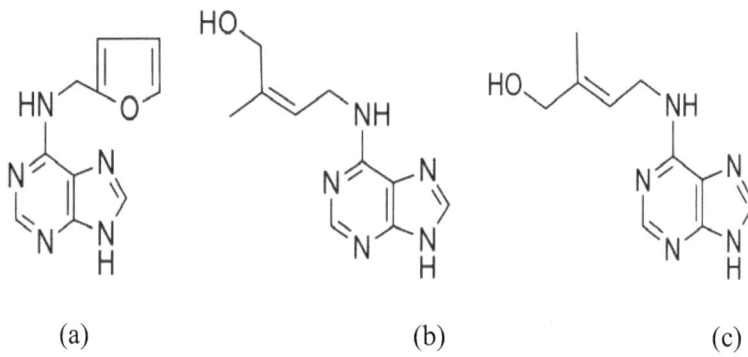

(a)   (b)   (c)

Fig. (5). Chemical structures of cytokinins (a) kinetin, (b) *cis*-zeatin and (c) *trans*-zeatin.

Ethylene

Ethylene (Fig. **6**) is an active biological gas, synthesized both enzymatically and photochemically, in all parts of the plants depending on the type and age of the plant tissue [58]. Ethylene is biosynthesized from methionine (amino acid) *via* different intermediates such as, S-adenosyl methionine (SAM) and 1-amino cyclopropane-1-carboxylic acid (ACC) where enzymes ACC synthase and ACC oxidase are involved [59]. Phytohormone ethylene plays a major role in growth and senescence of different plant parts such as, leaves, flowers and fruits [60]. Abiotic stressors in plants such as, chilling, drought and exposure to ozone cause enhanced production of ethylene in comparison to normal conditions [61]. Under stress conditions, various plant responses are produced due to signals processed by different ethylene receptors, which are mainly restricted to endoplasmic reticulum [62].

Hall *et al.* explained that in Arabidopsis, the family of ethylene receptors has five types of receptors with almost similar modular structure, where, transmembrane domains consist of the binding site for ethylene molecules near the N-terminus

[63]. This binding site is further followed by a GAF domain with unidentified function [63]. In C-terminal half, signal output domains are present [63]. Based on phylogenetic studies and similar structural properties, ethylene receptors can be categorized in two subfamilies: subfamily 1 (receptors ETR1 and ERS1) and subfamily 2 (receptors ETR2, ERS2, and EIN4) [63]. They also reported that receptors from the subfamily-1 transmit ethylene signals more efficiently than receptors from the subfamily-2 [63]. Raf-like kinase CTR1, a negative regulator of ethylene signalling functions as an integral part of the ethylene receptor signalling complex [64]. Ethylene helps in coping water stress by reducing water loss by increasing senescence and reducing growth, salt stress by enhancement of Na/K homeostasis, temperature stress by increasing abscission, *etc* [24].

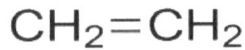

$$CH_2=CH_2$$

Fig. (6). Chemical structure of ethylene.

Gibberellins

Gibberellins (GAs) are a group of diterpenoid carboxylic acids (occur in plants, bacterial and fungi) act as endogenous growth regulators [65]. Gibberellins have important contribution in maintaining plant growth regulation of many developmental processes in plants such as, stem elongation, leaf expansion, leaf and fruit senescence, flowering induction, dormancy, seed germination, *etc* [66]. Gao *et al.*, explained that out of 136 fully characterized gibberellins (GA_1 to GA_{136}), most are inactive biologically with GA_1, GA_3, GA_4, and GA_7 being biologically active, where, GA_1 and GA_4 are most active in maximum plant species [67]. Structure of gibberellic acid (GA_3) has been presented in Fig. (7). Most of the biologically inactive gibberellins act as precursors and intermediates during the biosynthesis/ degradation of bioactive gibberellins [67].

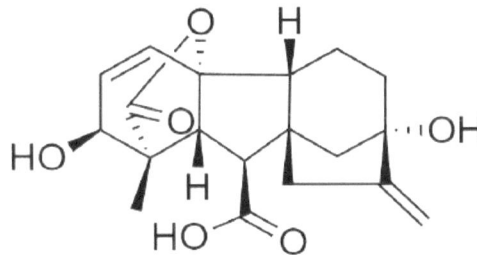

Fig. (7). Chemical structure of gibberellic acid.

Process of biosynthesis of bioactive gibberellins is a complex mechanism and a number of intermediates are produced, rendering it difficult to identify the exact site of GAs synthesis [68]. According to Upreti and Sharma, GAs are produced in plants *via* the methyl erythritol phosphate pathway from trans-geranylgeranyl diphosphate, formed in plastids in the presence of terpene cyclases [24]. This step is followed by oxidation by cytochrome P450 monooxygenases and soluble 2-oxoglutarate-dependent dioxygenases such as, GA_2-oxidase (GA2ox), GA_{20}-oxidase (GA20ox) and GA_3-oxidase (GA3ox) [24]. These dioxygenases aid metabolism of gibberellins in plants [24]. GA2ox genes are highly active during plant development and under abiotic stress [52]. In case of *Arabidopsis*, in GA response pathway, bioactive GA attaches to a receptor such as, GoGID1 [69]. According to Olszewski *et al.* (2002), in plants, binding of GA and its receptor, activates the G proteins (D1) and secondary messengers for nuclear localization gene PHOR1 and rapid destruction of proteins such as RGA/GAI [70]. Moreover, inside the nucleus, genes like KGM and SPY which alter different components of the response pathway to repress the responses of GlcNAc to GA. Later various transcription regulators modify the gene expression and cause increase/decrease in GA response [70].

Zhong *et al.* described that reduction of gibberellic acid or decreased sensitivity of plant towards gibberellic acid led to increase in drought resistance in plants [71]. Similarly, Plaza-Wüthrich *et al.* stated that inhibition of gibberellic acid with inhibitors such as, paclobutrazol (PBZ) enhanced the drought tolerance in plants [72]. Application of gibberellins on plants enhances salt tolerance in plants by inducing ion homeostasis and increasing nutrient uptake [73, 74].

Jasmonic Acid

Methyl jasmonate and jasmonic acid (free-acid of methyl jasmonate), collectively known as jasmonates, are important for regulating development in plants *e.g.*, seed germination, formation of gum and bulb, fruit ripening, senescence, primary root growth, flowering, senescence, fertility, *etc.*; and activation of plant defence system under environmental stressors such as, low temperature, drought and salinity [44, 45, 50, 75]. Jasmonic acid (Fig. **8**) is biosynthesized in leaves and roots of plants, with chloroplasts and peroxisomes being primary cellular organelles involved in synthesis [44, 76]. In LOX pathway for jasmonic acid synthesis, various enzymes participate in a series of reactions such as 13-lipoxygenase catalyzes oxygenation of α-linolenic acid, allene oxide synthase catalyses the metabolism of 13-hydroperoxy linolenic acid, allene oxide cyclase helps in production of 12-oxophytodienoic acid and a reductase alongwith three β-oxidation steps produce jasmonic acid [77]. According to Ruan *et al.* (2019), JA is converted to biologically active JA-Ile (JA, its methyl ester and isoleucine conjugate) by JAR1 (jasmonate resistant 1) [78]. JA-Ile is recognized by its COI1 (receptor coronatine insensitive1) leading to degradation of JAZ (jasmonate ZIM-domain) repressors, subsequent production of transcription factors and regulation of genes which are responsive to JAs in various processes in plant cells [78].

Jasmonic acid controls the plant response to water stress by regulating the expression of genes responsible for combating the stress, include genes encoding the vegetative storage proteins (VSP) [79]. Under stress, jasmonic acid accumulation in plants enhances levels of VSP mRNA in injured plant tissues [80]. Kang *et al.,* indicated that high concentration of jasmonates in salt tolerant plants in comparison to salt sensitive plants indicated the role of jasmonic acid in enhancing the capacity of plants to survive in saline soils and exogenous application of jasmonic acid can ameliorate the salt-sensitive rice seedlings under salt stress [81]. It may be due to alteration in balance of endogenous hormones in plants due to exogenous application of jasmonic acid [44]. Initiation of three genes responsive to jasmonic acid, *i.e.,* ribulose 1, 5-bisphosphate carboxylase/ oxygenase (Rubisco) activase, apoplastic invertase and arginine decarboxylase is important for plant response under salt stress [82].

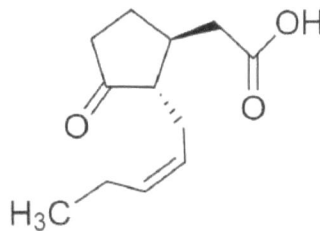

Fig. (8). Chemical structure of jasmonic acid.

Brassinosteroids

Brassinosteroids are polyhydroxylated steroidal phytohormones which have structural relationship with animal steroid hormones and have discrete properties to promote growth in plants [83]. Brassinosteroids (BRs) inherit pleiotropic effects influencing developmental and physiological mechanisms such as, cell division and elongation, rhizogenesis, photomorphogenesis, abscission, senescence, xylem differentiation, seed germination, and fruit ripening [52]. An active by-product of brassinolide biosynthesis called 24-epibrassnolide (EBL) is capable to stimulate various plant metabolic processes, *e.g.,* photosynthesis [84] (Fig. **9**). Content of BRs differs in different parts of the plants with highest in pollen, immature seeds, flowers and roots, however, lower amounts in leaves and shoots [83]. Upreti and Sharma explained that biosynthesis of brassinosteroids involves two steps [24]. First is a sterol-specific pathway converting squalene to campesterol and second step is a brassinosteroid specific pathway converting campesterol to brassinosteroids [24]. When BR couples with extracellular domain of BRI1 (a receptor kinase) on cell surface, BRI1 gets activated and forms a dimer with BAK1 (another receptor kinase) and these activated kinases control BIN2 kinase and BSU1 phosphatase, which regulate phosphorylation and nuclear accumulation of BZR1 and BES1 (transcription factors) [86]. These transcription factors regulate the expression of BR responsive genes by binding to them [87].

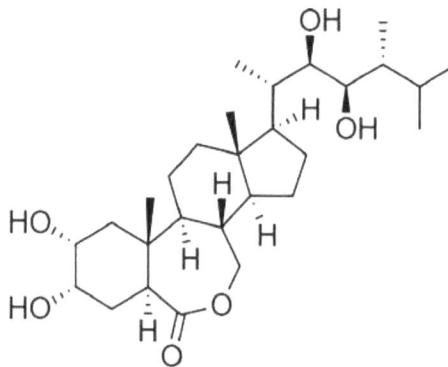

Fig. (9). Chemical structure of 24-epibrassnolide (EBL).

BRs have important role in enabling plants to tolerate different abiotic stressors, *e.g.*, salinity, drought and extreme temperatures [85]. Brassinosteroids control the plant responses to stressors through a sequence of biochemical reactions involving protein synthesis, initiation or termination of various enzymatic reactions and synthesis of defense compounds [88]. Brassinosteroids are capable of alleviating the inhibition of seed germination and seed growth induced due to salinity [50]. Increased levels of nucleic acids and soluble proteins in plants under salt stress become instrumental in raising the contents of brassinosteroids in plants [89]. Application of a brassinosteroid "Epibrassinolide" improves drought tolerance in seedlings by reducing the reactive oxygen species and increasing antioxidative enzyme activities producing antioxidant contents [90].

Salicylic Acid

Salicylic acid (SA) is a phenolic hormone in plants, which plays a major role as a growth regulator [91] (Fig. **10**). Exogenous application of SA has effect on seed germination, glycolysis, membrane permeability, stomatal movement, uptake and transport of ions, flowering, rate of photosynthesis, thermo-tolerance, rate of growth, fruit yield, senescence, transpiration, nodulation, *etc*., in plants [44, 75]. According to Chen *et al*. (2009), one of the two proposed pathways for SA biosynthesis suggest that SA in plants is synthesized from cinnamate formed by action of the enzyme phenylalanine ammonia lyase (PAL), whereas, other pathway as indicated by the genetic studies suggests that SA is synthesized from isochorismate [92]. There are two alternative mechanisms proposed for SA perception in plants, where according to the first model, SA is perceived by homologs of NPR_1 (Non-expressor of pathogenesis-related 1) *i.e.*, NPR_3 and NPR_4 and hence regulate the NPR_1 protein accumulation; whereas, according to the second model it is NPR_1 which perceives SA leading to alteration in conformational of NPR_1 and activation of SA-mediated transcription [93].

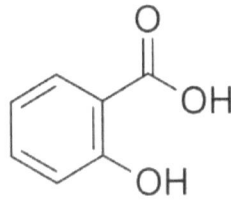

Fig. (10). Chemical structure of salicylic acid.

Salicylic acid plays an important role in increasing the tolerance in plants under stress conditions [94]. According to Jumali *et al.,* most of the genes in plants which respond to treatment with salicylic acid are active in different stress pathways and signalling mechanisms *e.g.*, genes which encode antioxidants, heat shock proteins, generation of secondary metabolites, *etc* [95]. Salicylic acid can protect against abiotic stressors posed by extreme temperature, drought and heavy metals [44]. In a study conducted by Nazar *et al.,* it was observed that treatment of plants with salicylic acid can alleviate the reduction in rate of photosynthesis due to salt stress by initiating the activity of enzymes involved in ascorbate-glutathione pathway and increased glutathione production [96]. Similarly, Khodary reported that treatment of maize plants with salicylic acid stimulated their salt tolerance by increase in their photosynthetic rate and carbohydrate metabolism [97].

Polyamines

Polyamines (PAs) are aliphatic nitrogenous bases with two or more attached amino groups and have strong biological activity [98]. Polyamines such as, putrescine, spermidine and spermine are involved in maintenance of different physiological processes, *e.g.* cell growth and development, embryogenesis, organogenesis, flowering, floral and fruit development, fruit ripening, root development and senescence [24]. According to Reddy, putrescine is the main product of the polyamine biosynthesis and acts as a precursor of spermidine and spermine [99] (Fig. **11**). Putrescine is synthesized in plants through three different pathways. First pathway is based on the removal of carbon number 8 from an arginine molecule to produce agmatine and carbon dioxide, in the presence of enzyme arginine decarboxylase [98, 100]. Then removal of number 2 nitrogen atom from agmatine leads to formation of N-carbamoylputrescine and ammonia [98]. Further, due to hydrolysation of N-carbamoylputrescine its carbamoyl group gets removed in the presence of enzyme N-carbamoylputrescine amidohydrolase to produce putrescine along with carbon dioxide and ammonia [98]. In the second pathway, arginine is converted to ornithine by enzyme arginase, then carboxyl group of first carbon atom of ornithine in the presence of enzyme ornithine decarboxylase produces putrescine and carbon dioxide [98, 101]. In the third pathway, arginine converts into citrulline and then decarboxylated to form putrescine by enzyme citrulline decarboxylase [98]. Polyamines activate kinases such as tyrosine kinases, causing signal transduction from the cell membrane to nucleus leading to activation of *ras,* MEKs and ERKs genes and at the end of process, various transcription factors and nuclear proto-oncogenes are expressed [102].

Polyamines play a vital role in inducing abiotic stress tolerance in plants [103]. Exogenous application of polyamines helps in elevating tolerance against the abiotic stressors such as, drought, flooding, low temperatures, excessive heat and salinity in plants [104]. Increase in concentration of polyamines under abiotic stress enhances stress tolerance in plants by stabilizing the structure and functioning of membranes, shifting hormonal balance and activating antioxidant enzymes [104].

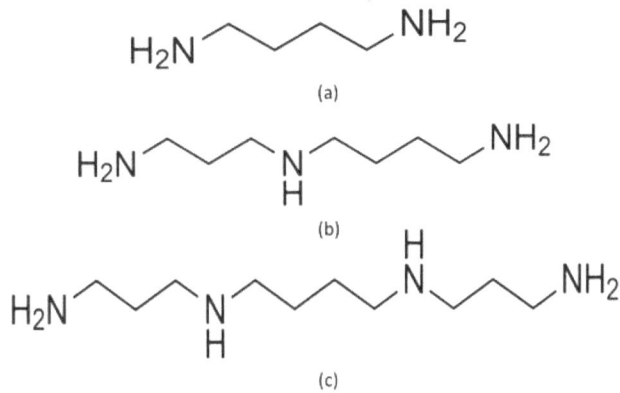

Fig. (11). Chemical structures of polyamines **(a)** putrescine, **(b)** spermidine and **(c)** spermine.

Strigolactones

Strigolactones (SLs) are a group of phytohormones derived from carotenoids and has a major role in plant growth and development [105]. These phytohormones are exuded by roots of plants in to the rhizosphere to strengthen the symbiotic relationships with the arbuscular mycorrhizal fungi in the soil [106]. Strigolactons play an endogenous role in signalling pathway for suppression of shoot branching [107]. Strigolactones also act as germination stimulants for various parasitic plant species such as witchweeds and broomrapes [108, 109]. Structure of strigolactones such as, strigol and sorgolactone, consists of a tricyclic lactone with three rings attached to a fourth ring *i.e.,* butenolide moiety *via* an enol-ether bond [105, 110] as represented in Fig. (**12**). According to Wang and Bouwmeester (2018), strigolactones, on the basis of difference in stereochemistry of third ring of tricyclic lactone, are of two canonical SL types *i.e.,* the strigol-type and orobanchol-type. The strigol-type SLs are derived from 5-deoxystrigol (5-DS), whereas, orobanchol-type SLs are derived from 4-deoxyorobanchol. On the other hand, third type which do not have a canonical part are called as non-canonical strigolactones.

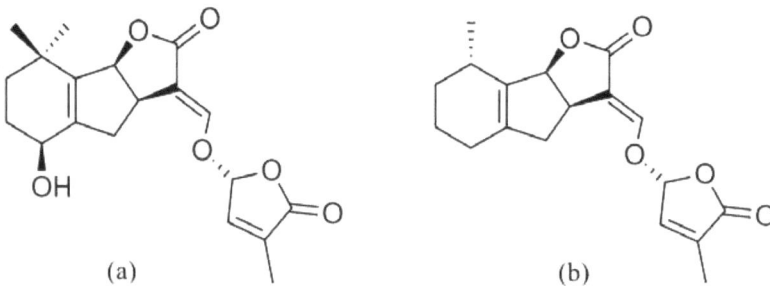

Fig. (12). Chemical structures of strigolactones **(a)** strigol and **(b)** sorgolactone.

Ruyter-Spira *et al.* (2013) explained that during strigolactone biosynthesis, tricyclic lactone and butenolide moieties are synthesized separately and then later attached to each other [108]. During biosynthesis of a strigolactone from a carotenoid such as an all trans-configured-β-carotene gets converted to a common intermediate in strigolactone biosynthesis *i.e.*, carlactone, under the effect of various enzymes such as carotenoid cleavage dioxygenase 7 (CCD7), and carotenoid cleavage dioxygenase 8 (CCD8) and DWARF 27 (D27) [108, 109]. According to Ruyter-Spira *et al.* (2013), carlactone undergoes various reactions such as dioxygenation, dehydrogenation and two ring closure steps catalyzed by a gene called More Axillary Growth 1 (MAX1) for production of 5-Deoxy strigol and its stereoisomers, which act as precursors of various strigolactones such as sorgolactone, strigol, orobanchol, orobacnhol, *etc* [108].

According to Ruyter-Spira *et al.*, (2013) and Morffy *et al.* (2016), SLs are perceived by a SL receptor called D14 leading to an association between D14, SCF^{MAX2} (SKP1-CULLIN-F box protein rich in C-terminal with Leucine rich repeats) and SMXL6,7,8 (SMAX1 Like genes) resulting in proteosomal degradation to initiate the responses to strigolactones [108, 111]. According to Liu *et al.* (2015), treatment of wild-type roots of *Lotus japonicas* with exogenous strigolactones inhibited the absicisic acid induced osmotic stress along with the suppression of an absicisic acid biosynthetic gene called *LjNCED2* [112]. Strigolactones through MAX2 (More Axillary Growth 2) help in plant stress tolerance against abiotic stressors such as excess salt and drought conditions [113]. Under abiotic stressors such as temperature, salinity, oxidative stresses, drought, light stress lead to accumulation of SLs in stress tolerance for maintenance of metabolic homeostasis in cells under stress conditions *via* expression induction of downstream osmolytes [114].

CONCLUSION

Abiotic stress limits the crop quality and productivity worldwide by disturbing the growth and development of plants. The plant growth regulators are chemical compounds that not only regulate physiological processes in the plants but also help in ameliorating the effects of abiotic stressors posed on them. These compounds increase the plant tolerance towards the abiotic stress by modulating the plant responses through complex mechanisms, such as, regulation of hormonal balance, maintenance of membrane fluidity, ion homeostasis, generation of antioxidant compounds, *etc.* at the cellular levels. An understanding of these stress tolerance mechanisms is instrumental in improving the crop yield and quality under abiotic stress. Therefore, exogenous application of plant growth regulators such as, auxins, abscisic acid, cytokinins, ethylene, gibberellins, jasmonic acid, brassinosteroids, salicylic acid, polyamines and strigolactones can prove to be a major tool for crop improvement.

CONSENT FOR PUBLICATION

Not applicable.

CONFLICT OF INTEREST

The author declares no conflict of interest, financial or otherwise.

ACKNOWLEDGEMENTS

Authors would like to thank Guru Nanak Dev University, Amritsar, Punjab and DAV College, Amritsar, Punjab for providing library and internet facilities for survey of literature during preparation of this chapter.

REFERENCES

[1]   Gull A, Lone AA, Wani NUI. Biotic and Abiotic Stresses in Plants. In: de Oliveira AB, Ed. Abiotic and Biotic Stress in Plants. London: IntechOpen 2019; pp. 1-6.
        [http://dx.doi.org/10.5772/intechopen.85832]

[2]   Lobell DB, Field CB. Global scale climate–crop yield relationships and the impacts of recent warming. Environ Res Lett 2007; 2(1): 014002.
        [http://dx.doi.org/10.1088/1748-9326/2/1/014002]

[3]   Dresselhaus T, Hückelhoven R. Biotic and abiotic stress responses in crop plants. Agron 2018; 8: 1-6.
        [http://dx.doi.org/10.3390/agronomy8110267]

[4]   Verma S, Nizam S, Verma PK. Biotic and abiotic stress signaling in plants. In: Sarwat M, Ahmad A, Abdin MZ, Ibrahim MM, Eds. Stress Signaling in Plants: Genomics and Proteomics Perspective. New York: Springer 2013; pp. 25-49.
        [http://dx.doi.org/10.1007/978-1-4614-6372-6_2]

[5]   Vijayalakshmi D. Abiotic stresses and its management in agriculture. Coimbatore: TNAU Agritech 2018.

[6]   Gupta A, Dixit SK, Senthil-Kumar M. Drought stress predominantly endures *Arabidopsis thaliana* to *Pseudomonas syringae* infection. Front Plant Sci 2016; 7: 808.
       [http://dx.doi.org/10.3389/fpls.2016.00808] [PMID: 27375661]

[7]   Fahad S, Bajwa AA, Nazir U, *et al.* Crop production under drought and heat stress: plant responses and management options. Front Plant Sci 2017; 8: 1147.
       [http://dx.doi.org/10.3389/fpls.2017.01147] [PMID: 28706531]

[8]   Kosová K, Vítámvás P, Urban MO, Klíma M, Roy A, Prášil IT. Biological networks underlying abiotic stress tolerance in temperate crops—a proteomic perspective. Int J Mol Sci 2015; 16(9): 20913-42.
       [http://dx.doi.org/10.3390/ijms160920913] [PMID: 26340626]

[9]   Pandey S, Fartyal D, Agarwal A, *et al.* Abiotic stress tolerance in plants: myriad roles of ascorbate peroxidase. Front Plant Sci 2017; 8: 581.
       [http://dx.doi.org/10.3389/fpls.2017.00581] [PMID: 28473838]

[10]   Khan MIR, Fatma M, Per TS, Anjum NA, Khan NA. Salicylic acid-induced abiotic stress tolerance and underlying mechanisms in plants. Front Plant Sci 2015; 6: 462.
       [http://dx.doi.org/10.3389/fpls.2015.00462] [PMID: 26175738]

[11]   Anwar A, Liu Y, Dong R, Bai L, Yu X, Li Y. The physiological and molecular mechanism of brassinosteroid in response to stress: a review. Biol Res 2018; 51(1): 46.
       [http://dx.doi.org/10.1186/s40659-018-0195-2] [PMID: 30419959]

[12]   Liao D, Wang S, Cui M, Liu J, Chen A, Xu G. Phytohormones regulate the development of arbuscular mycorrhizal symbiosis. Int J Mol Sci 2018; 19(10): 3146.
       [http://dx.doi.org/10.3390/ijms19103146] [PMID: 30322086]

[13]   Bhattacharya A. Effect of high-temperature stress on the metabolism of plant growth regulators. Effect of High Temperature on Crop Productivity and Metabolism of Macro Molecules. Cambridge: Academic Press 2019; pp. 485-591.
       [http://dx.doi.org/10.1016/B978-0-12-817562-0.00006-9]

[14]   Buchanan BB, Gruissem W, Jones RL, Eds. Biochemistry and molecular biology of plants. Hoboken: John Wiley & Sons 2015.

[15]   Onaga G, Wydra K. Advances in plant tolerance to biotic stresses. Plant Genome 2016; 14: 229-72.

[16]   Lesk C, Rowhani P, Ramankutty N. Influence of extreme weather disasters on global crop production. Nature 2016; 529(7584): 84-7.
       [http://dx.doi.org/10.1038/nature16467] [PMID: 26738594]

[17]   Gururani MA, Venkatesh J, Tran LS. Regulation of photosynthesis during abiotic stress-induced photoinhibition. Mol Plant 2015; 8(9): 1304-20.
       [http://dx.doi.org/10.1016/j.molp.2015.05.005] [PMID: 25997389]

[18]   Feller U. Drought stress and carbon assimilation in a warming climate: Reversible and irreversible impacts. J Plant Physiol 2016; 203: 84-94.
       [http://dx.doi.org/10.1016/j.jplph.2016.04.002] [PMID: 27083537]

[19]   Guo M, Liu JH, Ma X, Luo DX, Gong ZH, Lu MH. The plant heat stress transcription factors (HSFs): structure, regulation, and function in response to abiotic stresses. Front Plant Sci 2016; 7: 114.
       [http://dx.doi.org/10.3389/fpls.2016.00114] [PMID: 26904076]

[20]   Nadeem M, Li J, Wang M, *et al.* Unraveling field crops sensitivity to heat stress: Mechanisms, approaches, and future prospects. Agron 2018; 8: 128.
       [http://dx.doi.org/10.3390/agronomy8070128]

[21]   Tuteja N, Sopory SK. Chemical signaling under abiotic stress environment in plants. Plant Signal Behav 2008; 3(8): 525-36.
       [http://dx.doi.org/10.4161/psb.3.8.6186] [PMID: 19513246]

[22]   He M, He CQ, Ding NZ. Abiotic stresses: general defenses of land plants and chances for engineering multistress tolerance. Front Plant Sci 2018; 9: 1771.
       [http://dx.doi.org/10.3389/fpls.2018.01771] [PMID: 30581446]

[23]   Farooq M, Wahid A, Kobayashi N, Fujita D, Basra SMA. Plant drought stress: effects, mechanisms and management. In: Lichtfouse E, Navarrete M, Debaeke P, Véronique S, Alberola C, Eds. Sustainable agriculture. Dordrecht: Springer 2009; pp. 153-88.
       [http://dx.doi.org/10.1007/978-90-481-2666-8_12]

[24]   Upreti KK, Sharma M. Role of plant growth regulators in abiotic stress tolerance. In: Rao NKS, Shivashankara KS, Laxman RH, Eds. Abiotic stress physiology of horticultural crops New Delhi. New Delhi: Springer 2016; pp. 19-46.
       [http://dx.doi.org/10.1007/978-81-322-2725-0_2]

[25]   Waqas MA, Kaya C, Riaz A, *et al.* Potential mechanisms of abiotic stress tolerance in crop plants induced by thiourea. Front Plant Sci 2019; 10: 1336.
       [http://dx.doi.org/10.3389/fpls.2019.01336] [PMID: 31736993]

[26]   Roychoudhury A, Banerjee A. Abscisic acid signaling and involvement of mitogen activated protein kinases and calcium-dependent protein kinases during plant abiotic stress. In: Pandey GK, Ed. Mechanism of Plant Hormone Signaling Under Stress. Hoboken, NJ: Wiley 2017; pp. 197-241.
       [http://dx.doi.org/10.1002/9781118889022.ch9]

[27]   Fahad S, Hussain S, Saud S, *et al.* Exogenously applied plant growth regulators affect heat stressed rice pollens. J Agron Crop Sci 2016; 202: 139-50.
       [http://dx.doi.org/10.1111/jac.12148]

[28]   Chu J, Fang S, Xin P, Guo Z, Chen Y. Quantitative analysis of plant hormones based on LC-MS/MS. In: Li J, Li C, Smit SM, Eds. Authors Hormone metabolism and signaling in plants. Cambridge: Academic Press 2017; pp. 471-537.
       [http://dx.doi.org/10.1016/B978-0-12-811562-6.00014-1]

[29]   Korasick DA, Enders TA, Strader LC. Auxin biosynthesis and storage forms. J Exp Bot 2013; 64(9): 2541-55.
       [http://dx.doi.org/10.1093/jxb/ert080] [PMID: 23580748]

[30]   Minocha SC, Dibona S. Effect of auxin and abscisic acid on RNA and protein synthesis prior to the first cell division in Jerusalem artichoke tuber tissue cultured *in vitro.* Z Pflanzenphysiol 1979; 92: 367-74.
       [http://dx.doi.org/10.1016/S0044-328X(79)80019-7]

[31]   Zhao Y. Auxin biosynthesis and its role in plant development. Annu Rev Plant Biol 2010; 61: 49-64.
       [http://dx.doi.org/10.1146/annurev-arplant-042809-112308] [PMID: 20192736]

[32]   Bielach A, Hrtyan M, Tognetti VB. Plants under stress: involvement of auxin and cytokinin. Int J Mol Sci 2017; 18(7): 1427.
       [http://dx.doi.org/10.3390/ijms18071427] [PMID: 28677656]

[33]   Brumos J, Robles LM, Yun J, *et al.* Local auxin biosynthesis is a key regulator of plant development. Dev Cell 2018; 47(3): 306-318 e5.
       [http://dx.doi.org/10.1016/j.devcel.2018.09.022] [PMID: 30415657]

[34]   Zhao Y. Auxin biosynthesis. Arabidopsis Book 2014; 12: e0173.
       [http://dx.doi.org/10.1199/tab.0173] [PMID: 24955076]

[35]   Hagen G. Auxin signal transduction. Essays Biochem 2015; 58: 1-12.
       [http://dx.doi.org/10.1042/bse0580001] [PMID: 26374883]

[36]   Potters G, Pasternak TP, Guisez Y, Jansen MA. Different stresses, similar morphogenic responses: integrating a plethora of pathways. Plant Cell Environ 2009; 32(2): 158-69.
       [http://dx.doi.org/10.1111/j.1365-3040.2008.01908.x] [PMID: 19021890]

[37] Korver RA, Koevoets IT, Testerink C. Out of shape during stress: A key role for auxin. Trends Plant Sci 2018; 23(9): 783-93.
[http://dx.doi.org/10.1016/j.tplants.2018.05.011] [PMID: 29914722]

[38] Upreti KK, Srinivasa Rao NK, Jayaram HL. Floral abscission in capsicum under high temperature: role of endogenous hormones and polyamines. Indian J Plant Physiol 2012; 17: 207-14.

[39] Wareing PF. Abscisic acid as a natural growth regulator. Philosophical Transactions of the Royal Society of London. B. Biomed Sci 1978; 284: 483-98.

[40] Rücker g Sesquiterpenes. Angew Chem Int Ed Engl 1973; 12: 793-806.
[http://dx.doi.org/10.1002/anie.197307931]

[41] Finkelstein R. Abscisic Acid synthesis and response. Arabidopsis Book 2013; 11: e0166.
[http://dx.doi.org/10.1199/tab.0166] [PMID: 24273463]

[42] Ng LM, Melcher K, Teh BT, Xu HE. Abscisic acid perception and signaling: structural mechanisms and applications. Acta Pharmacol Sin 2014; 35(5): 567-84.
[http://dx.doi.org/10.1038/aps.2014.5] [PMID: 24786231]

[43] Hauser F, Li Z, Waadt R, Schroeder JI. SnapShot: abscisic acid signaling. Cell 2017; 171(7): 1708-8.
[http://dx.doi.org/10.1016/j.cell.2017.11.045] [PMID: 29245015]

[44] Fahad S, Nie L, Chen Y, *et al.* Crop plant hormones and environmental stress. In: Lichtfouse E, Ed. Sustainable Agriculture Reviews. Cham: Springer 2015; pp. 371-400.

[45] Zhu JK. Salt and drought stress signal transduction in plants. Annu Rev Plant Biol 2002; 53: 247-73.
[http://dx.doi.org/10.1146/annurev.arplant.53.091401.143329] [PMID: 12221975]

[46] Abobatta WF. Drought adaptive mechanisms of plants–a review. Adv Agr Environ Sci 2019; 2: 42-5.

[47] Zhou R, Cutler AJ, Ambrose SJ, *et al.* A new abscisic acid catabolic pathway. Plant Physiol 2004; 134(1): 361-9.
[http://dx.doi.org/10.1104/pp.103.030734] [PMID: 14671016]

[48] Verslues PE, Agarwal M, Katiyar-Agarwal S, Zhu J, Zhu JK. Methods and concepts in quantifying resistance to drought, salt and freezing, abiotic stresses that affect plant water status. Plant J 2006; 45(4): 523-39.
[http://dx.doi.org/10.1111/j.1365-313X.2005.02593.x] [PMID: 16441347]

[49] Marcińska I, Czyczyło-Mysza I, Skrzypek E, *et al.* Alleviation of osmotic stress effects by exogenous application of salicylic or abscisic acid on wheat seedlings. Int J Mol Sci 2013; 14(7): 13171-93.
[http://dx.doi.org/10.3390/ijms140713171] [PMID: 23803653]

[50] Javid MG, Sorooshzadeh A, Moradi F, Modarres Sanavy SA, Allahdadi I. The role of phytohormones in alleviating salt stress in crop plants. Aust J Crop Sci 2011; 5: 726.

[51] Kieber JJ, Schaller GE. Cytokinins. Arabidopsis Book 2014; 12: e0168.
[http://dx.doi.org/10.1199/tab.0168] [PMID: 24465173]

[52] Rao NS, Shivashankara KS, Laxman RH, Eds. Abiotic stress physiology of horticultural crops. New Delhi: Springer 2016.
[http://dx.doi.org/10.1007/978-81-322-2725-0]

[53] Brizzolari A, Marinello C, Carini M, Santaniello E, Biondi PA. Evaluation of the antioxidant activity and capacity of some natural N6-substituted adenine derivatives (cytokinins) by fluorimetric and spectrophotometric assays. J Chromatogr B Analyt Technol Biomed Life Sci 2016; 1019: 164-8.
[http://dx.doi.org/10.1016/j.jchromb.2015.12.047] [PMID: 26753810]

[54] Mok DW, Mok MC. Cytokinin metabolism and action. Annu Rev Plant Physiol Plant Mol Biol 2001; 52: 89-118.
[http://dx.doi.org/10.1146/annurev.arplant.52.1.89] [PMID: 11337393]

[55] Aoyama T, Oka A. Cytokinin signal transduction in plant cells. J Plant Res 2003; 116(3): 221-31.

[http://dx.doi.org/10.1007/s10265-003-0094-6] [PMID: 12836044]

[56] Kamínek M, Motyka V, Vaňková R. Regulation of cytokinin content in plant cells. Physiol Plant 1997; 101: 689-700.
[http://dx.doi.org/10.1111/j.1399-3054.1997.tb01053.x]

[57] Davies WJ, Kudoyarova G, Hartung W. Long-distance ABA signaling and its relation to other signaling pathways in the detection of soil drying and the mediation of the plant's response to drought. J Plant Growth Regul 2005; 24: 285.
[http://dx.doi.org/10.1007/s00344-005-0103-1]

[58] Abeles FB, Morgan PW, Saltveit ME Jr. Ethylene in Plant Biology. New York: Academic Press 1992.

[59] Kende H. Ethylene biosynthesis. Annu Rev Plant Biol 1993; 44: 283-307.
[http://dx.doi.org/10.1146/annurev.pp.44.060193.001435]

[60] Iqbal N, Khan NA, Ferrante A, Trivellini A, Francini A, Khan MIR. Ethylene role in plant growth, development and senescence: interaction with other phytohormones. Front Plant Sci 2017; 8: 475.
[http://dx.doi.org/10.3389/fpls.2017.00475] [PMID: 28421102]

[61] Wilkinson S, Davies WJ. Drought, ozone, ABA and ethylene: new insights from cell to plant to community. Plant Cell Environ 2010; 33(4): 510-25.
[http://dx.doi.org/10.1111/j.1365-3040.2009.02052.x] [PMID: 19843256]

[62] Gao Z, Wen CK, Binder BM, *et al.* Heteromeric interactions among ethylene receptors mediate signaling in Arabidopsis. J Biol Chem 2008; 283(35): 23801-10.
[http://dx.doi.org/10.1074/jbc.M800641200] [PMID: 18577522]

[63] Hall BP, Shakeel SN, Schaller GE. Ethylene receptors: ethylene perception and signal transduction. J Plant Growth Regul 2007; 26: 118-30.
[http://dx.doi.org/10.1007/s00344-007-9000-0]

[64] Kieber JJ, Rothenberg M, Roman G, Feldmann KA, Ecker JR. CTR1, a negative regulator of the ethylene response pathway in Arabidopsis, encodes a member of the raf family of protein kinases. Cell 1993; 72(3): 427-41.
[http://dx.doi.org/10.1016/0092-8674(93)90119-B] [PMID: 8431946]

[65] Urbanová T, Tarkowská D, Novák O, Hedden P, Strnad M. Analysis of gibberellins as free acids by ultra performance liquid chromatography-tandem mass spectrometry. Talanta 2013; 112: 85-94.
[http://dx.doi.org/10.1016/j.talanta.2013.03.068] [PMID: 23708542]

[66] Jini D, Joseph B. Use of Phytohormones in Improving Abiotic Stress Tolerance to Rice. In: Hasanuzzaman M, Fujita M, Nahar K, Biswas JK, Eds. Advances in Rice Research for Abiotic Stress Tolerance. Cambridge: Woodhead Publishing 2019; pp. 633-49.
[http://dx.doi.org/10.1016/B978-0-12-814332-2.00031-9]

[67] Gao X, Zhang Y, He Z, Fu X. Gibberellins. In: Li J, Li C, Smit SM, Eds. Authors Hormone metabolism and signaling in plants. Cambridge: Academic Press 2017; pp. 107-60.
[http://dx.doi.org/10.1016/B978-0-12-811562-6.00004-9]

[68] Binenbaum J, Weinstain R, Shani E. Gibberellin localization and transport in plants. Trends Plant Sci 2018; 23(5): 410-21.
[http://dx.doi.org/10.1016/j.tplants.2018.02.005] [PMID: 29530380]

[69] Wang X, Li J, Ban L, *et al.* Functional characterization of a gibberellin receptor and its application in alfalfa biomass improvement. Sci Rep 2017; 7: 41296.
[http://dx.doi.org/10.1038/srep41296] [PMID: 28128230]

[70] Olszewski N, Sun TP, Gubler F. Gibberellin signaling: biosynthesis, catabolism, and response pathways. Plant Cell 2002; 14 (Suppl.): S61-80.
[http://dx.doi.org/10.1105/tpc.010476] [PMID: 12045270]

[71] Zhong T, Zhang L, Sun S, Zeng H, Han L. Effect of localized reduction of gibberellins in different

tobacco organs on drought stress tolerance and recovery. Plant Biotechnol Rep 2014; 8: 399-408.
[http://dx.doi.org/10.1007/s11816-014-0330-7]

[72]   Plaza-Wüthrich S, Blösch R, Rindisbacher A, Cannarozzi G, Tadele Z. Gibberellin deficiency confers both lodging and drought tolerance in small cereals. Front Plant Sci 2016.
[http://dx.doi.org/10.3389/fpls.2016.00643]

[73]   Shomeili M, Nabipour M, Meskarbashee M, Memari HR. Effects of gibberellic acid on sugarcane plants exposed to salinity under a hydroponic system. Afr J Plant Sci 2011; 5: 609-16.

[74]   Iqbal M, Ashraf M. Gibberellic acid mediated induction of salt tolerance in wheat plants: Growth, ionic partitioning, photosynthesis, yield and hormonal homeostasis. Environ Exp Bot 2013; 86: 76-85.
[http://dx.doi.org/10.1016/j.envexpbot.2010.06.002]

[75]   Fahad S, Hussain S, Matloob A, *et al.* Phytohormones and plant responses to salinity stress: a review. Plant Growth Regul 2015; 75: 391-404.
[http://dx.doi.org/10.1007/s10725-014-0013-y]

[76]   Cheong JJ, Choi YD. Methyl jasmonate as a vital substance in plants. Trends Genet 2003; 19(7): 409-13.
[http://dx.doi.org/10.1016/S0168-9525(03)00138-0] [PMID: 12850447]

[77]   Wasternack C, Miersch O, Kramell R, *et al.* Jasmonic acid: biosynthesis, signal transduction, gene expression. Fett/Lipid 1998; 100(4-5): 139-46.
[http://dx.doi.org/10.1002/(SICI)1521-4133(19985)100:4/5<139::AID-LIPI139>3.0.CO;2-5]

[78]   Ruan J, Zhou Y, Zhou M, *et al.* Jasmonic acid signaling pathway in plants. Int J Mol Sci 2019; 20(10): 2479.
[http://dx.doi.org/10.3390/ijms20102479] [PMID: 31137463]

[79]   Mason HS, Mullet JE. Expression of two soybean vegetative storage protein genes during development and in response to water deficit, wounding, and jasmonic acid. Plant Cell 1990; 2(6): 569-79.
[PMID: 2152178]

[80]   Creelman RA, Tierney ML, Mullet JE. Jasmonic acid/methyl jasmonate accumulate in wounded soybean hypocotyls and modulate wound gene expression. Proc Natl Acad Sci USA 1992; 89(11): 4938-41.
[http://dx.doi.org/10.1073/pnas.89.11.4938] [PMID: 1594598]

[81]   Kang DJ, Seo YJ, Lee JD, *et al.* Jasmonic acid differentially affects growth, ion uptake and abscisic acid concentration in salt tolerant and salt sensitive rice cultivars. J Agron Crop Sci 2005; 191: 273-82.
[http://dx.doi.org/10.1111/j.1439-037X.2005.00153.x]

[82]   Walia H, Wilson C, Condamine P, Liu X, Ismail AM, Close TJ. Large-scale expression profiling and physiological characterization of jasmonic acid-mediated adaptation of barley to salinity stress. Plant Cell Environ 2007; 30(4): 410-21.
[http://dx.doi.org/10.1111/j.1365-3040.2006.01628.x] [PMID: 17324228]

[83]   Tang J, Han Z, Chai J. Q&A: what are brassinosteroids and how do they act in plants? BMC Biol 2016; 14(1): 113.
[http://dx.doi.org/10.1186/s12915-016-0340-8] [PMID: 28007032]

[84]   Tanveer M, Shahzad B, Sharma A, Biju S, Bhardwaj R. 24-Epibrassinolide; an active brassinolide and its role in salt stress tolerance in plants: A review. Plant Physiol Biochem 2018; 130: 69-79.
[http://dx.doi.org/10.1016/j.plaphy.2018.06.035] [PMID: 29966934]

[85]   Sharma I, Kaur N, Pati PK. Brassinosteroids: A promising option in deciphering remedial strategies for abiotic stress tolerance in rice. Front Plant Sci 2017; 8: 2151.
[http://dx.doi.org/10.3389/fpls.2017.02151] [PMID: 29326745]

[86]   Wang ZY, Wang Q, Chong K, *et al.* The brassinosteroid signal transduction pathway. Cell Res 2006; 16(5): 427-34.

[http://dx.doi.org/10.1038/sj.cr.7310054] [PMID: 16699538]

[87]   Clouse SD. Brassinosteroid signal transduction: from receptor kinase activation to transcriptional networks regulating plant development. Plant Cell 2011; 23(4): 1219-30.
[http://dx.doi.org/10.1105/tpc.111.084475] [PMID: 21505068]

[88]   Tran LSP, Pal S, Eds. Phytohormones: a window to metabolism, signaling and biotechnological applications. New York: Springer 2014.
[http://dx.doi.org/10.1007/978-1-4939-0491-4]

[89]   Anuradha S, Rao SS. Effect of brassinosteroids on salinity stress induced inhibition of seed germination and seedling growth of rice (*Oryza sativa* L.). Plant Growth Regul 2001; 33: 151-3.
[http://dx.doi.org/10.1023/A:1017590108484]

[90]   Kagale S, Divi UK, Krochko JE, Keller WA, Krishna P. Brassinosteroid confers tolerance in *Arabidopsis thaliana* and *Brassica napus* to a range of abiotic stresses. Planta 2007; 225(2): 353-64.
[http://dx.doi.org/10.1007/s00425-006-0361-6] [PMID: 16906434]

[91]   Rivas-San Vicente M, Plasencia J. Salicylic acid beyond defence: its role in plant growth and development. J Exp Bot 2011; 62(10): 3321-38.
[http://dx.doi.org/10.1093/jxb/err031] [PMID: 21357767]

[92]   Chen Z, Zheng Z, Huang J, Lai Z, Fan B. Biosynthesis of salicylic acid in plants. Plant Signal Behav 2009; 4(6): 493-6.
[http://dx.doi.org/10.4161/psb.4.6.8392] [PMID: 19816125]

[93]   Seyfferth C, Tsuda K. Salicylic acid signal transduction: the initiation of biosynthesis, perception and transcriptional reprogramming. Front Plant Sci 2014; 5: 697.
[http://dx.doi.org/10.3389/fpls.2014.00697] [PMID: 25538725]

[94]   Senaratna T, Touchell D, Bunn E, Dixon K. Acetyl salicylic acid (Aspirin) and salicylic acid induce multiple stress tolerance in bean and tomato plants. Plant Growth Regul 2000; 1(30): 157-61.
[http://dx.doi.org/10.1023/A:1006386800974]

[95]   Jumali SS, Said IM, Ismail I, Zainal Z. Genes induced by high concentration of salicylic acid in *Mitragyna speciosa*. Aust J Crop Sci 2011; 5: 296.

[96]   Nazar R, Umar S, Khan NA. Exogenous salicylic acid improves photosynthesis and growth through increase in ascorbate-glutathione metabolism and S assimilation in mustard under salt stress. Plant Signal Behav 2015; 10(3): e1003751.
[http://dx.doi.org/10.1080/15592324.2014.1003751] [PMID: 25730495]

[97]   Khodary SEA. Effect of salicylic acid on the growth, photosynthesis and carbohydrate metabolism in salt stressed maize plants. Int J Agric Biol 2004; 6: 5-8.

[98]   Chen D, Shao Q, Yin L, Younis A, Zheng B. Polyamine function in plants: metabolism, regulation on development, and roles in abiotic stress responses. Front Plant Sci 2019; 9: 1945.
[http://dx.doi.org/10.3389/fpls.2018.01945] [PMID: 30687350]

[99]   Reddy VP. Fluorinated Compounds in Enzyme-Catalyzed Reactions. In: Reddy VP, Ed. Organofluorine Compounds in Biology and Medicine. Amsterdam: Elsevier 2015; pp. 29-57.

[100] Docimo T, Reichelt M, Schneider B, *et al.* The first step in the biosynthesis of cocaine in Erythroxylum coca: the characterization of arginine and ornithine decarboxylases. Plant Mol Biol 2012; 78(6): 599-615.
[http://dx.doi.org/10.1007/s11103-012-9886-1] [PMID: 22311164]

[101] Lenis YY, Johnson GA, Wang X, *et al.* Functional roles of ornithine decarboxylase and arginine decarboxylase during the peri-implantation period of pregnancy in sheep. J Anim Sci Biotechnol 2018; 9: 10.
[http://dx.doi.org/10.1186/s40104-017-0225-x] [PMID: 29410783]

[102] Bachrach U, Wang YC, Tabib A. Polyamines: new cues in cellular signal transduction. News Physiol

Sci 2001; 16: 106-9.
[http://dx.doi.org/10.1152/physiologyonline.2001.16.3.106] [PMID: 11443226]

[103]   Groppa MD, Benavides MP. Polyamines and abiotic stress: recent advances. Amino Acids 2008; 34(1): 35-45.
[http://dx.doi.org/10.1007/s00726-007-0501-8] [PMID: 17356805]

[104]   Gill SS, Tuteja N. Polyamines and abiotic stress tolerance in plants. Plant Signal Behav 2010; 5(1): 26-33.
[http://dx.doi.org/10.4161/psb.5.1.10291] [PMID: 20592804]

[105]   Waters MT, Gutjahr C, Bennett T, Nelson DC. Strigolactone signaling and evolution. Annu Rev Plant Biol 2017; 68: 291-322.
[http://dx.doi.org/10.1146/annurev-arplant-042916-040925] [PMID: 28125281]

[106]   Mishra S, Upadhyay S, Shukla RK. The role of strigolactones and their potential cross-talk under hostile ecological conditions in plants. Front Physiol 2017; 7: 691.
[http://dx.doi.org/10.3389/fphys.2016.00691] [PMID: 28119634]

[107]   Gomez-Roldan V, Fermas S, Brewer PB, *et al.* Strigolactone inhibition of shoot branching. Nature 2008; 455(7210): 189-94.
[http://dx.doi.org/10.1038/nature07271] [PMID: 18690209]

[108]   Ruyter-Spira C, Al-Babili S, van der Krol S, Bouwmeester H. The biology of strigolactones. Trends Plant Sci 2013; 18(2): 72-83.
[http://dx.doi.org/10.1016/j.tplants.2012.10.003] [PMID: 23182342]

[109]   Wang Y, Bouwmeester HJ. Structural diversity in the strigolactones. J Exp Bot 2018; 69(9): 2219-30.
[http://dx.doi.org/10.1093/jxb/ery091] [PMID: 29522118]

[110]   Zwanenburg B, Pospíšil T. Structure and activity of strigolactones: new plant hormones with a rich future. Mol Plant 2013; 6(1): 38-62.
[http://dx.doi.org/10.1093/mp/sss141] [PMID: 23204499]

[111]   Morffy N, Faure L, Nelson DC. Smoke and hormone mirrors: action and evolution of karrikin and strigolactone signaling. Trends Genet 2016; 32(3): 176-88.
[http://dx.doi.org/10.1016/j.tig.2016.01.002] [PMID: 26851153]

[112]   Liu J, He H, Vitali M, *et al.* Osmotic stress represses strigolactone biosynthesis in *Lotus japonicus* roots: exploring the interaction between strigolactones and ABA under abiotic stress. Planta 2015; 241(6): 1435-51.
[http://dx.doi.org/10.1007/s00425-015-2266-8] [PMID: 25716094]

[113]   Li W, Tran LSP. Are karrikins involved in plant abiotic stress responses? Trends Plant Sci 2015; 20(9): 535-8.
[http://dx.doi.org/10.1016/j.tplants.2015.07.006] [PMID: 26255855]

[114]   Banerjee A, Roychoudhury A. Strigolactones: multi-level regulation of biosynthesis and diverse responses in plant abiotic stresses. Acta Physiol Plant 2018; 40: 86.
[http://dx.doi.org/10.1007/s11738-018-2660-5]

Genomics, Proteomics and Metabolic Approaches against Abiotic Stress

Harjit Kaur Bajwa[1] and **Hina Khan**[1,*]

[1] Department of Botany and Environmental Science, Sri Guru Granth Sahib World University, Fatehgarh Sahib, Punjab 140407, India

Abstract: Abiotic stressors such as drought, salination, flooding, cold, heat, ultraviolet radiation, heavy metals, *etc.*, are the paramount cause that reduce crop yield and weaken universal food security as they strongly affect plant growth, physiology, and metabolism. Plants frequently face a large number of environmental stressors and usually generate common responses to deal with these unfavorable conditions. However, crop improvement against abiotic stressors is one of the urgent priorities that need undivided attention, while a huge increase in demand for various plant-derived products will rise in the near future owing to the rising human population. As conventional methods for crop enhancement have limitations, therefore an epoch of omic research has shot up with new and encouraging perspectives in breeding to improve the crops against abiotic stress. In this light, the genomic, proteomic, and metabolomic approaches are emerging as powerful tools for the identification and description of cellular networks through which stress perception, signal transduction, and defensive responses are exhibited. Further advances in omic techniques have permitted a comprehensive investigation of crop genomes and have magnified the perception of convolution of the mechanisms controlling abiotic stressor tolerance and the adaptation to mitigate them. This chapter will give an overview of genomics, proteomics, and metabolic approaches and their usage to enhance the possibility of producing abiotic stressor tolerant crops.

Keywords: Abiotic stress, Genome, Metabolome, Proteome.

INTRODUCTION

Florae are often subjected to numerous abiotic stressors forever that limit their yield and productivity. Tolerance to abiotic stressors can halt yield loss in crops for maintaining agricultural productivity. Abiotic stressors such as high/low temperature, salinization or alkalinity, and drought are the major challenges fac-

---
* **Corresponding Author Dr. Hina Khan:** Department of Botany and Environmental Science, Sri Guru Granth Sahib World University, Fatehgarh Sahib, Punjab 140407, India; Tel: +91-7973287929; E-mail:hina786k786@gmail.com

Tajinder Kaur & Saroj Arora (Eds.)

ing agriculture. 50% of the total yield loss in essential crops worldwide is only due to these abiotic stressors [1]. Approximately 45% of the world's agricultural land is affected by drought, while 19% is considered saline [2].

Thus, the production of sustainable and high-yielding crops with improved tolerance to different abiotic stressors is a precondition to meet the increasing universal food supply. The conventional methods such as plant breeding and genetic engineering played a vital role during the last century for enhancing abiotic stress tolerance but have limited success due to the genetic complexity of stress responses [3]. Consequently, agriculture has to adopt novel approaches to produce plants with natural ability of enhanced level of tolerance to environmental stressors. Plant reaction to abiotic stressors is a complicated process because abiotic stressors could initiate multigene response which comprises a number of events including stress signaling, stress transduction, and gene expression. This whole process leads to the agglomeration of transcription factors, stress-related proteins, enzymes, and metabolites. Therefore, genomics, proteomics, and metabolomic techniques are playing a key role in studying plants' reactions to different stimuli [4]. Genomics had disclosed the gene and protein sequences while proteomics and metabolomics uncover the biological function of the gene product. The "omic" technologies are high-throughput innovations that allow a system biology approach toward understanding the compound interactions between genes, proteins, and metabolites within the resulting expression [5]. Furthermore, recent advances in molecular biology and omic technology allowed explaining functions of many key genes, proteins, and metabolites involved in plant responses to abiotic stressors [6]. The purpose of this book chapter is to provide an overview of genomics, proteomics, and metabolic approaches and their usage in developing abiotic stress-tolerant crops.

GENOMICS APPROACHES AGAINST ABIOTIC STRESS

Genomics studies the sequence, organization, and functions of DNA of an organism using a wide variety of experimental, computational and statistical methodologies. Genomics has been substantially expedited in the last few years with the emergence of next-generation sequencing technologies, high-density molecular assays, and advances in computational biology and biostatistics [7]. Sequencing the complete genome is the most effective method to accelerate molecular research and provided the ground for the production of simple sequence repeat (SSR) and single nucleotide polymorphism markers. This approach has already been successfully used in several crops and also has high scope in the future [8].

Arabidopsis thaliana was the initial plant to be completely sequenced. It was asserted that the *Arabidopsis* genome sequence helped in the extensive perception of plant development, environmental responses, genome organization, regulation, and evolution [9]. However, these days, crops like rice and maize have gained importance. Abiotic stressor reactions in plants are controlled by different transcription factors and signaling mechanisms. The relevant factors in each mechanism have been identified and addressed using responsive genes as markers [10]. Many stress-regulated genes are respondent to diverse stressors and stimuli; for instance, KNAT3, KNAT4, SEN1, DIN9, DIN10, and ACP4 are drought-regulated genes but also respond to light [11]. Similarly, ERD14, RD29B, and COR47 are responsive to duo stressors, water deprivation as well as cold. In *Arabidopsis*, 67 genes have been identified which exhibited responses to the cold, osmotic stressor, salinity, drought, genotoxic stressor, UV light, oxidative stressorosmotic, injury, and high temperature [12]. The approaches based on genomics provide ingress to agronomically desirable alleles present at quantitative trait loci (QTLs), thus enabling the improvement of abiotic stressor tolerant plants. Recently, marker-assisted selection (MAS), multi-parallel analysis of transcript levels, microarray analysis, RT-qPCR, serial analysis of gene expression (SAGE), massive parallel signature sequencing (MPSS), and oligoarray techniques are used to evaluate abiotic stressor responses in plants. Moreover, various bioinformatic tools, ESTs, and subtractive cDNA have added more to the area of genomics in order to get to the bottom of the genetics of stress tolerance [13]. The main goals of genomics is to assemble the genomic sequence of organisms, search out the position of the genes for analyzing the relationships, interpret the gene set, learn gene functions, and compare gene protein profiles among distinct organisms.

Structural Genomics

The main purpose of structural genomics is to identify the structure of the genome. It can be very useful in manipulating genes and DNA segments in particular plant species. The initial step in structural genomics is the assignment of genes and markers to individual chromosomes, then the mapping of these genes and markers within a chromosome, and finally the preparation of a physical map culminating in sequencing [14].

Molecular Markers

The first low-throughput markers were discovered using restriction fragment length polymorphism (RFLP) method. The restriction fragment length polymorphism subsequently improved by coupling PCR that gave rise to random amplified polymorphic DNA (RAPD), inter-simple sequence repeat (ISSR), amplified fragment length polymorphism (AFLP), selective amplification of polymorphic microsatellite loci (SAMPL), sequence-related amplified polymorphism (SRAP) and target region amplification polymorphism (TRAP) methods [15]. In order to identify the markers associated with cold stress-traits, a panel of 110 different *Citrus* accessions varying in reaction to cold stress was evaluated for eight physiologic traits and genomic markers at four stress temperatures. The outcomes revealed a significant relationship between several markers and the cold stress-related traits [16]. 17 calcium-dependent protein kinases (CDPKs) have been screened from drought and salt induced soybean transcriptome sequences using quantitative real-time PCR (qRT-PCR). *GmCDPK3* showed high relative expression and better drought and salt resistance in *Arabidopsis* and soybean. The results of this study indicated that *GmCDPK3* is crucial in resisting drought and salt stressors [17]. TaWRKY13, a nuclear protein also assists positively in salt stress. The *TaWRKY13* was characterized in a RNA-Seq data set from salt-treated wheat and additional functional identification was performed in *Arabidopsis thaliana* and *Oryza sativa* L [18].

The emergence of next-generation sequencing (NGS) enabled the identification of single nucleotide polymorphism (SNP) [19]. Afterwards, SNPs remained the preferred markers so far for large scale genotyping and genomic-assisted breeding. In recent times, microarrays with typically 10-50K markers were developed for several crops such as sorghum, rice and barley [20, 21]. Recently, the double digest restriction associated DNA sequencing is well liked method for the development of large-scale SNPs for genetic mapping using either linkage analysis or association mapping [22]. Su *et al.* [23] identified 179 DYW subgroup *PPR* genes in soybean genome using genome-wide analysis and found that these genes help in improving the drought tolerance in soybean. Similarly, the role of histone acetyltransferases (HATs) in response to developmental and environmental cues has been identified in cotton by genome-wide analysis. It was also reported that, various HATs were differentially regulated in response to different stressors such as salinity, drought, cold and heavy metal [24]. Mapping resolution of genome-wide association study (GWAS) is higher and it has also contributed to the characterization of favorable alleles for drought tolerance in several crops. Liu *et al.* [25], analyzed Dehydration Responsive Element Binding protein (DREB) genes in maize and also studied their relationship with the natural variation in drought tolerance among 368 maize varieties selected from different regions. Similarly, Thirunavukkarasu *et al.* [26] identified 61 significant SNPs from subtropical maize under water stress conditions and also found that 48% genes were stress tolerance. To characterize genetic variations underlying aluminum (Al) tolerance and 48 genomic regions associated with Al tolerance, Famoso *et al.* [27] conducted GWAS among 383 rice accessions in combination with bi-parental QTL mapping. Wehner *et al.* [28] investigated the genetic basis of drought resistance in *Hordeum vulgare*, *Cicer arietinum*, *Arabidopsis thaliana*, *Medicago truncatula* by GWAS. The disadvantage of GWAS and QTL is that former is sensible to the population structure that may lead to many false positives while latter is restricted to the genetic diversity existing into the parents of segregating population. However, "Meta-QTL analysis" is an advanced statistical tool that collates QTL data from distinct studies combined on the same linkage map for characterization of QTL region [29].

There are several methods that can bridge the advantages of QTL mapping and GWAS. For examples, nested association mapping (NAM) and Multi-parent advanced generation inter-cross (MAGIC) as they have a broader genetic diversity, and higher resolutions compared to bi-parental populations and the frequencies of alleles are quite well represented to reduce the problems of rare alleles. Vatter *et al.* [30] used SNP-based nested association mapping (NAM) to map QTL for *Pyrenophora teres* resistance in the barley population HEB-25 containing 1,420 lines obtained from BC_1S_3 generation. Thudi *et al.* [31] identified and introduced diverse drought tolerance traits in three chickpea varieties (JG 11, Chefe, KAK 2) by using marker-assisted backcrossing method and obtained various lines with 10-24% more yield than the respective recurrent parents. Similarly, MAGIC approach has been successfully used for mapping QTLs for blast and bacterial blight resistance, salt and submergence tolerance, and seed quality traits in rice [32]. A MAGIC population for cowpea (*Vigna unguiculata* L. Walp.) has been developed from eight genetically diverse parents possesses many abiotic and biotic stress resistance, grain quality and agronomic traits [33]. Despite its countless uses, the marker assisted selection has some major drawbacks such as high costs of application and the risk of recombination between the marker and the trait. Additionally, it requires the validation of QTLs when applied in distinct genetic background. Even so, marker assisted selection has been successfully applied to enhance crops for abiotic stress tolerance, including drouth, salt, and water logging [34, 35]. Even though the presence of highly advanced molecular genetic techniques, our goals are still away due to inaccurate phenotyping. These problems can be solved by using high-throughput phenotyping techniques using light, cameras, sensors and computers for the collection of very accurate phenotypic data [36].

Genome Sequencing and Mapping

The past decade has witnessed a revolution in DNA sequencing and mapping techniques that have made possible the generation of a lot of sequence information including whole genome sequence. In 1977, the first complete DNA genome sequenced was that of bateriophage ΦX174 and Frederick Sanger publishes "DNA sequencing with chain terminating inhibitors". In 2001, a draft sequence of the human genome is published and in 2006, era of next generation sequencing like Roche 454 GS FLX Titanium, Genome analyzer and solid Sequencer and third-generation sequencing (TGS) technologies such as Ion Torrent PGM/Proton, HiSeq from Illumina has acquired tremendous popularity in recent times because of its throughput, read lengths and low sequencing cost. Presently, next generation sequencing (NGS) has occupied a key position in breeding programmes [37] to increase the precision of trait mapping and trait transfer and now widely being used for de-novo sequencing, whole-genome sequencing (WGS), whole-genome resequencing (WGRS), quantitative trait mapping, Targeting induced local lesions in genome (TILLING) study, mutational map (MutMap), genotyping by sequencing (GBS) *etc.* Abe *et al*. [38] was the first researcher who used the MutMap technique in rice to characterize genomic regions controlling essential agronomic traits. MutMap technique involves development of a mutant population by chemical mutagen followed by the selection of line with desirable phenotype in M$_2$ which were then bulked and subjected to WGRS [39]. WGS provide detailed information on genomic features which can be exploited in functional studies such as microarray or tiling arrays. The WGS is highly efficient technique because it involves sequencing of known targeted region rather than the whole genome [40]. 454 GS-FLX Titanium technique yields around 1,000,000 sequences in a single 10-hour run with an average read length equal to 330 bp to 500 bp in shotgun libraries conditions. Goyal *et al*. [41] identified a total 453,882 reads from Kharchia Local wheat variety which is highly tolerant to salinity by using Roche 454-GS FLX Titanium sequencing technique. Illumina/Solexa is another method in which the DNA templates are sequenced in a massively parallel fashion using sequencing by synthesis approach. Gao *et al*. [42] characterized a batch of drought-responsive transcripts, including 36 transcription factors in leaves of *Ammopiptanthus monogolicus* using Illumina sequencing technology. However, initial cost of the instruments in this method is very high. Ion-torrent sequencing technique is based on the fact that the addition of one nucleotide in the template DNA chain always release on H$^+$ in the medium which alters the pH. This technique is appropriate for *de novo* sequencing of the genome with higher accuracy. Illumina genome analyzer and HiSeq instruments are presently the most widely used sequencing devices. However, based on the principle of nanotechnology there are a large number of "future-generation sequencing" technologies including Visigene and Oxford nanopore technology, which will soon be available for commercial use [43].

Functional Genomics

Functional genomics describes biological functions of genes and their products. The main purpose of functional genomics is to collect and use data from sequencing for describing gene and protein functions, functions of genes and non-gene sequences in genomes and relationship between gene and protein. The various techniques of functional genomics are genetic interaction mapping and encode project at the DNA level, expressed sequence tags, serial analysis of gene expression and DNA microarrays at gene expression profiling at the transcript level and protein microarray, 2D-PAGE during proteome analysis.

Sequencing-Based Approaches

Sequencing-based methods like Serial Analysis of Gene Expression (SAGE) and Massively Parallel Signature Sequencing (MPSS) do not require any pre compilation of an mRNA library of sequences rather they use type IIS restriction endonucleases to collect short tags of length typically 10–22 bases from each mRNA molecule, provided a relevant recognition site exists for an anchoring enzyme. Afterwards, through sequencing long concatamers of tags using conventional sequencer (SAGE), or by performing iterative parallel sequencing using a proprietary technique (MPSS), the characterization of a large amount of tags can be determined in an efficient manner. Expressed sequence tags (ESTs) have been proved to be efficient means to identify novel genes regulated by environmental stressors [44]. cDNA libraries from various tissues normally serve as the sources for EST sequencing to show differentially expressed genes. Presently, over a million ESTs are submitted in the EST database at National Center for Biotechnological Information (NCBI) for essential crops like maize, soybean, wheat, and rice, together with thousands of ESTs for other plants [45, 46]. SAGE and MPSS both techniques require large-scale sequencing of small sequence tags that permit comparatively unambiguous and efficient characterization of genes [47].

SAGE was first used for the quantification of global gene expression in rice seedling. This study analyzed total 10122 tags from 5921 expressed genes [48]. Similarly, the global gene expression patterns of *Arabidopsis* pollen has also been developed using SAGE. A total of 21,237 SAGE tags were sequenced and 4,211 unique tags were characterized [49]. MPSS enables analysis of millions of transcripts simultaneously. The potential of MPSS to apprehend rare transcripts is particularly useful in species that lack a whole genome sequence [50, 51]. 20 MPSS libraries established from various tissues of rice, including three from abiotic stress conditions *viz*. cold, salinity and dehydration [52]. The only snag of

MPSS is its high cost due to which, the full potential of this method in the universal expression profiling of abiotic stress response is still to be realized [53].

RNA-sequencing is a newly developed method applied to sequence cDNA in order to get information about the RNA content of a sample. It gives more accurate measurement of levels of transcripts and their isoforms than other methods. In this technique, a population of RNA is transformed to a library of cDNA fragments with adaptors attached to one or both ends and each molecule is then sequenced in a high-throughput manner to acquire short sequences from one end or both ends. The resulting reads are either aligned to a reference genome or assembled do novo without the genomic sequence to develop a genome-scale transcription map that consists of both the transcriptional structure and level of expression for each gene [54].

Similarly, ChIP-sequencing is another influential technique for characterizing genome-wide DNA binding sites for transcription factors and other proteins. This method uses a specific antibody for immune precipitating a DNA-bound protein followed by microarray (ChIP-chip). The bound DNA is then co precipitated, purified and sequenced. ChIP-seq enables through investigation of the inter-relationship between proteins and nucleic acids on a genome-wide scale and also does not require prior knowledge of probes derived from known sequences. ChIP-seq provides genome-wide profiling with massively parallel sequencing, generating millions of counts across numerous samples for economical, accurate, unbiased examination of epigenetic patterns. It provides higher resolution, less noise, higher genome coverage and wider dynamic range [55].

Hybridization-Based Approaches

Hybridization-based approaches include DNA microarrays which is based on simultaneous hybridization of mRNA extract from biological samples to a pre-selected mRNA library. Microarrays have revolutionized universal gene expression profiling by permitting the whole gene complement of the genome to be examined in a single experiment [56]. Kawasaki *et al.* [57] used microarray technique for the first time to investigate the universal gene expression profiling in rice in response to abiotic stressor using a cDNA microarray comprising 1728 cDNA clones developed from unstressed or salt-stressed roots of Pokkali. The gene expression profiles during pollination and fertilization under drought stressor has also been studied using microarray. Results of this study revealed that, exceeding one-half of the pollination and fertilization-related genes were regulated by dehydration stressor. It was concluded that water deficit may be a vital factor during pollination and fertilization. The global gene expression profiling in rice has also been studied using oligonucleotide microarrays [58, 59].

The suppression subtractive hybridization (SSH), microarray and Expressed Sequence Tags (ESTs) sequencing have been performed to characterize the abiotic stress-responsive transcripts in chickpea, tomato *etc* [60]. SSH is one of the powerful techniques for generating subtracted cDNA libraries [61]. In addition to the model species *Arabidopsis thaliana* and rice, the gene expression profiling in wheat, barley, maize, tomato, strawberry, petunia and lima bean has also been extensively studied by using cDNA and oligonucleotide microarrays [62 - 64]. Recently, tilling array technology has become a useful tool for the analysis of whole-genome transcriptome including mapping of transcript, characterization of splice sites, characterization of binding sites for the proteins, comparative genomic hybridization and mapping of histone modification sites [65]. Tilling array technology provides unbiased and more precise information about the transcriptome and transcriptional control at the chromosomal level. Hybridization-based approaches are relatively low cost but require prior knowledge of the transcript to be studied [66].

Among all genome-editing methods, CRISPR-Cas9 has appeared as very powerful technique for accurate genome editing study which can uncover therapeutic targets and mechanisms of neurological diseases. This technique enables editing of the human genome sequence and facilitate to correct disease related genetic variants [67]. It is comparatively simpler, precise and quicker than other genome editing technologies. CRISPR can be used to produce gene knockout mutant lines to examine the function of targeted genes. It involves designing of a guide RNA of 20 nucleotides complementary to the gene of interest and a Cas9 nuclease enzyme that cut 3-4 bases next to the protospacer adjacent motif which is later repaired. Annexin gene OsAnn3 knockout mutant lines developed *via* CRISPR-Cas9 technique showed the role of OsAnn3 gene in cold stressor tolerance in rice [68].

Comparative Genomics

Comparative genomics intent to characterize structural and functional genomic elements conserved across different species. It is a promising tool to obtain knowledge about the species with unexplored genomes by using the conservation between closely associated plant species. One of the main objectives of comparative genomics is to attempt prediction of gene function through sequence comparisons, gene order, and regulation. It also aims to differentiate conserved from divergent and functional from nonfunctional DNA [69]. It is therefore a strong area that becomes increasingly illuminative as genomic sequence data accumulate [70]. Comparative genomics has also appeared as an essential device for the characterization of micro-RNA (mi-RNA) targets that are conserved during evolution [71]. Several online comparative functional genomics analysis devices including gramene are also playing an important role in sequence comparison [72]. Comparative examination of RNA-seq expression profiling of watermelon resulted in the characterization of genes homologous to tomato controlling carotenoid synthesis [73]. Similarly, comparative examination of *Arabidopsis*, rice, barley and maize genomes allowed characterization of various important gene families such as *Sm* and *WAK* [74]. The future of crop improvement will be centered on comparison of individual plant genomes, and some of the best opportunities may lie in using combinations of new genetic mapping strategies and evolutionary analyses to direct and optimize the discovery and use of genetic variation.

PROTEOMICS APPROACHES AGAINST ABIOTIC STRESS

Proteomics deals with characterization of proteins, expression profile, post translation modifications and protein-protein interactions in a comparative manner under stress and non-stress conditions. Plant stressor proteomics is actually a dynamic area aimed at the study of plant proteome and enabled us to recognize possible target genes that can be used for genetic improvement and manipulation of plants against stressors. The role of proteins in the environmental stressor response is very important because proteins take part directly in the development of new plant phenotypes by regulating physiological characteristics, play a vital role in execution of cellular mechanisms and also act as key players in the maintenance of cellular homeostasis. Presently, much of our knowledge of plant responses to abiotic stressors is obtained through genomic and proteomic approaches.

The proteomics approach often take advantage of two-dimensional (2-D) gel electrophoresis, mass spectrometry (MS), matrix-assisted laser desorption ionization–time of flight (MALDI TOF), western blots, and ELISA in combination with bioinformatics tools to characterize proteins and map their interactions in a cellular context. Proteomics identify the useful protein markers whose changes in abundance can be related with quantitative alterations in some physiological parameters of the crops. The reaction to salinity stressor has been studied in many plant species such as rice, wheat, barley and maize using proteomics approach [75].

In *Arabidopsis*, nuclear proteome and alterations in the nuclear proteome in response to cold stressor has been studied by Bae *et al*. [76] using 2DGE and MALDI-TOF-MS technique. It was reported that 30% of the proteins altered in response to cold stressor. Similarly, Salekdeh *et al*. [77] analyzed proteome of rice to investigate alterations in response to drought stressor using 2DGE and found that expression of 42 proteins altered by stressors. When effect of salt stress was studied on rice plant root proteome and cold stress on anthers at young microspore stage using 2DGE method, it was observed that expression of 54 proteins was altered in the case of roots while 70 proteins showed differential expression in anthers [78]. Proteome analysis was performed on tobacco leaf apoplast to study effect of salt stress and found alterations in expression of 20 proteins [79]. Several proteomics studies are investigated in crops under various abiotic stressors such as drought, salt and extreme temperatures [80], proteomics of low temperature stress, dehydration stress [81, 82], heavy metal stress [83, 84], plant biotic stressors [85] with a special focus on fungal pathogens [86] and Fusarium head blight disease [87].

Moreover, specialized reviews were published on plant root proteome response to abiotic stress [88], plant post-translational modifications under abiotic stress [89], plant phosphoproteomics, redox proteomics [90 - 92], S-nitrosoproteomics [93], subcellular proteomics under stressors [94], chloroplast proteome under abiotic stress [95], crop proteomics [96], plant proteome responses to salinity [97 - 99], stress responses of major crops such as rice [100], maize [101], wheat and barley [102], soybean [103], common bean and solanaceae species [104, 105], stress proteomics of crops produced in temperate climate [106].

It has been reported that, amalgamation of genomics with proteomics and metabolomics may allow a better understanding of integrated plant reaction to abiotic stressors. Differences in proteins and metabolites will make possible the characterization of target genes and alleles for crop improvement. The target genes and proteins will decrypt the complex relationship between genes, proteins and metabolites. All these act synergistically and may provide valuable information for further study in molecular breeding for furtherance of crops (Fig. 1) .

METABOLOMIC APPROACHES AGAINST ABIOTIC STRESS

The knowledge of metabolites is critical to understand the biological processes involved in plant abiotic stress response along with the understanding of gene and proteins expression for a given plant species. Metabolomic technology is the most transversal as compared to the genomics and proteomics. It provides a better understanding of biochemical pathways and helps to characterize and quantify the complete range of primary and secondary metabolites involved in biological processes. Despite a big challenge due to the presence of several compounds with different physical and chemical properties in the plants, the presence or absence and relative accumulation of metabolites along with gene expression data provides accurate markers such as mQTL (metabotype quantitative trait locus) and MWAS (metabolome-wide association studies) for tolerant crop selection in breeding programs [107, 108].

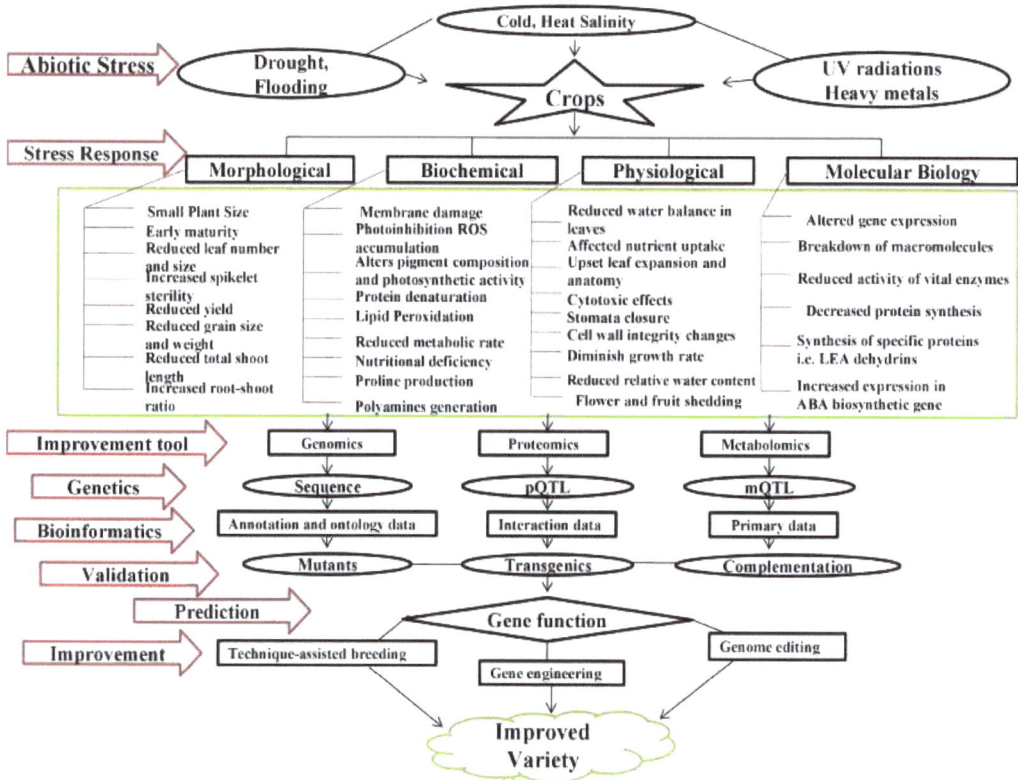

Fig. (1). Flowchart showing responses to various abiotic stressors and methodology for integrating the different omic approaches to identify target genes for crop improvement.

A single accession of *Arabidopsis* carries approximately more than 5000 metabolites, most of them yet unidentified. In order to characterize most of the compounds in a particular plant species, the adequate combination of extraction and detection technologies is paramount importance [107]. Presently, new chromatographic and mass spectrometric techniques like gas chromatography mass spectrometry (GC-MS), fourier transform ion cyclotron resonance mass spectrometry (FT-ICR-MS), liquid chromatography mass spectrometry (LC-MS), capillary electrophoresis mass spectrometry (CE-MS), and nuclear magnetic resonance (NMR) along with bioinformatics and statistical tools are commonly used in plant sciences [109]. Physiological responses of plants to environmental stress cause changes in proteins as well as metabolites leading to a particular physiological response. When plants are subjected to unfavorable environmental conditions, the most sensitive mechanism to abiotic stressor is photosynthesis, resulting in alteration in the concentration of primary metabolites such as carbohydrates, alcohols and amino acids in plant tissues. Contrarily, changes in the secondary metabolites are species specific and also specific to particular stress condition [110].

The main aim of investigating metabolic alterations during stressor responses is to characterize metabolites belonging to a particular group that are liable for stress tolerance. The accumulation of secondary metabolites is firmly dependent on a variety of environmental factors [111]. The accumulation of proline, raffinose family oligosaccharides, γ-amino butyrate (GABA) and tricarboxylic acid (TCA) cycle metabolites in leaves of *Arabidopsis* is reported under drought condition [112]. It has also been showed that stressor response in growing and mature leaves is largely distinct. The accumulation of proline, erythritol and putrecine was noticed only in mature leaves under drought condition [113]. The compounds like proline, sugars, glycine betaine, and sugar alcohols shows high concentration in drought and/or heat stress. On the other hand, total phenol, flavonoid, and saponin content decreases in response to drought and/or heat stress [114]. The accumulation of amino acids, alanine, proline and GABA, and the phosphoesters, glucose-6-phosphate and glycerol-3-phosphate, were recorded in *Arabidopsis* roots under anoxic conditions. It was also reported that, when oxygen is decreased to 4%, the levels of the intermediates both of sucrose degradation and the TCA cycle increased, however, levels of most amino acids declined when the oxygen further decreased to 1%. It indicates the inhibition and reactivation of metabolic activities. Similarly, high temperature induces the biosynthesis of alkaloids. Contrarily, in the roots of *Glycine max*, level of phenolic acids and isoflavonoid improved after the treatment at low temperature for 24 hr [115]. Metabolomic changes during cold acclimation was also studied and compared in two ecotypes of *Arabidopsis thaliana*, Wassilewskija-2 (Ws-2) and Cape verde islands-1 (Cvi-1). In cold acclimation process, plants respond to freezing environment.

Wassilewskija-2 (Ws-2) is freezing tolerant while Cape verde islands-1 (Cvi-1) is a freezing sensitive. Results revealed that, metabolome of Ws-2 species were extensively altered in response to low temperature [116, 117]. Regarding light, its quantity as well as quality both affects plant metabolism. When metabolic profiling of leaves of *Arabidopsis* were studied under high light, maximum metabolites of the glycolysis, TCA cycle and oxidative pentose phosphate pathway were changed in their content. It indicates that plants exposed to high light undergo a metabolic shift and improve the Calvin-Benson cycle to fix more carbon [118]. When *Arabidopsis* is exposed to ultraviolet (UV) light, most of the changes in the levels of primary metabolites, including ascorbate derivatives and secondary metabolites, such as flavonoids and phenolics were reported.

Salinity is one of the most brutal abiotic stressors that prompt complex interconnections among numerous morphological, physiological and biochemical processes [119]. It also causes the oxidative stress due to high production of reactive oxygen species (ROS) so as to change plant metabolism. When *Aegiceras corniculatum* treated by 250 mM NaCl, polyphenol content considerably increased to almost double as compared to control plants [120]. Similarly, salinization was studied in *Rosmarinus officinalis* and reported alteration in the accumulation of secondary metabolites, mainly inducing a pronounced effect on monoterpenes composition [121]. Ghasemi *et al.* [122] reported the affect of UV light on the expression of key genes involved in the biosynthesis of Phenylalanine ammonia lyase (PAL), Hydroxymethylglutaryl-CoA reductase (HMG-CoA reductase), GPP synthases, Deoxyribonino heptulosinate 7-phosphate synthase (DAHP), and Deoxy Xylose Phosphate Synthase (DXS)) and reported a remarkable increase in total levels of terpenoids, phenols, flavonoids, anthocyanins, alkaloids, beta-carotene, and lycopene. Although, the field of metabolomics has approached a long way in the past, but it still has a long way to go. Several metabolites have very short turnovers, which explain their diurnal changes and also the content vary even in response to short term alteration in the environment. Hence, the environmental variables must be recorded over the course of the trial for metabolomics. Finally, when combined with genomics and proteomics, metabolomics represents a critical puzzle piece in the understanding of biological systems—and with that understanding, much more will be possible [123].

CONCLUSION

Abiotic stressor is one of the crucial prohibitive factors in plant growth and productivity. Consequently, understanding mechanisms underlying stressor tolerance in plants is not only of an academic interest, but also has great socioeconomic importance. Different '-omic' techniques are being used in a system biology approach to better understand plant processes to aquire a comprehensive knowledge for the identification and description of cellular networks through which stress perception, signal transduction and defensive responses are mediated, which is essential to sustainably increase food production under adverse environmental conditions.

CONSENT FOR PUBLICATION

Not applicable.

CONFLICT OF INTEREST

The author declares no conflict of interest, financial or otherwise.

ACKNOWLEDGEMENTS

Declared none.

REFERENCES

[1]   Pandey P, Irulappan V, Bagavathiannan MV, Senthil-Kumar M. Impact of combined abiotic and biotic stressors on plant growth and avenues for crop improvement by exploiting physio-morphological traits. Front Plant Sci 2017; 8: 537.
       [http://dx.doi.org/10.3389/fpls.2017.00537] [PMID: 28458674]

[2]   Dos Reis SP, Lima AM, de Souza CR. Recent molecular advances on downstream plant responses to abiotic stress. Int J Mol Sci 2012; 13(7): 8628-47.
       [http://dx.doi.org/10.3390/ijms13078628] [PMID: 22942725]

[3]   Meena KK, Sorty AM, Bitla UM, *et al.* Abiotic stress responses and microbe-mediated mitigation in plants: the omics strategies. Front Plant Sci 2017; 8: 172.
       [http://dx.doi.org/10.3389/fpls.2017.00172] [PMID: 28232845]

[4]   Rodziewicz P, Swarcewicz B, Chmielewska K, *et al.* Influence of abiotic stresses on plant proteome and metabolome changes. Acta Physiol Plant 2014; 36: 1-19.
       [http://dx.doi.org/10.1007/s11738-013-1402-y]

[5]   Yuan JS, Galbraith DW, Dai SY, Griffin P, Stewart CN Jr. Plant systems biology comes of age. Trends Plant Sci 2008; 13(4): 165-71.
       [http://dx.doi.org/10.1016/j.tplants.2008.02.003] [PMID: 18329321]

[6]   Munns R, James RA, Läuchli A. Approaches to increasing the salt tolerance of wheat and other cereals. J Exp Bot 2006; 57(5): 1025-43.
       [http://dx.doi.org/10.1093/jxb/erj100] [PMID: 16510517]

[7]   Akpınar BA, Lucas SJ, Budak H. Genomics approaches for crop improvement against abiotic stress. ScientificWorldJournal 2013; 2013: 361921.
       [http://dx.doi.org/10.1155/2013/361921] [PMID: 23844392]

[8]   Schmutz J, Cannon SB, Schlueter J, *et al.* Genome sequence of the palaeopolyploid soybean. Nature 2010; 463(7278): 178-83.
       [http://dx.doi.org/10.1038/nature08670] [PMID: 20075913]

[9]   Feldmann KA, Goff SA. The first plant genome sequence-*Arabidopsis thaliana.* Advances in botanical

research. Academic press 2014; Vol. 69: pp. 91-117.

[10]   Ahanger MA, Akram NA, Ashraf M, Alyemeni MN, Wijaya L, Ahmad P. Signal transduction and biotechnology in response to environmental stresses. Biol Plant 2017; 61(3): 401-16.
[http://dx.doi.org/10.1007/s10535-016-0683-6]

[11]   Huang B, Xu C. Identification and characterization of proteins associated with plant tolerance to heat stress. J Integr Plant Biol 2008; 50(10): 1230-7.
[http://dx.doi.org/10.1111/j.1744-7909.2008.00735.x] [PMID: 19017110]

[12]   Swindell WR. The association among gene expression responses to nine abiotic stress treatments in *Arabidopsis thaliana*. Genetics 2006; 174(4): 1811-24.
[http://dx.doi.org/10.1534/genetics.106.061374] [PMID: 17028338]

[13]   Roychoudhury A, Datta K, Datta SK. Abiotic stress in plants: from genomics to metabolomics Omics and plant abiotic stress tolerance. Chicago: Bentham Science Publishers 2011; pp. 91-120.

[14]   Griffiths AJF, Miller JH, Suzuki DT, *et al.* An introduction to genetic analysis. Structural Genomics. 7th edition., New York: W. H. Freeman 2000. Available from: https://www.ncbi.nlm.nih.gov/books/NBK21841/

[15]   Muthamilarasan M, Venkata Suresh B, Pandey G, Kumari K, Parida SK, Prasad M. Development of 5123 intron-length polymorphic markers for large-scale genotyping applications in foxtail millet. DNA Res 2014; 21(1): 41-52.
[http://dx.doi.org/10.1093/dnares/dst039] [PMID: 24086082]

[16]   Abouzari A, Solouki M, Golein B, Fakheri BA, Sabouri A, Dadras AR. Screening of molecular markers associated to cold tolerance- related traits in Citrus. Sci Hortic (Amsterdam) 2020; 263: 109-45.
[http://dx.doi.org/10.1016/j.scienta.2019.109145]

[17]   Wang D, Liu YX, Yu Q, *et al.* Functional analysis of the soybean *GmCDPK3* gene responding to drought and salt stresses. Int J Mol Sci 2019; 20(23): 5909.
[http://dx.doi.org/10.3390/ijms20235909] [PMID: 31775269]

[18]   Zhou S, Zheng WJ, Liu BH, *et al.* Characterizing the Role of *TaWRKY13* in Salt Tolerance. Int J Mol Sci 2019; 20(22): 5712.
[http://dx.doi.org/10.3390/ijms20225712] [PMID: 31739570]

[19]   Appleby N, Edwards D, Batley J. New technologies for ultra-high throughput genotyping in plants. Plant genomics. Humana Press 2009; pp. 19-39.
[http://dx.doi.org/10.1007/978-1-59745-427-8_2]

[20]   McCouch SR, Wright MH, Tung CW, *et al.* Open access resources for genome-wide association mapping in rice. Nat Commun 2016; 7(1): 1-4.

[21]   Bayer MM, Rapazote-Flores P, Ganal M, *et al.* Development and evaluation of a barley 50k iSelectSNP array. Front Plant Sci 2017; 8: 1792.
[http://dx.doi.org/10.3389/fpls.2017.01792] [PMID: 29089957]

[22]   Peterson BK, Weber JN, Kay EH, Fisher HS, Hoekstra HE. Double digest RADseq: an inexpensive method for de novo SNP discovery and genotyping in model and non-model species. PLoS One 2012; 7(5): e37135.
[http://dx.doi.org/10.1371/journal.pone.0037135] [PMID: 22675423]

[23]   Su HG, Li B, Song XY, *et al.* Genome-wide analysis of the DYW subgroup PPR gene family and identification of *GmPPR4* responses to drought stress. Int J Mol Sci 2019; 20(22): 5667.
[http://dx.doi.org/10.3390/ijms20225667] [PMID: 31726763]

[24]   Imran M, Shafiq S, Farooq MA, *et al.* Comparative genome-wide analysis and expression profiling of Histone acetyltransferase (HAT) gene family in response to hormonal applications, metal and abiotic stresses in cotton. Int J Mol Sci 2019; 20(21): 5311.
[http://dx.doi.org/10.3390/ijms20215311] [PMID: 31731441]

[25]   Liu S, Wang X, Wang H, *et al.* Genome-wide analysis of ZmDREB genes and their association with natural variation in drought tolerance at seedling stage of *Zea mays* L. PLoS Genet 2013; 9(9): e1003790.
[http://dx.doi.org/10.1371/journal.pgen.1003790] [PMID: 24086146]

[26]   Thirunavukkarasu N, Hossain F, Arora K, *et al.* Functional mechanisms of drought tolerance in subtropical maize (*Zea mays* L.) identified using genome-wide association mapping. BMC Genomics 2014; 15(1): 1182.
[http://dx.doi.org/10.1186/1471-2164-15-1182] [PMID: 25539911]

[27]   Famoso AN, Zhao K, Clark RT, *et al.* Genetic architecture of aluminum tolerance in rice (*Oryza sativa*) determined through genome-wide association analysis and QTL mapping. PLoS Genet 2011; 7(8): e1002221.
[http://dx.doi.org/10.1371/journal.pgen.1002221] [PMID: 21829395]

[28]   Wehner GG, Balko CC, Enders MM, Humbeck KK, Ordon FF. Identification of genomic regions involved in tolerance to drought stress and drought stress induced leaf senescence in juvenile barley. BMC Plant Biol 2015; 15(1): 125.
[http://dx.doi.org/10.1186/s12870-015-0524-3] [PMID: 25998066]

[29]   Deshmukh R, Tomar NS, Tripathi N, Tiwari S. Identification of RAPD and ISSR markers for drought tolerance in wheat (*Triticum aestivum* L.). Physiol Mol Biol Plants 2012; 18(1): 101-4.
[http://dx.doi.org/10.1007/s12298-011-0096-0] [PMID: 23573046]

[30]   Vatter T, Maurer A, Kopahnke D, Perovic D, Ordon F, Pillen K. A nested association mapping population identifies multiple small effect QTL conferring resistance against net blotch (*Pyrenophora teres* f. teres) in wild barley. PLoS One 2017; 12(10): e0186803.
[http://dx.doi.org/10.1371/journal.pone.0186803] [PMID: 29073176]

[31]   Thudi M, Gaur PM, Krishnamurthy L, *et al.* Genomics-assisted breeding for drought tolerance in chickpea. Funct Plant Biol 2014; 41(11): 1178-90.
[http://dx.doi.org/10.1071/FP13318] [PMID: 32481067]

[32]   Bandillo N, Raghavan C, Muyco PA, *et al.* Multi-parent advanced generation inter-cross (MAGIC) populations in rice: progress and potential for genetics research and breeding. Rice (N Y) 2013; 6(1): 11.
[http://dx.doi.org/10.1186/1939-8433-6-11] [PMID: 24280183]

[33]   Huynh BL, Ehlers JD, Huang BE, *et al.* A multi-parent advanced generation inter-cross (MAGIC) population for genetic analysis and improvement of cowpea (*Vigna unguiculata* L. Walp.). Plant J 2018; 93(6): 1129-42.
[http://dx.doi.org/10.1111/tpj.13827] [PMID: 29356213]

[34]   Edmeades GO, McMaster GS, White JW, Campos H. Genomics and the physiologist: bridging the gap between genes and crop response. Field Crops Res 2004; 90(1): 5-18.
[http://dx.doi.org/10.1016/j.fcr.2004.07.002]

[35]   Ahmed F, Rafii MY, Ismail MR, *et al.* Waterlogging tolerance of crops: breeding, mechanism of tolerance, molecular approaches, and future prospects. BioMed Res Int 2013; 2013: 963525.
[http://dx.doi.org/10.1155/2013/963525] [PMID: 23484164]

[36]   Nadeem MA, Nawaz MA, Shahid MQ, *et al.* DNA molecular markers in plant breeding: current status and recent advancements in genomic selection and genome editing. Biotechnol Biotechnol Equip 2018; 32(2): 261-85.
[http://dx.doi.org/10.1080/13102818.2017.1400401]

[37]   Sehgal D, Baliyan N, Kaur P. Progress towards identification and validation of candidate genes for abiotic stress tolerance in wheat. In: Rajpal V, Sehgal D, Kumar A, Raina S, Eds. Genomics Assisted Breeding of Crops for Abiotic Stress Tolerance. Cham: Springer 2019; Vol. II: pp. 31-48.

[38]   Abe A, Kosugi S, Yoshida K, *et al.* Genome sequencing reveals agronomically important loci in rice

using MutMap. Nat Biotechnol 2012; 30(2): 174-8.
[http://dx.doi.org/10.1038/nbt.2095] [PMID: 22267009]

[39]   Jaganathan D, Bohra A, Thudi M, Varshney RK. Fine mapping and gene cloning in the post-NGS era: advances and prospects. Theor Appl Genet 2020; 133(5): 1791-810.
[http://dx.doi.org/10.1007/s00122-020-03560-w] [PMID: 32040676]

[40]   Warr A, Robert C, Hume D, Archibald A, Deeb N, Watson M. Exome sequencing: current and future perspective. G3: Genes, Genomes. G3 (Bethesda) 2015; 5(8): 1543-50.
[http://dx.doi.org/10.1534/g3.115.018564] [PMID: 26139844]

[41]   Goyal E, Amit SK, Singh RS, Mahato AK, Chand S, Kanika K. Transcriptome profiling of the salt-stress response in *Triticum aestivum* cv. Kharchia Local. Sci Rep 2016; 6(1): 27752.
[http://dx.doi.org/10.1038/srep27752] [PMID: 27293111]

[42]   Gao L, Gonda I, Sun H, *et al.* The tomato pan-genome uncovers new genes and a rare allele regulating fruit flavor. Nat Genet 2019; 51(6): 1044-51.
[http://dx.doi.org/10.1038/s41588-019-0410-2] [PMID: 31086351]

[43]   Mondal TK, Sutoh K. Application of next-generation sequencing for abiotic stress tolerance OMICS: Applications in Biomedical, Agricultural, and Environmental sciences. CRC Press 2013; pp. 347-60.

[44]   Li C, Yue J, Wu X, Xu C, Yu J. An ABA-responsive DRE-binding protein gene from *Setaria italica*, SiARDP, the target gene of SiAREB, plays a critical role under drought stress. J Exp Bot 2014; 65(18): 5415-27.
[http://dx.doi.org/10.1093/jxb/eru302] [PMID: 25071221]

[45]   Bohnert HJ, Ayoubi P, Borchert C, *et al.* A genomic approach towards salt stress tolerance. Plant Physiol Biochem 2001; 39(3-4): 295-311.
[http://dx.doi.org/10.1016/S0981-9428(00)01237-7]

[46]   Houde M, Belcaid M, Ouellet F, *et al.* Wheat EST resources for functional genomics of abiotic stress. BMC Genomics 2006; 7(1): 149.
[http://dx.doi.org/10.1186/1471-2164-7-149] [PMID: 16772040]

[47]   Velculescu VE, Vogelstein B, Kinzler KW. Analysing uncharted transcriptomes with SAGE. Trends Genet 2000; 16(10): 423-5.
[http://dx.doi.org/10.1016/S0168-9525(00)02114-4] [PMID: 11050322]

[48]   Matsumura H, Nirasawa S, Terauchi R. Technical advance: transcript profiling in rice (*Oryza sativa* L.) seedlings using serial analysis of gene expression (SAGE). Plant J 1999; 20(6): 719-26.
[http://dx.doi.org/10.1046/j.1365-313X.1999.00640.x] [PMID: 10652144]

[49]   Lee JY, Lee DH. Use of serial analysis of gene expression technology to reveal changes in gene expression in *Arabidopsis* pollen undergoing cold stress. Plant Physiol 2003; 132(2): 517-29.
[http://dx.doi.org/10.1104/pp.103.020511] [PMID: 12805584]

[50]   Reinartz J, Bruyns E, Lin JZ, *et al.* Massively parallel signature sequencing (MPSS) as a tool for in-depth quantitative gene expression profiling in all organisms. Brief Funct Genomics Proteomics 2002; 1(1): 95-104.
[http://dx.doi.org/10.1093/bfgp/1.1.95] [PMID: 15251069]

[51]   Brenner S, Johnson M, Bridgham J, *et al.* Gene expression analysis by massively parallel signature sequencing (MPSS) on microbead arrays. Nat Biotechnol 2000; 18(6): 630-4.
[http://dx.doi.org/10.1038/76469] [PMID: 10835600]

[52]   Nakano M, Nobuta K, Vemaraju K, Tej SS, Skogen JW, Meyers BC. Plant MPSS databases: signature-based transcriptional resources for analyses of mRNA and small RNA. Nucleic Acids Res 2006; 34(Database issue): D731-5.
[http://dx.doi.org/10.1093/nar/gkj077] [PMID: 16381968]

[53]   Vij S, Tyagi AK. Emerging trends in the functional genomics of the abiotic stress response in crop plants. Plant Biotechnol J 2007; 5(3): 361-80.

[http://dx.doi.org/10.1111/j.1467-7652.2007.00239.x] [PMID: 17430544]

[54]   Wang Z, Gerstein M, Snyder M. RNA-Seq: a revolutionary tool for transcriptomics. Nat Rev Genet 2009; 10(1): 57-63.
[http://dx.doi.org/10.1038/nrg2484] [PMID: 19015660]

[55]   Park PJ. ChIP-seq: advantages and challenges of a maturing technology. Nat Rev Genet 2009; 10(10): 669-80.
[http://dx.doi.org/10.1038/nrg2641] [PMID: 19736561]

[56]   Jiao Y, Jia P, Wang X, *et al.* A tiling microarray expression analysis of rice chromosome 4 suggests a chromosome-level regulation of transcription. Plant Cell 2005; 17(6): 1641-57.
[http://dx.doi.org/10.1105/tpc.105.031575] [PMID: 15863518]

[57]   Kawasaki S, Borchert C, Deyholos M, *et al.* Gene expression profiles during the initial phase of salt stress in rice. Plant Cell 2001; 13(4): 889-905.
[http://dx.doi.org/10.1105/tpc.13.4.889] [PMID: 11283343]

[58]   Lan L, Li M, Lai Y, *et al.* Microarray analysis reveals similarities and variations in genetic programs controlling pollination/fertilization and stress responses in rice (*Oryza sativa* L.). Plant Mol Biol 2005; 59(1): 151-64.
[http://dx.doi.org/10.1007/s11103-005-3958-4] [PMID: 16217609]

[59]   Cooper JL, Henikoff S, Comai L, Till BJ. Tilling and ecotilling for rice. Methods Mol Biol 2013; 956: 39-56.
[http://dx.doi.org/10.1007/978-1-62703-194-3_4] [PMID: 23135843]

[60]   Garg B, Lata C, Prasad M. A study of the role of gene TaMYB2 and an associated SNP in dehydration tolerance in common wheat. Mol Biol Rep 2012; 39(12): 10865-71.
[http://dx.doi.org/10.1007/s11033-012-1983-3] [PMID: 23065204]

[61]   Rebrikov DV, Desai SM, Siebert PD, Lukyanov SA. Suppression subtractive hybridization. Methods Mol Biol 2004; 258: 107-34.
[PMID: 14970460]

[62]   Luo M, Liu J, Lee RD, Scully BT, Guo B. Monitoring the expression of maize genes in developing kernels under drought stress using oligo-microarray. J Integr Plant Biol 2010; 52(12): 1059-74.
[http://dx.doi.org/10.1111/j.1744-7909.2010.01000.x] [PMID: 21106005]

[63]   Loukehaich R, Wang T, Ouyang B, *et al.* SpUSP, an annexin-interacting universal stress protein, enhances drought tolerance in tomato. J Exp Bot 2012; 63(15): 5593-606.
[http://dx.doi.org/10.1093/jxb/ers220] [PMID: 22915741]

[64]   Singh A, Pandey A, Baranwal V, Kapoor S, Pandey GK. Comprehensive expression analysis of rice phospholipase D gene family during abiotic stresses and development. Plant Signal Behav 2012; 7(7): 847-55.
[http://dx.doi.org/10.4161/psb.20385] [PMID: 22751320]

[65]   Matsui A, Ishida J, Morosawa T, *et al. Arabidopsis* tiling array analysis to identify the stress-responsive genes. Methods Mol Biol 2010; 639. 141-55.
[http://dx.doi.org/10.1007/978-1-60761-702-0_8] [PMID: 20387044]

[66]   Rensink WA, Buell CR. Microarray expression profiling resources for plant genomics. Trends Plant Sci 2005; 10(12): 603-9.
[http://dx.doi.org/10.1016/j.tplants.2005.10.003] [PMID: 16275051]

[67]   Kampmann M. CRISPR-based functional genomics for neurological disease. Nat Rev Neurol 2020; 16(9): 465-80.
[http://dx.doi.org/10.1038/s41582-020-0373-z] [PMID: 32641861]

[68]   Zafar SA, Hameed A, Nawaz MA, *et al.* Mechanisms and molecular approaches for heat tolerance in rice (*Oryza sativa* L.) under climate change scenario. J Integr Agric 2018; 17(4): 726-38.
[http://dx.doi.org/10.1016/S2095-3119(17)61718-0]

[69]   Hardison RC. Comparative genomics. PLoS Biol 2003; 1(2): E58.
[http://dx.doi.org/10.1371/journal.pbio.0000058] [PMID: 14624258]

[70]   Mochida K, Shinozaki K. Genomics and bioinformatics resources for crop improvement. Plant Cell Physiol 2010; 51(4): 497-523.
[http://dx.doi.org/10.1093/pcp/pcq027] [PMID: 20208064]

[71]   Friedman RC, Burge CB. MicroRNA target finding by comparative genomics. Methods Mol Biol 2014; 1097: 457-76.
[http://dx.doi.org/10.1007/978-1-62703-709-9_21] [PMID: 24639172]

[72]   Monaco MK, Stein J, Naithani S, *et al.* Gramene 2013: comparative plant genomics resources. Nucleic Acids Res 2014; 42(Database issue): D1193-9.
[http://dx.doi.org/10.1093/nar/gkt1110] [PMID: 24217918]

[73]   Grassi S, Piro G, Lee JM, *et al.* Comparative genomics reveals candidate carotenoid pathway regulators of ripening watermelon fruit. BMC Genomics 2013; 14: 781.
[http://dx.doi.org/10.1186/1471-2164-14-781] [PMID: 24219562]

[74]   Chen Y, Cao J. Comparative genomic analysis of the Sm gene family in rice and maize. Gene 2014; 539(2): 238-49.
[http://dx.doi.org/10.1016/j.gene.2014.02.006] [PMID: 24525402]

[75]   Zhang H, Han B, Wang T, *et al.* Mechanisms of plant salt response: insights from proteomics. J Proteome Res 2012; 11(1): 49-67.
[http://dx.doi.org/10.1021/pr200861w] [PMID: 22017755]

[76]   Bae MS, Cho EJ, Choi EY, Park OK. Analysis of the *Arabidopsis* nuclear proteome and its response to cold stress. Plant J 2003; 36(5): 652-63.
[http://dx.doi.org/10.1046/j.1365-313X.2003.01907.x] [PMID: 14617066]

[77]   Salekdeh GH, Siopongco J, Wade LJ, Ghareyazie B, Bennett J. Proteomic analysis of rice leaves during drought stress and recovery. Proteomics 2002; 2(9): 1131-45.
[http://dx.doi.org/10.1002/1615-9861(200209)2:9<1131::AID-PROT1131>3.0.CO;2-1]   [PMID: 12362332]

[78]   Imin N, De Jong F, Mathesius U, *et al.* Proteome reference maps of *Medicago truncatula* embryogenic cell cultures generated from single protoplasts. Proteomics 2004; 4(7): 1883-96.
[http://dx.doi.org/10.1002/pmic.200300803] [PMID: 15221745]

[79]   Dani V, Simon WJ, Duranti M, Croy RR. Changes in the tobacco leaf apoplast proteome in response to salt stress. Proteomics 2005; 5(3): 737-45.
[http://dx.doi.org/10.1002/pmic.200401119] [PMID: 15682462]

[80]   Ahmad P, Abdel Latef AAH, Rasool S, Akram NA, Ashraf M, Gucel S. Role of proteomics in crop stress tolerance. Front Plant Sci 2016; 7: 1336.
[http://dx.doi.org/10.3389/fpls.2016.01336] [PMID: 27660631]

[81]   Janmohammadi M, Zolla L, Rinalducci S. Low temperature tolerance in plants: Changes at the protein level. Phytochemistry 2015; 117: 76-89.
[http://dx.doi.org/10.1016/j.phytochem.2015.06.003] [PMID: 26068669]

[82]   Johnová P, Skalák J, Saiz-Fernández I, Brzobohatý B. Plant responses to ambient temperature fluctuations and water-limiting conditions: a proteome-wide perspective. J Biochem Biophys Acta Prot Proteo 2016; 1864: 31-916.

[83]   Ahsan N, Renaut J, Komatsu S. Recent developments in the application of proteomics to the analysis of plant responses to heavy metals. Proteomics 2009; 9(10): 2602-21.
[http://dx.doi.org/10.1002/pmic.200800935] [PMID: 19405030]

[84]   Hossain Z, Komatsu S. Contribution of proteomic studies towards understanding plant heavy metal stress response. Front Plant Sci 2013; 3: 310.

[http://dx.doi.org/10.3389/fpls.2012.00310] [PMID: 23355841]

[85]   Sergeant K, Renaut J. Plant biotic stress and proteomics. J Curr Proteo 2010; 7: 275-97.
       [http://dx.doi.org/10.2174/157016410793611765]

[86]   Rampitsch C, Bykova NV. Proteomics and plant disease: advances in combating a major threat to the
       global food supply. Proteomics 2012; 12(4-5): 673-90.
       [http://dx.doi.org/10.1002/pmic.201100359] [PMID: 22246663]

[87]   Yang F, Jacobsen S, Jørgensen HJL, Collinge DB, Svensson B, Finnie C. *Fusarium graminearum* and
       its interactions with cereal heads: studies in the proteomics era. Front Plant Sci 2013; 4: 37.
       [http://dx.doi.org/10.3389/fpls.2013.00037] [PMID: 23450732]

[88]   Ghosh D, Xu J. Abiotic stress responses in plant roots: a proteomics perspective. Front Plant Sci 2014;
       5: 6.
       [http://dx.doi.org/10.3389/fpls.2014.00006] [PMID: 24478786]

[89]   Wu X, Gong F, Cao D, Hu X, Wang W. Advances in crop proteomics: PTMs of proteins under abiotic
       stress. Proteomics 2016; 16(5): 847-65.
       [http://dx.doi.org/10.1002/pmic.201500301] [PMID: 26616472]

[90]   Rampitsch C, Bykova NV. The beginnings of crop phosphoproteomics: exploring early warning
       systems of stress. Front Plant Sci 2012; 3: 144.
       [http://dx.doi.org/10.3389/fpls.2012.00144] [PMID: 22783265]

[91]   Rinalducci S, Murgiano L, Zolla L. Redox proteomics: basic principles and future perspectives for the
       detection of protein oxidation in plants. J Exp Bot 2008; 59(14): 3781-801.
       [http://dx.doi.org/10.1093/jxb/ern252] [PMID: 18977746]

[92]   Mock HP, Dietz KJ. Redox proteomics for the assessment of redox-related posttranslational regulation
       in plants. Biochim Biophys Acta 2016; 1864(8): 967-73.
       [http://dx.doi.org/10.1016/j.bbapap.2016.01.005] [PMID: 26784836]

[93]   Hossain Z, Nouri MZ, Komatsu S. Plant cell organelle proteomics in response to abiotic stress. J
       Proteome Res 2012; 11(1): 37-48.
       [http://dx.doi.org/10.1021/pr200863r] [PMID: 22029473]

[94]   Ning F, Wang W. The response of chloroplast proteome to abiotic stress. In: MA Hossain SH, Wani S,
       Bhattacharjee DJ, Burritt LSP, Eds. Drought Stress Tolerance in Plants. Verlag: Springer 2016; pp.
       232-49.
       [http://dx.doi.org/10.1007/978-3-319-32423-4_9]

[95]   Salekdeh GH, Komatsu S. Crop proteomics: aim at sustainable agriculture of tomorrow. Proteomics
       2007; 7(16): 2976-96.
       [http://dx.doi.org/10.1002/pmic.200700181] [PMID: 17639607]

[96]   Zhang J, Liang S, Duan J, *et al.* *De novo* assembly and characterization of the transcriptome during
       seed development and generation of genic-SSR markers in Peanut (*Arachis hypogaea* L.). BMC
       Genomics 2012; 13: 13-90.
       [http://dx.doi.org/10.1186/1471-2164-13-90]

[97]   Kosová K, Práil IT, Vítámvás P. Protein contribution to plant salinity response and tolerance
       acquisition. Int J Mol Sci 2013; 14(4): 6757-89.
       [http://dx.doi.org/10.3390/ijms14046757] [PMID: 23531537]

[98]   Kosov KR, V T Mv S P, Urban MOI, Pr Il IT. Plant proteome responses to salinity stress - comparison
       of glycophytes and halophytes. Funct Plant Biol 2013; 40(9): 775-86.
       [http://dx.doi.org/10.1071/FP12375] [PMID: 32481150]

[99]   Agrawal GK, Jwa NS, Rakwal R. Rice proteomics: ending phase I and the beginning of phase II.
       Proteomics 2009; 9(4): 935-63.
       [http://dx.doi.org/10.1002/pmic.200800594] [PMID: 19212951]

[100] Pechanova O, Takáč T, Samaj J, Pechan T. Maize proteomics: an insight into the biology of an important cereal crop. Proteomics 2013; 13(3-4): 637-62.
[http://dx.doi.org/10.1002/pmic.201200275] [PMID: 23197376]

[101] Komatsu S, Kamal AH, Hossain Z. Wheat proteomics: proteome modulation and abiotic stress acclimation. Front Plant Sci 2014; 5: 684.
[http://dx.doi.org/10.3389/fpls.2014.00684] [PMID: 25538718]

[102] Yin X, Komatsu S. Comprehensive analysis of response and tolerant mechanisms in early-stage soybean at initial-flooding stress. J Proteomics 2017; 169: 225-32.
[http://dx.doi.org/10.1016/j.jprot.2017.01.014] [PMID: 28137666]

[103] Zargar SM, Mahajan R, Nazir M, *et al.* Common bean proteomics: Present status and future strategies. J Proteomics 2017; 169: 239-48.
[http://dx.doi.org/10.1016/j.jprot.2017.03.019] [PMID: 28347863]

[104] Ghatak A, Chaturvedi P, Paul P, *et al.* Proteomics survey of Solanaceae family: Current status and challenges ahead. J Proteomics 2017; 169: 41-57.
[http://dx.doi.org/10.1016/j.jprot.2017.05.016] [PMID: 28528990]

[105] Kosová K, Vítámvás P, Urban MO, Klíma M, Roy A, Prášil IT. Biological networks underlying abiotic stress tolerance in temperate crops - a proteomic perspective. Int J Mol Sci 2015; 16(9): 20913-42.
[http://dx.doi.org/10.3390/ijms160920913] [PMID: 26340626]

[106] Arbona V, Manzi M. de OC, Gómez-Cadenas A. Int J Mol Sci 2013; 14: 4885-911.
[http://dx.doi.org/10.3390/ijms14034885] [PMID: 23455464]

[107] Beckles D, Roessner U. Plant metabolomics – applications and opportunities for agricultural biotechnology. In: Altmann A, Hasegawa PM, Hasegawa PM, Eds. Plant Biotechnology and Agriculture: Prospects for the 21st Century. London: Elsevier 2012; pp. 67-78.

[108] Putri SP, Yamamoto S, Tsugawa H, Fukusaki E. Current metabolomics: technological advances. J Biosci Bioeng 2013; 116(1): 9-16.
[http://dx.doi.org/10.1016/j.jbiosc.2013.01.004] [PMID: 23466298]

[109] Verslues PE, Agarwal M, Katiyar-Agarwal S, Zhu J, Zhu JK. Methods and concepts in quantifying resistance to drought, salt and freezing, abiotic stresses that affect plant water status. Plant J 2006; 45(4): 523-39.
[http://dx.doi.org/10.1111/j.1365-313X.2005.02593.x] [PMID: 16441347]

[110] Valerio C, Costa A, Marri L, *et al.* Thioredoxin-regulated beta-amylase (BAM1) triggers diurnal starch degradation in guard cells, and in mesophyll cells under osmotic stress. J Exp Bot 2011; 62(2): 545-55.
[http://dx.doi.org/10.1093/jxb/erq288] [PMID: 20876336]

[111] Yang L, Wen KS, Ruan X, Zhao YX, Wei F, Wang Q. Response of plant secondary metabolites to environmental factors. Molecules 2018; 23(4): 762.
[http://dx.doi.org/10.3390/molecules23040762] [PMID: 29584636]

[112] Urano K, Maruyama K, Ogata Y, *et al.* Characterization of the ABA-regulated global responses to dehydration in *Arabidopsis* by metabolomics. Plant J 2009; 57(6): 1065-78.
[http://dx.doi.org/10.1111/j.1365-313X.2008.03748.x] [PMID: 19036030]

[113] Skirycz A, De Bodt S, Obata T, *et al.* Developmental stage specificity and the role of mitochondrial metabolism in the response of *Arabidopsis* leaves to prolonged mild osmotic stress. Plant Physiol 2010; 152(1): 226-44.
[http://dx.doi.org/10.1104/pp.109.148965] [PMID: 19906889]

[114] Alhaithloul HA, Soliman MH, Ameta KL, *et al.* Changes in ecophysiology, osmolytes, and secondary metabolites of the medicinal plants of *Mentha piperita* and *Catharanthus roseus* subjected to drought and heat stress. Biomolecules 2019; 10: 43.

[115] Janas KM, Cvikrová M, Pałagiewicz A, Szafranska K, Posmyk MM. Constitutive elevated accumulation of phenylpropanoids in soybean roots at low temperature. Plant Sci 2002; 163: 369-73.
[http://dx.doi.org/10.1016/S0168-9452(02)00136-X]

[116] Rocha M, Licausi F, Araújo WL, *et al.* Glycolysis and the tricarboxylic acid cycle are linked by alanine aminotransferase during hypoxia induced by waterlogging of *Lotus japonicus*. Plant Physiol 2010; 152(3): 1501-13.
[http://dx.doi.org/10.1104/pp.109.150045] [PMID: 20089769]

[117] Cook D, Fowler S, Fiehn O, Thomashow MF. A prominent role for the CBF cold response pathway in configuring the low-temperature metabolome of *Arabidopsis*. Proc Natl Acad Sci USA 2004; 101(42): 15243-8.
[http://dx.doi.org/10.1073/pnas.0406069101] [PMID: 15383661]

[118] Wulff-Zottele C, Gatzke N, Kopka J, *et al.* Photosynthesis and metabolism interact during acclimation of *Arabidopsis thaliana* to high irradiance and sulphur depletion. Plant Cell Environ 2010; 33(11): 1974-88.
[http://dx.doi.org/10.1111/j.1365-3040.2010.02199.x] [PMID: 20573050]

[119] Cuartero J, Bolarín MC, Asíns MJ, Moreno V. Increasing salt tolerance in the tomato. J Exp Bot 2006; 57(5): 1045-58.
[http://dx.doi.org/10.1093/jxb/erj102] [PMID: 16520333]

[120] Parida AK, Das AB, Sanada Y, Mohanty P. Effects of salinity on biochemical components of the mangrove. *Aegiceras corniculatum*. Aquat Bot 2004; 80: 77-87.
[http://dx.doi.org/10.1016/j.aquabot.2004.07.005]

[121] Tounekti T, Vadel AM, Ennajeh M, Khemira H, Munné-Bosch S. Ionic interactions and salinity affect monoterpene and phenolic diterpene composition in rosemary (*Rosmarinus officinalis*). J Plant Nutr Soil Sci 2011; 174: 504-14.
[http://dx.doi.org/10.1002/jpln.201000213]

[122] Ghasemi S, Kumleh HH, Kordrostami M. Changes in the expression of some genes involved in the biosynthesis of secondary metabolites in *Cuminum cyminum* L. under UV stress. Protoplasma 2019; 256(1): 279-90.
[http://dx.doi.org/10.1007/s00709-018-1297-y] [PMID: 30083789]

[123] Sulpice R. Closing the yield gap: can metabolomics be of help? J Exp Bot 2020; 71(2): 461-4.
[http://dx.doi.org/10.1093/jxb/erz322] [PMID: 31425582]

CHAPTER 12

CRISPR/CAS9 Technologies to Enhance Tolerance to Abiotic Stress in Crop Plants

Ravinder Singh[1,*], **Rajani Gupta**[2], **Renu Bhardwaj**[1] and **Rattandeep Singh**[3,*]

[1] *Department of Botanical & Envrionmental Sciences, Guru Nanak Dev University, Amritsar, Punjab 143005, India*

[2] *Department of Botanical & Environmental Sciences, Hindu College, Amritsar, Punjab 143005, India*

[3] *School of Bioengineering and Biosciences, Lovely Professional University, Jalandar, Punjab 144001, India*

Abstract: Plants can identify and cope up with biotic and abiotic stressors, thus resulting in a reduction in agricultural production significantly. Currently, the primary goal of plant breeders is to develop the ability to tolerate multiple stress conditions without lowering the productivity of the crop. However, numerous attempts have failed to release these plants due to persistent divergence between growth and resistance to stressors. Such strategies are not appropriate to effectively enhance characteristics and ensure potential environmental impact. Genome editing approaches and RNA interference (RNAi) technique has been used to develop plants resistant to different environmental changes. Newly developed approaches such as CRISPR are used to grow new varieties that have external stressors tolerance and reduce a vital yield loss. This chapter addresses the use of CRISPR/CAS techniques to improve the stress tolerating ability of the plants

Keywords: Abiotic, CRISPR, Plant breeding, RNA interference (RNAi), Stress.

INTRODUCTION

The global reduction in crop yield is caused by abiotic stress, which further restricts the average cultivation of all agricultural crops [1]. It also restricts plant efficiency, growth [2], and overproduction of toxic reactive oxygen (ROS) species [3]. These plants immediately evolve different mechanisms for reducing environmental stress-related toxic effects [1]. Plants establish specific immune responses during growth and development with metabolic alterations [4]. Besides

---

[*] **Corresponding Author Ravinder Singh & Rattandeep Singh:** Department of Botanical & Envrionmental Sciences, Guru Nanak Dev University, Amritsar, Punjab 143005, India; E-mail:ravinderbali@gmail.com and School of Bioengineering and Biosciences, Lovely Professional University, Jalandar, Punjab 144001, India; Tel: +91-6006083110; E-mail:rattan.19383@lpu.co.in

Tajinder Kaur & Saroj Arora (Eds.)

the management of genetic strategies in plant growth, development, and productivity [5], the mechanism for plant defense needs in-depth investigations of stress-sensitive messengers and signals, including protein interaction, transcription factors (TFs), and promoters. Also, a broad spectrum of abiotic stressors, including temperatures, light, heavy metals, osmotic stressors, and a range of biotic factors, can over-produce highly toxic ROS [1]. The plant may establish a defensive mechanism by which the damage is either prevented or even alleviated. This process triggers the highly stressors-responsive, antioxidant gene expressions [1]. Besides this, these genes will function through metabolic modifications in the plant cell.

IMPORTANCE OF BREEDING TOWARDS ABIOTIC STRESS TOLERANCE

Drought, salinity, heavy metals, and extreme temperatures are significant abiotic stress factors that thrive in modern farming. The decrease in yield in the primary cultivation of plants is approximately 50 percent, although the degree of yield losses depends on the nature and intensity of stressors during plants' growing cycle [6]. Different abiotic stressors-triggered gene reduced expression of genes responsible for biochemical and phenotypic modifications, thereby reducing photosynthetic capacity, lowered nitrogen assimilation capability, altered behaviors of several core molecules, plasma membrane properties, and cellular biochemical changes. Plants are affected in many ways, *e.g.*, by altering growth and development of plants, membrane diffusion, contraction of stomata, chlorophyll loss, osmotic fluctuations, photosynthetic activity declined and decreased the CO_2 assimilation. In contrast, the level of abscisic acid (ABA), osmolytes, proline, sorbitol, mannitol, radical scavenging molecules (glutathione, ascorbate) enhanced [7]. Plant breeders should implement new genetic engineering approaches to improve the breeding system. Breeders might concentrate on strengthening the resistance of plants towards environmental stressors [8]. Research and development of crop varieties with improved response to stressors *via* the use of genetic techniques helps in the development of several specific characteristics and reduces time and complexity under stressors conditions. Based on the continuously rising world population, it is very difficult to meet the demands of the increasing supply of food production. The biological response of crops may help to develop systems or prototypes that can better understand a range of plant responses to diverse abiotic stressors. These strategies help to identify the adequate technology for the development of enhanced abiotic-resistant crop species. The advancement of plant science in systems biology is inevitable. It would help to coordinate different high-level "omic" networks such as transcriptome, proteome, metabolome, and genomics to recognize plant responses and regulatory mechanisms and their interactions [9].

GENETIC APPROACHES TO ENHANCE ABIOTIC STRESS TOLERANCES IN PLANTS

Classic breeding techniques have become a standard protocol in the study of the genetic diversity of crop varieties. Most of the crop cultivars introduced several conventional breeding programmes, such as wheat [10, 11,], rice [12, 13], soybeans [14], and maize [15]. Hence, traditional breeding technology has strengthened crop quality at anatomical, biochemical, and molecular levels against numerous abiotic stressors. However, the limitations of this technology are noteworthy because it is arduous, time-consuming, complicated, and expensive. Therefore, the development of a new cultivar with improved features by this technology takes several years. Besides, several of the genes with inconsequential or undesired traits might also be introduced, which may be challenging to eliminate. The availability of germplasm or low genetic variations can also be a significant selection factor [16]. The complexity of adapting the plant under stress makes breeding difficult. This suggests that several traits can combine and help to improve the tolerance of plants towards various abiotic stressors. Therefore, the tremendous progress in genomics science led to a deeper understanding, *e.g.*, gene expression, communication, and the interaction toward stressors tolerance. Qualitative biotechnology and strategies, particularly, the high-throughput study of expressed sequence tags, targeted or random mutagenesis, large-scale simultaneous genomic study, the identification of innovative genes, and their trends of expression in response to abiotic stressors help in improving knowledge of their functions in stressors tolerance [17].

RNA INTERFERENCE (RNAI) APPROACHES TO RESIST ABIOTIC STRESSORS

The next-generation sequence has quickly resolved the weaknesses of DNA-mutagen dependent screening by mapping mutations using transposons and phages, rather than synthetic mutagens, to recognise sequences of insertions that are conducive to sequencing-based research [18, 19]. Besides, new genetic disruption methods for gene knock-down using the RNAi pathway, and forward genetic predictions were revolutionized [20]. Small RNAs (sRNAs) emerged as outstanding alternatives for promoting crop development and quantitative biotechnology experiments these days. siRNA and miRNA are core regulators in the growth and development of both the post-transcriptional and transcriptional stages. Different studies have indicated RNA silencing technology as an epigenetic regulation, which results in double-stranded post-transcriptional gene silencing (dsRNAs) to hinder the expression of specific genes [21]. RNAi technology is reliable, competent and is one of the main strategies for the improvement of crops, as well as for modifications relevant molecular engineering organisms [22]. The RNAi technique was used successfully in rice to inhibit Squalene Synthase (SQS) enzyme and to enhance drought resistance during plant growth stages in vegetative and reproductive period [23]. This result in increase in water restoration, reduced stomatal conductance and reduced water retention in growing plants. Hu *et al*. [24], have generated better rice transgenic plants for dry Stressors through RNAi-mediated silencing of sGRXS17 (Glutaredoxins) and transgenic plants have demonstrated ROS modulation and stomatal closure.

Although the RNAi technique provided a lot of knowledge regarding gene regulation but, it also has shortcomings of incomplete knockdown of genes and substantial off-target results, making it challenging to understand phenotypic modifications [25]. In addition, some of the off-target genes may potentially be disrupted [26]. Moreover, siRNA or shRNA may interact with endogenous cell machinery, which may even lead to cell death. Current genetic methods to genome editing have arisen recently as effective methods that allow for the addition, deletion or replacement of DNA into the genome to overcome these restrictions [27]. Genome editing is the powerful techniqe to describe the gene structure and to construct new and beneficial variants of phenotype. ZFN (Zinc Finger nucleases), TALEN's (transcription activator-like effector nucleases) and CRISPR-Cas9 are few distinct approaches for creating a sequence which can be used to improve a variety of engineered nucleases [28]. However use of ZFNs and TALENs are costly, time consuming and give poor results especially in high-performance experiments [29]. As a result, CRISPR-Cas9 has been used in a variety of species, including both model plants and crops, due to the simple and efficient application of short RNA guides to almost all genomic locations [30].

CRISPR/CAS9-SINGLE GUIDE RNA (SGRNA) SCHEME FOR GENETIC ENGINEERING

The development of new regulatory systems (genes, proponents, cis-regulatory elements, small RNAs, epigenetic changes) from ordinarily existing components can encourage the design of signaling / regulatory and metabolic methods to reduce abiotic Stressors tolerance of plants. Several studies show that the assembly and steady transmission of homozygous transgenic plants within one generation has been introduced to progressive generations [31 - 33]. In 2002, Rudd Jansen and co-workers introduced Clustered Regularly Interspaced short Palindrome Repeats (CRISPR) at the University of Utrecht, Netherlands [34]. The CRISPRs are clustered in internal transcribed regions with multiple short direct repeats (21–37 bp), alternated with non-repetitive sequences surrounded by a universal leader sequence (300–500 bp) and the existence of a numerous CRISPR chromosome implies that these are moving elements [34]. While CRISPRs seem to be the most scattered group of tandem repeats present in archaeal and bacteria chromosomes [35]. Kira Makarova (2011–2017) continued to contribute an improved epigenetic description of CRISPR-Cas systems at the National Center for Biotechnology Information, National Institutes of Health (USA), currently classified into multi-subunit protein complexes (Class 1) including Types I, III and IV, and single-effector protein (Class 2) with Types II, V and VI [36 - 41]. Cas9 is the dominant single abundant multidomain intracellular enzyme with 3822 orthology [42, 43]. For advanced viability, enthusiastic identification, and benefit attempting to make effectiveness, genomic guidance widely recognized for economically important crops [44]. A CRISPR-Cas system is already transforming researchers, to understand the importance of genes as well as their function. In recent years, genome editing technology (as shown in the Fig. **1**) involved in almost every area of the biological field (including the plant sciences) has been transformed the level of research [45]. Genetic editing ensued certainly by enhancing the synchronous processing of nucleases and templates, and by eliminating non-homologous end joining (NHEJ) [46]. Consequently, addition to its potential not to leave a footprint, it is difficult to determine whether the transition has been done deliberately or occured by natural mutation [47]. The genes have been successfully modified for the phenotypical expression of different traits: rice amylose [48], wheat Stressors-reactivity [49], sodium hairy roots [50], grape- and blight-powdery mildew [51], *etc.* Moreover, the application of the Crisper-Cas9 method of altering abiotic tolarances in plants has successfully demonstrated by Osakabe *et al.* [52]. The CRISPR-Cas system is utilized to produce point mutations, rearrangements and transcriptional control [28] in order to identify the biological history of plant's abiotic and biotic Stressors immunity. Our review focuses on the development of Cas9 based genetic screening tools to enhance plant production and abiotic Stressors response using CRISPR-Cas methods.

CRISPR/CAS9 TECHNOLOGIES TO ENHANCE ABIOTIC TOLERANCE TO STRESSORS IN CROP PLANTS

Abiotic Stress is one of the applauded challenges for agriculture and the predicted environment change expected to further deteriorate yield [53]. The CRISPR / Cas9 practice relies here on specific DNA/RNA hybrids which contain gene-specific characteristics and can modify almost any genome sequence to expose their role [54]. DNA target tolerance can be improved by manipulating the single-guided RNA chimer as sgRNA sequence, and various sgRNAs can interact with similar Cas9 protein on multiple targets [55]. Nowadays, the CRISPRCas9 method used frequently to alter crop genomes by selective gene modifications. It is crucial to develop the genome editing technologies to produce new and specific varieties of plant genomes using CRISPR-Cas9. There is an effort to reduce

Cas9's off-target effects to produce new mutant alleles of the Stress response. A number of software resources for the design of sgRNA's engineered in multiple genes have been developed for reducing off-target problems [56]. CRISPR-P, a new software platform for manipulating sgRNA in over 20 plant species, has recently been created [28]. CRISPR-P uses different vectors and toolkits and has been built up till now to reconcile plant genome edition using CRISPR-Cas9 [57].

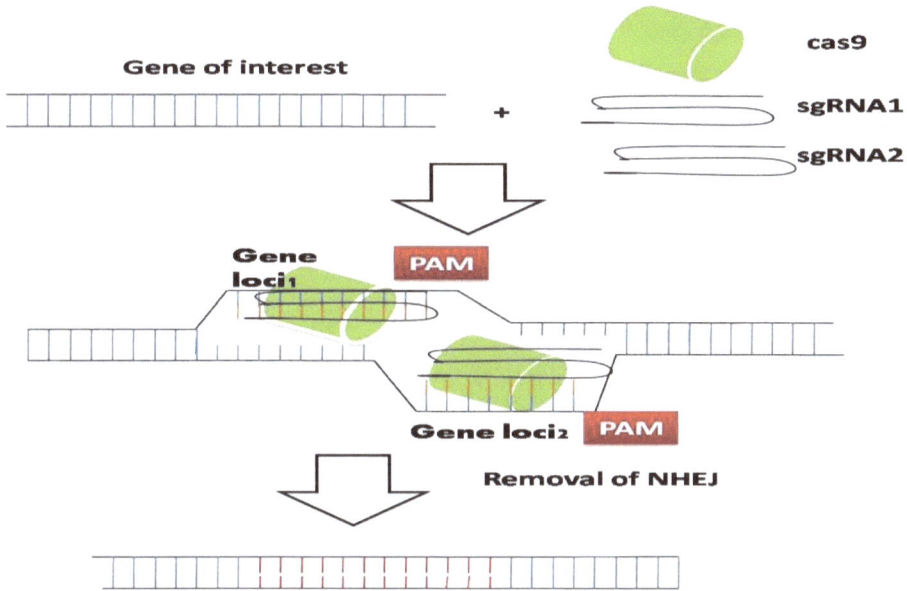

Fig. (1). CRISPR-Cas9 Technology. Cas9 is an endonuclease-guided RNA enzyme. Cas9 nuclease induced guidance sequence (20nt) of the short guidance RNA (sgRNA) at the target site which can be repaired either by NHEJ method or by te HR homologous recombination sequences which resulted in the addition/modification of the gene.

The invention of the CRISPR-Cas technique has profoundly improved genome editing and brought considerable success. CRISPR-Cas9 RNA-guided technique is now widely used for crop enhancement. Shi *et al.* [58] reported a corn strain modified with CRISPR being more impregnable to drought. They found that CRISPR-cas9 provoked ARGOS8 mutations reported an increased maize grain production during dry Stressors. These results showed that the process CRISPR / cas9 have been used to develop new changes successfully and efficiently in the cultivation of drought-resistant crop varieties. Similarly, a powerful CRISPR / Cas9 method using tru-gRNAs and Cas9 controlled by an AtEF1 tissue regulator reported triggering off-targ*et al*terations to abiotic-responsive gene(OST2 / AHA1) [59]. The OST2 / AHA1 alleles have been produced by standardizing CRISPR / Cas9 and intensifying stomata responses in Arabidopsis. Himatani *et al.* [60] examined nickase CRISPR / Cas9 (nCas9) or nuclease deficiency Cas9 (dCas9) with Petromyzon marinus cytidine deaminase (PmCDA1) to induce target genes point mutation in *Oryza sativa*, resulting in mutant herbicide tolerance phenotypes. Similar, CRISPR / Cas9 have also been used to develop maize (*Zea mays*) which overexpressed auxin-regulating organ-sized gene (ARGOS8) plants by replacing the GOS 2 promoter with respected ARGOS8 promoter [58]. Thus, the findings of the above experiments enabled prospects and methods of CRISPR / Cas9 directed genetic modifications to improve crop production and abiotic Stressors resistance in crops.

FUTURE APPROACHES

Although, Crop production is among the earliest human practices but to meet the ever increasing demand of plant based foods it is imperative to search for the technologies for developing improved crop varieties. New gene-editing techniques, especially CRISPR / Cas9, have benefits over traditional breeding because they can reduce obstacles to incompatibility between specific organisms and introduce artificial/synthetic genes into crops. The CRISPR / Cas9 for gene editing are exceptionally reliable and consistent in which the genetic sequence target site is specially picked to prevent off-target mutations. Due to the extreme lack of gene sequence data, extensive uses of CRISPR / Cas9 in plant crop genetic engineering remain limited. Therefore, the critical challenge for CRISPR / Cas9 techniques has been extended to crop enhancements. However, these experiments showed that this method would potentially enhance abiotic Stressors resistance for future applications in plant breeding. In comparison, metabolic and regulatory pathways may be modified to modulate abiotic Stressors tolerance in crop plants with the natural component, genes, promoters, cis-regulating elements, and epigenetic modifications, which may serve as the basis for the development of new biological components. CRISPR-Cas is required for both potential development and substantial genetic evolution of the target gene's diverse abiotic Stressors-related characteristics.

CONSENT FOR PUBLICATION

Not applicable.

CONFLICT OF INTEREST

The authors declare no conflict of interest, financial or otherwise.

ACKNOWLEDGEMENTS

Declared none.

REFERENCES

[1]   Ciarmiello LF, Woodrow P, Fuggi A, Pontecorvo G, Carillo P. Plant genes for abiotic stress. Abiotic stress in plants–mechanisms and adaptations 2011; 22: 283-308.

[2]   El-Esawi M, Alayafi A. Overexpression of StDREB2 transcription factor enhances drought stressors tolerance in cotton (*Gossypium barbadense* L.). Genes 2019; 14: 10(2): 142.

[3]   Consentino L, Lambert S, Martino C, *et al.* Blue-light dependent reactive oxygen species formation by *Arabidopsis cryptochrome* may define a novel evolutionarily conserve signaling mechanism. New Phytologist 2015; 26:206(4): 1450-62.

[4]   Elansary HO, Szopa A, Kubica P, Ekiert H, Ali HM, Elshikh MS, *et al.* Bioactivities of traditional medicinal plants in Alexandria. Evidence-Based Complementary and Alternative Medicine 2018; 1-13.
[http://dx.doi.org/10.1155/2018/1463579]

[5]   El-Esawi MA, Elansary HO, El-Shanhorey NA, Abdel-Hamid AME, Ali HM, Elshikh MS. Salicylic acid-regulated antioxidant mechanisms and gene expression enhance rosemary performance under saline conditions. Frontiers in Physiology 2017; 21: 8.
[http://dx.doi.org/10.3389/fphys.2017.00716]

[6]   Jaleel CA, Gopi R, Panneerselvam R. Growth and photosynthetic pigments responses of two varieties of catharanthus roseus to triadimefon treatment. Frontiers in Physiology 2008; 331(4): 272-7.
[http://dx.doi.org/10.1016/j.crvi.2008.01.004]

[7]   Osakabe Y, Osakabe K, Shinozaki K, Tran L-SP. Response of plants to water stressors. Front Plant Sci 2014; 13; 5

[8]   Tester M, Langridge P. Breeding technologies to increase crop production in a changing world. Science 2010; 327(5967): 818-22.
[http://dx.doi.org/10.1126/science.1183700]

[9]   Cramer GR, Urano K, Delrot S, Pezzotti M, Shinozaki K. Effects of abiotic Stressors on plants: a systems biology perspective. BMC Plant Biology 2011; 11(1): 163.

[10]   Valkoun J. Wheat pre-breeding using wild progenitors. Wheat in a global environment developments in plant breeding. 2001 May; 699–707.
[http://dx.doi.org/10.1007/978-94-017-3674-9_94]

[11]   Zaharieva M, Gaulin E, Havaux M, Acevedo E, Monneveux P. Drought and heat responses in the wild wheat relative aegilops geniculata roth: potential interest for wheat improvement. Crop Science. 2001 Jul1; 41(4): 1321–9

[12]   Mackill DJ, Amante MM, Vergara BS, Sarkarung S. Improved semidwarf rice lines with tolerance to submergence of seedlings. Crop Science 1993; 33(4): 749-53.
[http://dx.doi.org/10.2135/cropsci1993.0011183X003300040023x]

[13]   Mishra S, Senadhira D, Manigbas N. Genetics of submergence tolerance in rice (*Oryza sativa* L.). Field Crops Research. 1995; 46 (1-3): 177–81.

[14]   Vantoai TT, Beuerlein AF, Schmitthenner SK, Martin SKS. Genetic variability for flooding tolerance in soybeans. Crop Science 1994; 34(4): 1112–5.
[http://dx.doi.org/10.2135/cropsci1994.0011183X003400040051x]

[15]   Bänziger M, Setimela PS, Hodson D, Vivek B. Breeding for improved abiotic stressors tolerance in maize adapted to Southern Africa. Agric Water Manag 2006; 80(1-3): 212-4.

[16]   Ashraf M. Inducing drought tolerance in plants: recent advances. Biotechnol Adv 2010; 28(1): 169-83.
[http://dx.doi.org/10.1016/j.biotechadv.2009.11.005] [PMID: 19914371]

[17]   Cushman JC, Bohnert HJ. Genomic approaches to plant Stressors tolerance. Current Opinion in Plant Biology 2000; 3(2): 117–24.

[18] Copeland NG, Jenkins NA. Harnessing transposons for cancer gene discovery. Nature Reviews Cancer 2010; 10(10): 696–706.
[http://dx.doi.org/10.1038/nrc2916]

[19] Rad R, Rad L, Wang W, Cadinanos J, Vassiliou G, Rice S, *et al.* PiggyBac transposon mutagenesis: a tool for cancer gene discovery in mice. Science 2010; 330(6007): 1104–7.
[http://dx.doi.org/10.1126/science.1193004]

[20] Hammond SM, Bernstein E, Beach D, Hannon GJ. An RNA-directed nuclease mediates post-transcriptional gene silencing in Drosophila cells. Nature 2000; 404(6775): 293-6.

[21] Younis A, Siddique MI, Kim C-K, Lim K-B. RNA Interference (RNAi) Induced Gene Silencing: A Promising Approach of Hi-Tech Plant Breeding. Int J Biol Sci 2014; 10(10): 1150-8.

[22] Saurabh S, Vidyarthi AS, Prasad D. RNA interference: concept to reality in crop improvement. Planta 2014; 239(3): 543-64.
[http://dx.doi.org/10.1007/s00425-013-2019-5]

[23] Manavalan LP, Chen X, Clarke J, Salmeron J, Nguyen HT. RNAi-mediated disruption of squalene synthase improves drought tolerance and yield in rice. J Exp Bot 2011; 63(1): 163-75.
[http://dx.doi.org/10.1093/jxb/err258]

[24] Hu Y, Wu Q, Peng Z, *et al.* Silencing of OsGRXS17 in rice improves drought Stressors tolerance by modulating ROS accumulation and stomatal closure. Scientific Reports 2017; 7(1).

[25] Guan B, Hu Y, Zeng Y, Wang Y, Zhang F. Molecular characterization and functional analysis of a vacuolar Na/H antiporter gene (HcNHX1) from Halostachys caspica. Molecular Biology Reports 2010; 38(3): 1889-99.

[26] Jackson AL, Bartz SR, Schelter J, *et al.* Expression profiling reveals off-target gene regulation by RNAi. Nature Biotechnology 2003; 21(6): 635–7.
[http://dx.doi.org/10.1038/nbt831]

[27] Ali Z, Abulfaraj A, Idris A, Ali S, Tashkandi M, Mahfouz MM. Function genomics of abiotic Stressors tolerance in plants: a CRISPR approach. Front Plant Sci 2015; 27(6).

[28] Jain M. Function genomics of abiotic Stressors tolerance in plants:a CRISPR approach. Front Plant Sci 2015; 6(375).

[29] Choudhary PK, Mushtaq MA. Genome editing using Crispr/Cas System: new era genetic technology in agriculture to boost crop output. Europ. J Exper Biol 2017; 07(03).

[30] Ceasar SA, Rajan V, Prykhozhij SV, Berman JN, Ignacimuthu S. Insert, remove or replace: A highly advanced genome editing system using CRISPR/Cas9. Biochimica et Biophysica Acta (BBA) - Molecular Cell Research 2016; 1863(9): 2333-44.

[31] Feng Z, Zhang B, Ding W, *et al.* Efficient genome editing in plants using a CRISPR/Cas system. Cell Res 2013; 23(10): 1229-32.
[http://dx.doi.org/10.1038/cr.2013.114] [PMID: 23958582]

[32] Brooks C, Nekrasov V, Lippman ZB, Eck JV. Efficient gene editing in tomato in the first generation using the clustered regularly interspaced short palindromic repeats/CRISPR-Associated9 system. Plant Physiology 2014; 166(3): 1292-7.

[33] Zhou H, Liu B, Weeks DP, Spalding MH, Yang B. Large chromosomal deletions and heritable small genetic changes induced by CRISPR/Cas9 in rice. Nucleic Acids Res 2014; 42(17): 10903-4.

[34] Jansen R, Embden JDAV, Gaastra W, Schouls LM. Identification of genes that are associated with DNA repeats in prokaryotes. Mol Microbiol 2002; 43(6): 1565-75.
[http://dx.doi.org/10.1046/j.1365-2958.2002.02839.x]

[35] Ishino Y, Krupovic M, Forterre P. History of CRISPR-cas from encounter with a mysterious repeated sequence to genome editing technology. J Bacteriol 2018; 200(7).

[36]   Makarova KS, Aravind L, Wolf YI, Koonin EV. Unification of cas protein families and a simple scenario for the origin and evolution of CRISPR-Cas systems. Biol Direct 2011; 6(1): 38.

[37]   Makarova KS, Koonin EV. Evolution and classification of CRISPR-Cas systems and cas protein families. CRISPR-Cas Systems 2012; 61-91.

[38]   Makarova KS, Wolf YI, Koonin EV. The basic building blocks and evolution of CRISPR–cas systems. Biochemical Soc Trans 2013; 41(6): 1392-400.

[39]   Makarova KS, Koonin EV. Annotation and classification of CRISPR-Cas systems. Methods in Mol Biol CRISPR 2015; 47-75.

[40]   Makarova KS, Wolf YI, Alkhnbashi OS, *et al.* An updated evolutionary classification of CRISPR-Cas systems. Nat Rev Microbiol 2015; 13(11): 722-36.
[http://dx.doi.org/10.1038/nrmicro3569] [PMID: 26411297]

[41]   Koonin EV, Makarova KS, Zhang F. Diversity, classification and evolution of CRISPR-Cas systems. Curr Opin Microbiol 2017; 37: 67-78.

[42]   Shmakov S, Smargon A, Scott D, *et al.* Diversity and evolution of class 2 CRISPR–cas systems. Nat Rev Microbiol 2017; 15(3): 169-82.

[43]   Pyzocha NK, Chen S. Diverse Class 2 CRISPR-cas effector proteins for genome engineering applications. ACS Chem Biol 2017; 13(2): 347-56.

[44]   Noman A, Aqeel M, He S. CRISPR-Cas9: tool for qualitative and quantitative plant genome editing. Front Plant Sci 2016; 7.

[45]   Ma H, Marti-Gutierrez N, Park SW. Correction of a pathogenic gene mutation in human embryos. Nature 2017; 548(7668): 413-9.

[46]   Maggio I, Gonçalves MA. Genome editing at the crossroads of delivery, specificity, and fidelity. Trends in Biotechnology 2015; 33(5): 280-91.
[http://dx.doi.org/10.1016/j.tibtech.2015.02.011]

[47]   Rodríguez-Leal D, Lemmon ZH, Man J, Bartlett ME, Lippman ZB. Engineering quantitative trait variation for crop improvement by genome editing. Cell 2017; 171(2).
[http://dx.doi.org/10.1016/j.cell.2017.08.030]

[48]   Sun Y, Jiao G, Liu Z, Zhang X, Li J, Guo X, *et al.* Generation of high-amylose rice through CRISPR/Cas9-mediated targeted mutagenesis of starch branching enzymes. Front Plant Sci 2017; 8.

[49]   Kim D, Alptekin B, Budak H. CRISPR/Cas9 genome editing in wheat. Funct Integ Genom 2017; 18(1): 31-41.

[50]   Ueta R, Abe C, Watanabe T, *et al.* Rapid breeding of parthenocarpic tomato plants using CRISPR/Cas9. Sci Rep 2017; 7(1).

[51]   Malnoy M, Viola R, Jung M-H, *et al.* DNA-free genetically edited grapevine and apple protoplast using CRISPR/Cas9 ribonucleoproteins. Front Plant Sci 2016; 7.

[52]   Osakabe Y, Watanabe T, Sugano SS, *et al.* Optimization of CRISPR/Cas9 genome editing to modify abiotic Stressors responses in plants. Sci Rep 2016; 6(1).

[53]   Pereira A. Plant abiotic stressors challenges from the changing environment. Front Plant Sci 2016; 7.

[54]   Sander JD, Joung JK. CRISPR-Cas systems for editing, regulating and targeting genomes. Nat. Biotechn 2014; 32(4): 347–55.

[55]   Wu X, Kriz AJ, Sharp PA. Target specificity of the CRISPR-Cas9 system. Quantitative Biology 2014; 2(2): 59–70.

[56]   Montagne L, Raimondo A, Delobel B. Identification of two novel loss-of-functionSIM1mutations in two overweight children with developmental delay. Obesity 2014; 22(12): 2621-4.
[http://dx.doi.org/10.1002/oby.20886]

[57]   Xing H-L, Dong L, Wang Z-P, *et al.* A CRISPR/Cas9 toolkit for multiplex genome editing in plants. BMC Plant Biol 2014; 14(1).

[58]   Shi J, Gao H, Wang H, *et al.* ARGOS8 variants generated by CRISPR-Cas9 improve maize grain yield under field drought Stressors conditions. Plant Biotechn J 2016; 15(2): 207–16.

[59]   Osakabe Y, Osakabe K, *et al.* Genome Editing with Engineered Nucleases in Plants. Plant and Cell Physiol 2014; 56(3): 389–400.
[http://dx.doi.org/10.1093/pcp/pcu170]

[60]   Shimatani Z, Kashojiya S, Takayama M, *et al.* Targeted base editing in rice and tomato using a CRISPR-Cas9 cytidine deaminase fusion. Nat. Biotechn 2017; 35(5): 441–3.

SUBJECT INDEX

A

Ability 49, 206
  osmotic 49
  stress tolerating 206
Abiotic stress
  combating 159
  damages 6
  factors 1, 81, 84
  suppressing 91
Abiotic stressor(s) 2, 3, 4, 5, 6, 44, 158, 159,
    160, 161, 162, 163, 164, 166, 174, 183,
    184, 185, 193, 207, 209, 212
  combat 164
  combating 162
  diverse 207, 212
  environmental 44
  in plants 166
  reactions in plants 185
  responses 185
  resistance 164
  tolerance 4, 183, 209
Accumulation 41, 42, 43, 45, 82, 83, 91, 92,
    93, 116, 132, 144, 145, 149, 151, 152,
    170, 195
  anthocyanin 149
  ferritin 93
  nuclear 170
  of flavonoids in cytoplasm 148
  osmolytes 96
  rapid 132
Acid(s) 7, 27, 28, 45, 77, 78, 91, 93, 108, 111,
    112, 114, 124, 127, 144, 145, 146, 147,
    148, 149, 158, 159, 162, 164, 167, 168,
    169, 170, 171, 174, 175, 190, 195
  absicisic 174
  benzoic 148
  boric 78
  caffeic 148, 149
  dihydrophaseic 164
  fatty 145, 148
  ferulic 146, 148

  gallic 148
  gibberellic 167, 168
  homovanillic 148
  hydroxycinnamic 147
  jasmonic 112, 127, 158, 159, 162, 168, 169,
    175
  nucleic 45, 77, 91, 108, 145, 170, 190
  phenolic 114, 144, 195
  phenylacetic 162
  phosphatidic 7
  salicylic 27, 28, 111, 124, 127, 148, 158,
    159, 162, 171, 175
  sinapic 146
  thiobarbituric 93
  trans-cinnamic 146, 148
  tricarboxylic 195
  vanillic 148
Activities 22, 23, 27, 28, 76, 80, 82, 89, 42,
    56, 93, 94, 95, 98, 99, 100, 124, 125,
    128, 129, 132, 136, 144, 146, 149, 162,
    207
  anthropogenic 42, 56
  antibacterial 149
  antioxidant 27, 28, 89, 125, 128, 129, 136
  antioxidase 124
  anti-stress 128
  boosting antioxidant enzyme 132
  broad-spectrum antimicrobial 144
  enhancing antioxidant enzyme 132
  enzymatic 76, 146, 162
  of antioxidant enzymes 93, 95, 100
  of nitrate reductase 22, 80, 98
  of peroxidase 94, 99
  photochemical 23
  photosynthetic 23, 80, 82, 207
  plant nutrient uptake 22
  reduced enzymatic 80
  synergistic 149
Agricultural 1, 40, 44, 59, 61, 64, 76, 83, 90,
    159
  productivity 1, 40, 44, 59, 61, 90, 159
  products 64, 159

fundamental living 4
geographical information 45
immune 125
non-enzymatic 113
plant defence 168
salt stress photosynthetic 50
water drainage 42

T

Target 186, 212
    region amplification polymorphism
        (TRAP) 186
    site, genetic sequence 212
Techniques 66, 183, 184, 185, 187, 188, 189,
        190, 191, 195, 197, 206
    high-throughput phenotyping 187
    mass spectrometric 195
    metabolomic 184
    oligoarray 185
    omic 183
Technologies 20, 21, 45, 153, 184, 188, 191,
        207, 208, 210, 212
    genome editing 191, 210
    next-generation sequencing 184
    omic 184
    remote sensing 45
    tilling array 191
    traditional breeding 208
Third-generation sequencing (TGS) 188
Tolerance 25, 26, 27, 28, 89, 91, 92, 158, 159,
        163, 172, 183, 184, 186, 187
    elevating 172
    submergence 187
Toxicity 40, 41, 76, 81, 159
    metabolic 40
    mineral 159
Transaminases 163
Transcription 160
    factor binding proteins 160
Transcription factors (TFs) 26, 30, 31, 125,
        135, 169, 170, 172, 184, 185, 188, 190,
        207
    regulating 125

stress-associated 135
Transcriptome 191, 207
    whole-genome 191

V

Vegetative 58, 169
    growth period 58
    storage proteins (VSP) 169

W

Water 2, 8, 9, 10, 24, 28, 29, 30, 41, 42, 43,
        45, 49, 50, 75, 80, 81, 82, 162
    conserving 82
    excess 10
    excessive 162
    irrigation 28
    poor quality 45
    saline 43
    sterile 29
Water deficiency 2, 3, 8, 10, 22, 43
    stresses 3
Water deficit 9, 10, 132, 148, 190
    pressure 9
Water stress 1, 9, 23, 24, 41, 48, 128, 132,
        145, 164, 166, 167, 169
    coping 167
    response genes 132
Wheat 9, 22, 27, 28, 30, 44, 45, 61, 64, 65, 91,
        95, 100, 101, 186, 189, 191, 192, 193,
        210
    crops 44, 45, 91
    drought-tolerant 9
    genotypes 44, 64, 65
    salt-treated 186
    spring 61
    stressors-reactivity 210
Whole-genome sequencing (WGS) 188

www.ingramcontent.com/pod-product-compliance
Lightning Source LLC
Chambersburg PA
CBHW050830220326
41598CB00006B/345